Frontiers of Computational Science

Yukio Kaneda, Hiroshi Kawamura
and Masaki Sasai (Eds.)

Frontiers of Computational Science

Proceedings of the International Symposium on Frontiers of Computational Science 2005

With 157 Figures and 23 Tables

Prof. Yukio Kaneda
Department of Computational Science
Graduate School of Engineering
Nagoya University
Chikusa-ku
Nagoya 464-8603
Japan

E-mail: kaneda@cse.nagoya-u.ac.jp

Prof. Masaki Sasai
Department of Computational
Science and Engineering
Nagoya University
2-203-2, Omoteyama
Tempaku-Ku
Nagoya 468-0069
Japan

E-mail: sasai@cse.nagoya-u.ac.jp

Prof. Hiroshi Kawamura
Director
Research Center for Holistic Computational Science
Tokyo University of Science
1-56-90, Sakae-Cho
Ushiku
Ibaraki 300-1233
Japan

E-mail: kawa@rs.noda.tus.ac.jp

Library of Congress Control Number: 2006933559

ISBN-10 3-540-46373-9 Springer Berlin Heidelberg New York
ISBN-13 978-3-540-46373-3 Springer Berlin Heidelberg New York

This work is subject to copyright. All rights are reserved, whether the whole or part of the material is concerned, specifically the rights of translation, reprinting, reuse of illustrations, recitation, broadcasting, reproduction on microfilm or in any other way, and storage in data banks. Duplication of this publication or parts thereof is permitted only under the provisions of the German Copyright Law of September 9, 1965, in its current version, and permission for use must always be obtained from Springer. Violations are liable to prosecution under the German Copyright Law.

Springer is a part of Springer Science+Business Media.

springer.com

© Springer-Verlag Berlin Heidelberg 2007

The use of general descriptive names, registered names, trademarks, etc. in this publication does not imply, even in the absence of a specific statement, that such names are exempt from the relevant protective laws and regulations and therefore free for general use.

Typesetting by the SPi using a Springer LATEX macro package
Cover design: Erich Kirchner, Heidelberg

Printed on acid-free paper   SPIN 11661238   89/3100/SPi   5 4 3 2 1 0

Preface

This volume contains papers presented at the International Symposium on Frontiers of Computational Science 2005 held on December 12–13, 2005, at Noyori-Conference Hall, Nagoya University, Nagoya, Japan.

Computational science based on the high-level use of computers has been called the third major method in the field of scientific research, standing alongside the conventional methods of theory and experiment. Computational science covers wide-ranging scientific areas, from basic research fields such as algorithms and soft-computing to diverse applied fields targeting structures and dynamics in fluid systems (Macro Systems), electronic and atomic scale description of materials and chemical reactions (Micro Systems), bioinformatics and biomolecular structures and functions (Genome Systems), and optimization and visualization of complex systems (Complex systems). The aim of this symposium is to provide an opportunity for researchers from different fields of computational science to come together and discuss the latest research findings, so as to stimulate collaboration and cooperation among researchers from different fields, as well as to stimulate and promote the research activities of young computational scientists.

The symposium catalyzed interactions among basic and applied researches and we are sure that readers of this book will feel the atmosphere of lively discussions at the conference among participants from a wide variety of fields.

The program includes both invited lectures by distinguished scientists and poster sessions by young researchers. It consisted of seven plenary lectures, two key note lectures, 19 invited lectures, four oral presentations, and 57 poster presentations. These cover a wide range of Frontiers of Computational Science.

The symposium was organized under collaboration between the "21st Century COE Program: Frontiers of Computational Science (FCS)" at Nagoya University, and the "Academic Frontier: New Development of Computational Science through Holistic Approach (HOLCS)" at Tokyo University of Science (TUS). The FCS program is one of the 5-year "21st Century Center of Excellence (COE) Programs" in innovative academic fields, which were selected in 2004 by the Ministry of Education, Culture, Sports, Science and Technology (MEXT) in Japan to assist the formation of worldwide renowned centers. The project HOLCS started in 2005 as a follow-up of its predecessor, the Frontier Research for Computational Science (1996–2003) at TUS, and also is supported by MEXT to stimulate distinguished research activity in private universities.

We are grateful to the participants, the authors of written contributions to this volume, and the reviewers, whose valuable work helped to improve the quality of these proceedings. We wish to acknowledge, as well, the members of Local Organizing Committee, Drs. T. Furuhashi, S. Mitaku, M. Sonoyama, I. Ueno, K. Watanabe, M. Yamamoto, T. Yoshikawa, and Shao-Liang Zhang. We express our gratitude to Springer, Berlin, for making possible the publication of these proceedings.

Nagoya, August 2006

Yukio Kaneda
Hiroshi Kawamura
Masaki Sasai

Contents

Plenary & Keynote Papers

Computational Methods Combining Large Eddy Simulation
with Approximate Wall-Layer Models for Predicting
Separated Turbulent Near-Wall Flows
Michael A. Leschziner, Ning Li, and Fabrizio Tessicini 3

Numerical Experimentation: A Third Way
to Study Nature
Marie Farge 15

Recent Computational Developments in Quantum
Conductance and Field Emission
of Carbon Nanotubes
Jisoon Ihm 31

Energy Landscapes of Protein Self-Assembly: Lessons from
Native Topology-Based Models
Yaakov Levy and José N. Onuchic 37

A Brief Introduction to Krylov Space Methods
for Solving Linear Systems
Martin H. Gutknecht 53

Recent Developments in DNS and Modeling of Rotating
Wall-Turbulent Shear Flows with Heat Transfer
Yasutaka Nagano and Hirofumi Hattori 63

Contribution of Computational Biology
and Structural Genomics to Understand Genome
and Transcriptome
Mitiko Go, Kei Yura, and Masafumi Shionyu 75

Invited Papers

Achromatic and Chromatic Visual Information Processing and Discrete Wavelets
Hitoshi Arai 83

On Decaying Two-Dimensional Turbulence in a Circular Container
Kai Schneider and Marie Farge 91

A Study Toward Improving Accuracy of Large Scale Industrial Turbulent Flow Computations
Kazuhiko Suga 99

Direct Numerical Simulations of Turbulent Premixed Flames with Realistic Kinetic Mechanisms
Mamoru Tanahashi, Yuzuru Nada, Nobuhiro Shiwaku, and Toshio Miyauchi 107

Variational Monte Carlo Studies of Superconductivity in κ-BEDT-TTF Salts
Tsutomu Watanabe, Hisatoshi Yokoyama, Yukio Tanaka, and Jun-ichiro Inoue 115

Simulation for Measurements of Electric Properties of Surface Nanostructures
Satoshi Watanabe, Ryoji Suzuki, Tomofumi Tada, and Hideomi Totsuka 119

Ultra-Fast Dynamics in Nanocarbon Explored by TDDFT-MD Simulations
Yoshiyuki Miyamoto 125

Heat Transport in Nanoscale Objects: Classical to Quantum
Takahiro Yamamoto, Naoaki Kondo, Kazuyuki Watanabe, Yingwen Song, and Satoshi Watanabe 131

Aromaticity Driven Rupture of CN and CC Multiple Bonds
Cherumuttathu H. Suresh and Nobuaki Koga 137

Protein Function Prediction in Proteomics Era
Daisuke Kihara, Troy Hawkins, Stan Luban, Bin Li, Karthik Ramani, and Manish Agrawal 143

What We can Learn about Protein Folding from Recent Progress in Structure Prediction
George Chikenji, Yoshimi Fujitsuka, and Shoji Takada 149

Brownian Dynamics Approach to Protein Folding
Tadashi Ando and Ichiro Yamato 157

A Fourth Type of Secondary Structure Breaker
Kenichiro Imai, Masashi Sonoyama, and Shigeki Mitaku 165

Neighborhood Parallel Simulated Annealing
Keiko Ando, Mitsunori Miki, and Tomoyuki Hiroyasu 171

New Computing System Architecture for Simulation of Biological Signal Transduction Networks
Hitoshi Hemmi, Tetsuya Maeshiro, and Katsunori Shimohara 177

Convergence Analysis of GMRES Methods for Least Squares Problems
Ken Hayami and Tokushi Ito 181

A Numerical Method for Calculating the Green's Function Arising from Electronic Structure Theory
*Tomohiro Sogabe, Takeo Hoshi, Shao-Liang Zhang,
and Takeo Fujiwara* 189

General Contribution

Direct Numerical Simulation of Turbulent Channel Flow Imposed with Spanwise Couette Flow
Yohji Seki, Kaoru Iwamoto, and Hiroshi Kawamura 199

Reynolds-Number Dependence of Transport Barriers in Chaotic Mixing
Yoshinori Mizuno and Mitsuaki Funakoshi 205

On the Treatment of Long-range Electrostatic Interactions in Biomolecular Simulations
Yoshiteru Yonetani 209

DNS of Turbulent Channel Flow Obstructed by Rectangular Ribs
*Fusao Kawamura, Yohji Seki, Kaoru Iwamoto,
and Hiroshi Kawamura* 215

Source Models of Gravity Waves in an F-plane Shallow Water System
Norihiko Sugimoto, Keiichi Ishioka, and Katsuya Ishii 221

Direct Numerical Simulation of Neutrally Stratified Ekman Boundary Layer
Katsuhiro Miyashita, Kaoru Iwamoto, and Hiroshi Kawamura 227

Application of the Renormalization Group Analysis
to a Noisy Kuramoto–Sivashinsky Equation
and its Numerical Simulation
Kazuto Ueno 231

Wavelet-Based Statistics in High-Resolution Direct Numerical
Simulations of Three-Dimensional Homogeneous
Isotropic Turbulence
Naoya Okamoto, Katsunori Yoshimatsu, and Yukio Kaneda 235

Hydrophobic Hydration on Nanometer Length Scale
Takeshi Hotta and Masaki Sasai 239

Ultra-Fast Genome Wide Simulation of Biological Signal
Transduction Networks: Starpack
Tetsuya Maeshiro, Hitoshi Hemmi, and Katsunori Shimohara 243

Molecular Dynamics Study on Interaction of C-Component
Dislocation Loop and Edge Dislocation in α-Zirconium
Kenji Yamaura, Akiyuki Takahashi, and Masanori Kikuchi 247

Theoretical Analysis of Intramolecular Interaction
Kenta Yamada and Nobuaki Koga 253

Effect of Nanoscale Structure of Solid Surface
on Moving Droplet
Takuma Osawa and Ichiro Ueno 259

Proximity Effect in NS Junctions in the Presence
of Anderson Impurities
Takehito Yokoyama, Yukio Tanaka, and Alexander A. Golubov 265

Wiedemann–Franz Law in Diffusive Normal Metal/p-Wave
Superconductor Junctions
*Takehito Yokoyama, Yukio Tanaka, Alexander A. Golubov,
and Yasuhiro Asano* 269

Nodal Structures of Wave Functions in
Chaotic Billiards
Hiromu Ishio 273

Elongation of Water Residence Time at the Protein Interior
in Aqueous Solution with Ectoine
Isseki Yu and Masataka Nagaoka 277

Analytical Model for Protein Folding
Yoko Suzuki and José N. Onuchic 283

Theoretical Research on Dynamics of the
Genetic Toggle Switch
*Tomohiro Ushikubo, Wataru Inoue, Mitsumasa Yoda,
and Masaki Sasai* 289

Non-Condon Theory for the Energy Gap Dependence of
Electron Transfer Rate
*Hirotaka Nishioka, Akihiro Kimura, Takahisa Yamato,
and Toshiaki Kakitani* 293

Density Functional Molecular Orbital Calculations on Longer
DNA–DNA and PNA–DNA Double Strands
*Takayuki Natsume, Yasuyuki Ishikawa, Kenichi Dedachi,
and Noriyuki Kurita* 299

A Parallel Computing Framework for Nonexperts of
Computers: Easy Installation, Programming
and Execution of Master–Worker Applications Using
Spare Computing Power of PCs
*Takashi Noda, Hisaya Mine, Noriyuki Fujimoto,
and Kenichi Hagihara* 305

Derivation of TIS for Search Performance Quantification
on HC Method
Naotoshi Hoshiya and Tomohiro Yoshikawa 309

Improvement of Search Performance of Genetic Algorithm
Through Visualization of Search Process
*Daisuke Yamashiro, Tomohiro Yoshikawa,
and Takeshi Furuhashi* 313

A Proposal for Analysis of SD Evaluation Data by Using
Clustering Method Focused on Data Distribution
*Shou Kuroda, Kosuke Yamamoto, Tomohiro Yoshikawa,
and Takeshi Furuhashi* 317

Multiobjective Evolutionary RBF Networks and Its
Application to Ensemble Learning
Nobuhiko Kondo, Toshiharu Hatanaka, and Katsuji Uosaki 321

A New Concept of Cerebral Autoregulation Using
Mathematical Expressions
Tomoki Youkey Shiozawa, Hiroki Takada, and Masaru Miyao 327

Evolution Strategies on Desktop Grid Environment Using
Scheduling Algorithms RR for Same Length Tasks
*Yoshiyuki Matsumura, Noriyuki Fujimoto, Xin Yao, Yoshikazu
Murayama, and Kenichi Hagihara* 333

Solution of Black-Scholes Equation By Using RBF Approximation
Zhai Fei, Yumi Goto, and Eisuke Kita 339

Visualization of Differences on Two Subjects' Impression Words Based on Correlation of *SD Data*
Teruyuki Kojima, Takeshi Furuhashi, Tomohiro Yoshikawa, and Kosuke Yamamoto 345

Evaluation of High Imaging Quality LCDs Which Features Moving Picture: Comparison using Stabilometers
Kazuhiro Fujikake, Koichi Miura, Takahiro Sakurai, Hiroki Takada, Satoshi Hasegawa, Masako Omori, Ryumon Honda, and Masaru Miyao 349

Template Translation for Multilingual Disaster Information System
Shozo Tanaka, Masaru Miyao, Kohei Okamoto, and Satoshi Hasegawa 353

An Application of Double-Wayland Algorithm to Detect Anomalous Signals
Hiroki Takada, Hitoshi Tsunashima, Daishi Yamazaki, Tomoki Shiozawa, Meiho Nakayama, Takayuki Morimoto, and Masaru Miyao 357

List of Contributors

Manish Agrawal
Mechanical Engineering
Purdue University
West Lafayette, IN
USA
magrawal@purdue.edu

Keiko Ando
Graduate School of Knowledge
Engineering and Computer Science
Doshisha University
Kyoto
Japan
ando@mikilab.doshisha.ac.jp

Tadashi Ando
Department of Biological Science
and Technology
Tokyo University of Science
2641 Yamazaki
Noda-shi Chiba 278-8510
Japan
tando@rs.noda.tus.ac.jp

Hitoshi Arai
Graduate School of Mathematical
Sciences
University of Tokyo
3-8-1 Komaba, Meguro-ku
Tokyo 153-8914, Japan
h-arai@ms.u-tokyo.ac.jp

Yasuhito Asano
Department of Applied Physics
Hokkaido University
Sapporo 060-8628, Japan
asano@eng.hokudai.ac.jp

George Chikenji
Department of Chemistry
Faculty of Science
Kobe University
Nada Kobe 657-8501, Japan

and

Department of Computational
Science and Engineering
Graduate School of Engineering
Nagoya University
Nagoya 464-8603, Japan
chikenji@tbp.cse.nagoya-u.ac.jp

Kenichi Dedachi
Department of Knowledge-based
Information Engineering
Toyohashi University of Technology
Tenpaku-cho, Toyohashi 441-8580
Japan

Marie Farge
LMD–CNRS
Ecole Normale Supérieure
Paris, France
farge@lmd.ens.fr

Zhai Fei
Graduate School
of Information Sciences
Nagoya University
Nagoya 464-8301, Japan

Kazuhiro Fujikake
Nagoya University
Chikusa-ku
Nagoya 464-8601, Japan
fujikake@nagoya-u.jp

Noriyuki Fujimoto
Graduate School of Information
Science and Technology
Osaka University
Machikaneyama, Toyonaka
Osaka 560-8531, Japan
fujimoto@ist.osaka-u.ac.jp

Yoshimi Fujitsuka
Department of Chemistry
Faculty of Science
Kobe University
Nada Kobe 657-8501, Japan
yoshimi@theory.chem.sci.
kobe-u.ac.jp

Takeo Fujiwara
Department of Applied Physics
The University of Tokyo
Hongo, 7-3-1, Bunkyo-ku
Tokyo 113-8656, Japan

and

Core Research for Evolutional
Science and Technology
Japan Science and Technology
Agency (CREST-JST)
4-1-8 Honcho, Kawaguchi-shi
Saitama 332-0012, Japan
fujiwara@coral.t.u-tokyo.ac.jp

Mitsuaki Funakoshi
Department of Applied Analysis
and Complex Dynamical Systems

Kyoto University
Kyoto, Japan
mitsu@acs.i.kyoto-u.ac.jp

Takeshi Furuhashi
Department of Computational
Science and Engineering
Nagoya University
Nagoya-shi, Chikusa-ku furo-cho
Nagoya University
Nagoya 464-8603, Japan
furuhashi@cse.nagoya-u.ac.jp

Mitiko Go
Ochanomizu University
2-1-1 Otsuka
Bunkyo-ku
Tokyo 112-8610, Japan

and

Department of Bio-Science
Faculty of Bio-Science
Nagahama Institute of Bio-Science
and Technology
1266, Tamura-cho Nagahama
Shiga 526-0829, Japan
m_go@cc.ocha.ac.jp

Alexander A. Golubov
Faculty of Science and Technology
University of Twente
7500 AE, Enschede
The Netherlands
A.Golubov@tnw.utwente.nl

Yumi Goto
Graduate School
of Information Sciences
Nagoya University
Nagoya 464-8301, Japan

Martin H. Gutknecht
ETH Zurich
Seminar for Applied Mathematics
Germany
mhg@math.ethz.ch

Kenichi Hagihara
Graduate School
of Information Science
and Technology
Osaka University
Machikaneyama, Toyonaka
Osaka 560-8531, Japan
hagihara@ist.osaka-u.ac.jp

Satoshi Hasegawa
Nagoya Bunri University
365 Maeda, Inazawa-cho
Inazawa 492-8520, Japan

and

Graduate School
of Information Science
Nagoya University
Chikusa-ku
Nagoya 464-8601, Japan
hasegawa@nagoya-bunri.ac.jp

Toshiharu Hatanaka
Department of Information
and Physical Sciences
Osaka University
Suita 565-0871, Japan
hatanaka@ist.osaka-u.ac.jp

Hirofumi Hattori
Department of Mechanical
Engineering
Nagoya Institute of Technology
Japan
hattori@nitech.ac.jp

Troy Hawkins
Department of Biological Sciences
Purdue University
West Lafayette, IN
USA
thawkins@purdue.edu

Ken Hayami
National Institute of Informatics
2-1-2, Hitotsubashi
Chiyoda-ku
Tokyo 101-8430, Japan

and

The Graduate University
for Advanced Studies (Sokendai)
Department of Informatics
hayami@nii.ac.jp

Hitoshi Hemmi
ATR Network Informatics
Laboratories
Kyoto 691-0288, Japan
hemmi@atr.jp

Tomoyuki Hiroyasu
Department of Knowledge
Engineering and Computer Science
Doshisha University
Kyoto, Japan
tomo@is.doshisha.ac.jp

Ryumon Honda
Kanazawa Medical University
1-1 Daigaku
Unchinada-machi, Kahoku-gun
Ishikawa-ken 920-0293, Japan

Takeo Hoshi
Department of Applied Physics
The University of Tokyo
Hongo 7-3-1, Bunkyo-ku
Tokyo 113-8656
Japan
hoshi@coral.t.u-tokyo.ac.jp

Naotoshi Hoshiya
Mie University, Japan
hoshiya@ip.elec.mie-u.ac.jp

Takeshi Hotta
Department of Complex
Systems Science
Graduate School
of Information Science

Nagoya University
Nagoya 464-8603
Japan
hotta@tbp.cse.nagoya-u.ac.jp

Jisoon Ihm
School of Physics
Seoul National University
Seoul 151-747
Korea
jihm@snu.ac.kr

Kenichiro Imai
Department of Applied Physics
Graduate School of Engineering
Nagoya University
Furocho, Chikusa-ku
Nagoya 464-8603
Japan
imai@bp.nuap.nagoya-u.ac.jp

Jun-Ichiro Inoue
Department of Applied Physics
Nagoya University
Japan
inoue@nuap.nagoya-u.ac.jp

Wataru Inoue
Department of Complex
Systems Science
Graduate School
of Information Science
Nagoya University
Furocho, Chikusa
Nagoya 464-8601
Japan
wataru@sasai.human.
nagoya-u.ac.jp

Katsuya Ishii
Information Technology Center
Nagoya University
Furo-cho Chikusa-ku
Nagoya 464-8603
Japan
ishii@itc.nagoya-u.ac.jp

Yasuyuki Ishikawa
Department of Chemistry
University of Puerto
Rico, P.O. Box 23346
UPR Station
San Juan, PR 00931-3346
USA

Hiromu Ishio
Departamento de Química C-IX
Universidad Autónoma de Madrid
Cantoblanco, 28049
Madrid, Spain
h.ishio@uam.es

Keiichi Ishioka
Graduate School of Science
Kyoto University
Kyoto, Japan
ishioka@gfd-dennou.org

Tokushi Ito
Business Design Laboratory Co. Ltd.
Tokyo, Japan
ito@business-design.co.jp

Kaoru Iwamoto
Department of Mechanical
Engineering
Tokyo University of Science
2641 Yamazaki, Noda-shi
Chiba 278-8510, Japan
iwamoto@rs.noda.tus.ac.jp

Toshiaki Kakitani
Graduate School of Environmental
and Human Science
Meijo University
Tempaku-ku
Nagoya 468-8502, Japan
kakitani@ccmfs.meijo-u.ac.jp

Yukio Kaneda
Department of Computational
Science and Engineering
Graduate School of Engineering
Nagoya University
Nagoya 464-8603, Japan
kaneda@cse.nagoya-u.ac.jp

Fusao Kawamura
Department of
Mechanical Engineering
Tokyo University of Science
2641 Yamazaki
Noda-shi Chiba 278-8510, Japan
a7501034@rs.noda.tus.ac.jp

Hiroshi Kawamura
Department of
Mechanical Engineering
Tokyo University of Science
2641 Yamazaki, Noda-shi
Chiba 278-8510, Japan
kawa@rs.noda.tus.ac.jp

Daisuke Kihara
Department of Biological
Sciences
Computer Science
Purdue University
West Lafayette, IN
USA
dkihara@purdue.edu

Masanori Kikuchi
Department of
Mechanical Engineering
Faculty of Science and
Technology
Tokyo University of Science
Tokyo, Japan
kik@me.noda.tus.ac.jp

Akihiro Kimura
Department of Physics
Graduate School of Science
Nagoya University
Furo-cho, Chikusa-ku
Nagoya 464-8602, Japan

Eisuke Kita
Graduate School
of Information Sciences
Nagoya University
Nagoya 464-8301, Japan
kita@is.nagoya-u.ac.jp

Nobuaki Koga
Graduate School
of Information Science
Nagoya University
Chikusa-ku
Nagoya 464-8601
Japan
koga@is.nagoya-u.ac.jp

Teruyuki Kojima
Department of Computational
Science and Engineering
Nagoya-shi Chikusa-ku furo-cho
Nagoya University
Japan
kojima@cmplx.cse.nagoya-u.ac.jp

Naoaki Kondo
Department of Physics
Tokyo University of Science
1-3 Kagurazaka, Shinjuku-ku
Tokyo 162-8601
Japan

and

CREST
Japan Science
and Technology Agency
4-1-8 Honcho Kawaguchi
Saitama 332-0012
Japan

Nobuhiko Kondo
Department of Information
and Physical Sciences
Osaka University
Suita 565-0871
Japan
nobuhiko@ist.osaka-u.ac.jp

Noriyuki Kurita
Department of Knowledge-based
Information Engineering
Toyohashi University of Technology
Tenpaku-cho
Toyohashi 441-8580, Japan

Shou Kuroda
Nagoya University
Furo-cho, Chikusa-ku
Nagoya 464-8603, Japan
kuroda@cmplx.cse.nagoya-u.ac.jp

Michael A. Leschziner
Department of Aeronautics
Imperial College London
Prince Consort Road
South Kensington, London
SW7 2AZ
UK
mike.leschziner@imperial.ac.uk

Yaakov Levy
Center for Theoretical
Biological Physics
Department of Physics
University of California
at San Diego
9500 Gilman Drive
La Jolla, CA 92093
USA

Present address

Department of Structural Biology
Weizmann Institute of Science
Rehovot 76100, Israel
klevy@physics.ucsd.edu

Bin Li
Department of Computer Science
Purdue University
West Lafayette, IN
USA
lib@purdue.edu

Ning Li
Department of Aeronautics
Imperial College London
Prince Consort Road
South Kensington, London
SW7 2AZ, UK

Stan Luban
Department of Biological Sciences
Purdue University
West Lafayette, IN
USA
sluban@purdue.edu

Tetsuya Maeshiro
University of Tsukuba
1-1-1, Tennodai
Tsukuba Ibaraki, Japan

and

ATR Network Informatics
Laboratories
Kyoto 691-0288, Japan
maeshiro@slis.tsukuba.ac.jp

Yoshiyuki Matsumura
Shinshu University
Tokida, Ueda
Nagano 386-8567, Japan
matsumu@shinshu-u.ac.jp

Mitsunori Miki
Department of Knowledge
Engineering and Computer Science
Doshisha University
Kyoto, Japan
mmiki@mail.doshisha.ac.jp

Hisaya Mine
Department of Information and
Computer Sciences
Faculty of Engineering Science
Osaka University
Osaka, Japan
h-mine@ics.es.osaka-u.ac.jp

Shigeki Mitaku
Department of Applied Physics
Graduate School of Engineering
Nagoya University
Furocho, Chikusa-ku
Nagoya 464-8603, Japan
imai@bp.nuap.nagoya-u.ac.jp

Koichi Miura
Nagoya University
Chikusa-ku
Nagoya 464-8601, Japan

Yoshiyuki Miyamoto
Fundamental and Environmental
Research Laboratories
NEC, 34 Miyukigaoka
Tsukuba 305-8501, Japan
y-miyamoto@ce.jp.nec.com

Masaru Miyao
Information Technology Center
Nagoya University
Chikusa-ku
Nagoya 464-8601, Japan

Katsuhiro Miyashita
Department of
Mechanical Engineering
Tokyo University of Science
2641 Yamazaki, Noda-shi
Chiba 278-8510
Japan

Toshio Miyauchi
Tokyo Institute of Technology
Japan
tmiyauch@mes.titech.ac.jp

Yoshinori Mizuno
Department of Computational
Science and Engineering
Nagoya University
Chikusa-ku
Nagoya 464-8602, Japan
mizuno@fcs.coe.nagoya-u.ac.jp

Takayuki Morimoto
Nagoya University
Chikusa-ku
Nagoya 464-8601, Japan
morimot@fcs.coe.nagoya-u.ac.jp

Yoshikazu Murayama
University of Birmingham
Edgbaston, Birmingham
B15 2TT, UK
Y.Matsumura@cs.bham.ac.uk

Yuzuru Nada
Chiba Institute of Science
Japan
ynada@cis.ac.jp

Yasutaka Nagano
Department of
Mechanical Engineering
Nagoya Institute of Technology
Japan
nagano@nitech.ac.jp

Masataka Nagaoka
Graduate School of
Information Science
Nagoya University
Nagoya 464-8603
Japan
mnagaoka@is.nagoya-u.ac.jp

Meiho Nakayama
Aichi Medical University
Nagakute
Aichi 480-1195
Japan
nakayama@aichi-med-u.ac.jp

Takayuki Natsume
Department of Knowledge-based
Information Engineering
Toyohashi University of Technology
Tenpaku-cho
Toyohashi 441-8580, Japan
natsume@theo.tutkie.tut.ac.jp

Hirotaka Nishioka
Department of Physics
Graduate School of Science
Nagoya University
Furo-cho, Chikusa-ku
Nagoya 464-8602, Japan

Takashi Noda
Graduate School of Information
Science and Technology
Osaka University
Japan
ntakashi@ist.osaka-u.ac.jp

Kohei Okamoto
Graduate School of
Environment Studies
Nagoya University
Chikusa-ku
Nagoya 464-8601, Japan
h44540a@cc.nagoya-u.ac.jp

Naoya Okamoto
Department of Computational
Science and Engineering
Graduate School of Engineering
Nagoya University
Nagoya 464-8603, Japan
okamoto@fluid.cse.nagoya-u.ac.jp

José N. Onuchic
Center for Theoretical Biological
Physics
Department of Physics
University of California at San Diego
9500 Gilman Drive
La Jolla, CA 92093
USA
jonuchic@ucsd.edu

Takuma Osawa
Division of Mechanical Engineering
School of Science and Technology
Tokyo University of Science
Tokyo, Japan
a7501020@rs.noda.tus.ac.jp

Karthik Ramani
Mechanical Engineering
Purdue University
West Lafayette, IN
USA
ramani@purdue.edu

Takahiro Sakurai
Nagoya University
Chikusa-ku
Nagoya 464-8601, Japan

Masaki Sasai
Department of Computational
Science and Engineering
Graduate School of Engineering
Nagoya University
Furocho, Chikusa
Nagoya 464-8603
Japan
sasai@cse.nagoya-u.ac.jp;
sasai@tbp.cse.nagoya-u.ac.jp

Kai Schneider
MSNM–CNRS and CMI
Université de Provence
Marseille, France
kschneid@cmi.univ-mrs.fr

Yohji Seki
Department of
Mechanical Engineering
Tokyo University of Science
2641 Yamazaki, Noda-shi
Chiba 278-8510
Japan
a7598074@rs.noda.tus.ac.jp

Katsunori Shimohara
ATR Network Informatics
Laboratories
Kyoto 691-0288
Japan
katsu@atr.jp

Masafumi Shionyu
Department of Bio-Science
Faculty of Bio-Science
Nagahama Institute of
Bio-Science and Technology
1266, Tamura-cho, Nagahama
Shiga 526-0829, Japan
m_shionyu@nagahama-i.bio.ac.jp

Tomoki Shiozawa
Nihon University
1-2-1 Izumicho
Narashino-shi
Chiba 275-8575, Japan
tsuna@cit.nihon-u.ac.jp

Tomoki Youkey Shiozawa
Department of Environmental
Health and Preventive Medicine
Nihon University School of Medicine
3-51-7 Daita, Setagaya-ku
Tokyo 155-0033, Japan
shiozawa@med.nihon-u.ac.jp

Nobuhiro Shiwaku
Tokyo Institute of Technology
Japan
nshiwaku@navier.mes.titech.ac.jp

Tomohiro Sogabe
Department of Computational
Science and Engineering
Nagoya University
Furo-cho, Chikusa-ku
Nagoya 464-8603, Japan
sogabe@na.cse.nagoya-u.ac.jp

Yingwen Song
Department of Materials Engineering
The University of Tokyo
7-3-1 Hongo, Bunkyo-ku
Tokyo 113-8656, Japan

and

CREST
Japan Science
and Technology Agency
4-1-8 Honcho Kawaguchi
Saitama 332-0012, Japan

Masashi Sonoyama
Department of Applied Physics
Graduate School of Engineering
Nagoya University
Furocho, Chikusa-ku
Nagoya 464-8603
Japan
imai@bp.nuap.nagoya-u.ac.jp

Cherumuttathu H. Suresh
Computational Modeling
and Simulation Section
Regional Research
Laboratory (CSIR)
India
sureshch@csrrltrd.ren.nic.in

Ryoji Suzuki
Department of Materials Engineering
The University of Tokyo
Tokyo 113-8656
Japan

and

CREST
Japan Science and Technology
Agency
Saitama 332-0012, Japan

Kazuhiko Suga
Toyota Central R&D
Laboratories Inc.
Nagakute
Aichi 480-1192
Japan
suga@me.osakafer-u.ac.jp

Norihiko Sugimoto
Graduate School of Engineering
Nagoya University
Furo-cho, Chikusa-ku
Nagoya 464-8603, Japan
nori2@hpc.itc.nagoya-u.ac.jp

Yoko Suzuki
Department of Physics
Faculty of Physical Sciences
and Engineering
Meisei University
2-1-1 Hodokubo, Hino-shi
Tokyo 191-8506
Japan
youko@phys.meisei-u.ac.jp

Tomofumi Tada
Department of Materials Engineering
The University of Tokyo
Tokyo 113-8656
Japan

and

CREST
Japan Science and Technology
Agency
Saitama 332-0012
Japan

Hiroki Takada
Information Technology Center
Nagoya University
Chikusa-Ku, Nagoya-shi
Aichi 464-8601
Japan
takada@fcs.coe.nagoya-u.ac.jp

Shoji Takada
Department of Chemistry
Faculty of Science
Kobe University
Nada Kobe 657-8501, Japan

and

Core Research for Evolutionary
Science and Technology
Japan Science and
Technology Agency
Nada
Kobe 657-8501, Japan
stakada@kobe-u.ac.jp

Akiyuki Takahashi
Department of
Mechanical Engineering
Faculty of Science and Technology
Tokyo University of Science
Tokyo, Japan
takahash@me.noda.tus.ac.jp

Mamoru Tanahashi
Tokyo Institute of Technology
Japan
mtanahas@mes.titech.ac.jp

Shozo Tanaka
Graduate School
of Information Science
Nagoya University
Chikusa-ku
Nagoya 464-8601, Japan
s.tanaka@tc.kawai-juku.ac.jp

Yukio Tanaka
Department of Applied Physics
Nagoya University
Nagoya 464-8603, Japan

and

CREST
Japan Science and
Technology Agency
4-1-8 Honcho Kawaguchi
Saitama 332-0012, Japan
ytanaka@nuap.nagoya-u.ac.jp

Fabrizio Tessicini
Department of Aeronautics
Imperial College London
Prince Consort Road
South Kensington, London
SW7 2AZ, UK

Hideomi Totsuka
Department of Physics
Nihon University
Tokyo 101-8308, Japan

and

CREST
Japan Science and Technology
Agency
Saitama 332-0012, Japan

Hitoshi Tsunashima
Nihon University
1-2-1 Izumicho
Narashino-shi
Chiba 275-8575
Japan
tsuna@cit.nihon-u.ac.jp

Ichiro Ueno
Department of
Mechanical of Engineering
Faculty of Science and Technology
Tokyo University of Science
Tokyo, Japan
ich@rs.noda.tus.ac.jp

Kazuto Ueno
Graduate School of Engineering
Nagoya University
Nagoya, Japan
ueno@fcs.coe.nagoya-u.ac.jp

Katsuji Uosaki
Department of Information
and Physical Sciences
Osaka University
Suita 565-0871
Japan
uosaki@ist.osaka-u.ac.jp

Tomohiro Ushikubo
Department of Computational
Science and Engineering
Graduate School of Engineering
Nagoya University
Furocho, Chikusa
Nagoya 464-8603
Japan
ushikubo@tbp.cse.nagoya-u.ac.jp

Kazuyuki Watanabe
Department of Physics
Tokyo University of Science
1-3 Kagurazaka, Shinjuku-ku
Tokyo 162-8601, Japan

and

CREST
Japan Science and Technology
Agency
4-1-8 Honcho Kawaguchi
Saitama 332-0012, Japan

Satoshi Watanabe
Department of Materials Engineering
The University of Tokyo
Tokyo 113-8656
Japan

and

CREST
Japan Science and Technology
Agency
Saitama 332-0012
Japan
watanabe@cello.t.u-tokyo.ac.jp

Tsutomu Watanabe
Department of Applied Physics
Nagoya University
Japan
h042203d@mbox.nagoya-u.ac.jp

Kenta Yamada
Graduate School
of Information Science
Nagoya University
Chikusa-ku
Nagoya 464-8601
Japan
ykenta@nagoya-u.jp

Kosuke Yamamoto
Department of Computational
Science and Engineering
Nagoya University
Nagoya-shi, Chikusa-ku furo-cho
Nagoya University
yamamoto@cmplx.cse.
nagoya-u.ac.jp

Takahiro Yamamoto
Department of Physics
Tokyo University of Science
1-3 Kagurazaka, Shinjuku-ku
Tokyo 162-8601, Japan

and

CREST
Japan Science and Technology
Agency
4-1-8 Honcho Kawaguchi
Saitama 332-0012, Japan
takahiro@rs.kagu.tus.ac.jp

Daisuke Yamashiro
Department of Computational
Science and Engineering
Nagoya University
Furo-cho, Chikusa-ku
Nagoya 464-8603, Japan
daisuke@cmplx.cse.nagoya-u.ac.jp

Ichiro Yamato
Department of Biological Science
and Technology
Tokyo University of Science
2641 Yamazaki
Noda-shi Chiba 278-8510
Japan
iyamato@rs.noda.tus.ac.jp

Takahisa Yamato
Department of Physics
Graduate School of Science
Nagoya University
Furo-cho, Chikusa-ku
Nagoya 464-8602, Japan

Kenji Yamaura
Department of
Mechanical Engineering
Faculty of Science and Technology
Tokyo University of Science
Tokyo, Japan
chivo@me.noda.tus.ac.jp

Daishi Yamazaki
Nihon University
1-2-1 Izumicho
Narashino-shi
Chiba 275-8575
Japan
c61350@cit.nihon-u.ac.jp

Xin Yao
University of Birmingham
Edgbaston, Birmingham
B15 2TT, UK
Y.Matsumura@cs.bham.ac.uk

Mitsumasa Yoda
Department of Computational
Science and Engineering
Graduate School of Engineering
Nagoya University
Furo-cho, Chikusa
Nagoya 464-8603
Japan
yoda@tbp.cse.nagoya-u.ac.jp

Hisatoshi Yokoyama
Department of Physics
Tohoku University
Japan
yoko@cmpt.phys.tohoku.ac.jp

Takehito Yokoyama
Department of Applied Physics
Nagoya University
Nagoya 464-8603, Japan

and

CREST
Japan Science
and Technology Agency
4-1-8 Honcho Kawaguchi
Saitama 332-0012
Japan
h042224m@mbox.nagoya-u.ac.jp

Yoshiteru Yonetani
CREST JST
Japan Atomic Energy Agency
8-1 Umemidai, Kizu-cho
Soraku-gun
Kyoto 619-0215
Japan
yonetani.yoshiteru@jaea.go.jp

Tomohiro Yoshikawa
Department of Computational
Science and Engineering
Nagoya-shi Chikusa-ku furo-cho
Nagoya University
Japan
tom@fcs.coe.nagoya-u.ac.jp

Katsunori Yoshimatsu
Department of Computational
Science and Engineering
Graduate School of Engineering
Nagoya University
Nagoya 464-8603, Japan
yosimatu@fluid.cse.
nagoya-u.ac.jp

Isseki Yu
Graduate School of Information
Science
Nagoya University
Nagoya 464-8603, Japan
yu@ncube.human.nagoya-u.ac.jp

Kei Yura
Quantum Bioinformatics Team
Center for Computational Science
and Engineering
Japan Atomic Energy Agency
8-1 Umemidai
Kizu-cho, Souraku-gun
Kyoto 619-0215, Japan
yura.kei@jaea.go.jp

Shao-Liang Zhang,
Department of Computational
Science and Engineering
Nagoya University
Furo-cho, Chikusa-ku
Nagoya 464-8603, Japan
zhang@na.cse.nagoya-u.ac.jp

Plenary & Keynote Papers

Computational Methods Combining Large Eddy Simulation with Approximate Wall-Layer Models for Predicting Separated Turbulent Near-Wall Flows

Michael A. Leschziner, Ning Li, and Fabrizio Tessicini

Summary. Two approximate RANS-type near-wall treatments are implemented within LES strategies to the simulation of flow separation from curved surfaces at high Reynolds numbers. One is a hybrid RANS-LES scheme in which the LES field is interfaced, dynamically, with a full RANS solution in the near-wall layer; and the other is a zonal (two-layer) scheme in which the state of the near-wall layer is described by parabolized Navier–Stokes equations which only return the wall shear stress to the LES domain as a wall boundary condition. In both cases, the location of the interface can be chosen freely. The performance of the two methods is illustrated by reference to fully-developed channel flow and two separated flows, one developing over the trailing edge of a hydrofoil and the other around a 3d hill. Comparisons are made against fine-grid LES, experimental data and simulations in which the near-wall layer is treated with log-law-based wall functions.

1 Introduction

After many years of turbulence-model developments and numerous extensive validation campaigns, it is now generally accepted that Reynolds-averaged Navier–Stokes methods do not, as a rule, provide quantitatively reliable predictions of massively separated flows over a broad range of conditions. In recent years, much effort has been directed, in particular, toward second-moment closure and nonlinear eddy-viscosity models, and while some perform well in statistically two-dimensional conditions, most fail to return quantitatively correct predictions in much more complex three-dimensional cases, as demonstrated by Wang et al. [1], among others.

The turbulence state in separated flows is strongly affected by the dynamics of large-scale, energetic motions and associated highly nonlocal processes. This so regardless of whether or not the flow contains periodically shed vortices and Görtler-Taylor-type instabilities – although these add complexity to the turbulence dynamics. Such conditions are fundamentally at odds with the closure assumptions in virtually all (one-point) turbulence models, which are based, in one way or another, on the notion that turbulence is not far from a state of local equilibrium and that inhomogeneities are relatively modest. In all cases, turbulence-model constants are calibrated by reference to simple shear flows, and only in a few cases is the variability of the constants taken into account as the flow departs significantly from local equilibrium.

Against the above background, large eddy simulation (LES) is held to offer the principal route to a reliable prediction of separated flows, especially if the resolution of unsteady features is a specific objective. However, there is an increasing recognition that LES is no panacea and that it faces some major practical challenges. In general, LES is frustratingly expensive, despite advances in computing hardware, and demands high numerical accuracy and high grid quality (low skewness, low cell aspect ratio, low gradation). The problem of resolution and expense is especially acute in near-wall flows, because the range and size of the dynamically important scales diminishes rapidly as the wall is approached. As a result, the resolution requirements of near-wall turbulence rise in proportion to $Re^{1.8}$, relative to $Re^{0.4}$ away from walls. Thus, at a mean Reynolds number of order 10^6, the presence of a wall at which viscous effects are important demands grids of order 10^8–10^9 nodes. This is untenable in a practical environment, and alternative, more economical strategies must be sought.

A particularly challenging type of separation is that from curved surfaces. This is characterized by intermittent, rapidly varying patches of reverse flow and the ejection of large-scale vortices over an area that can extend to several boundary-layer thicknesses in the streamwise direction. The correct resolution of this area can be very important to the overall predictive accuracy of the simulation. Temmerman et al. [2] show, for the case of separation from a ducted 2d hill, that a 1% error in the prediction of the time–mean separation line results, approximately, in a 7% error in the length of the recirculation region. This sensitivity, associated with the representation of the near-wall physics, can be a serious obstacle to the effective utilization of LES in the prediction of separation from continuous(-derivative) walls, and presents a challenging testing ground for any approximate near-wall treatment designed to circumvent a full resolution of the near-wall region.

Approximate near-wall models are based on wall functions or the use of some RANS-type near-wall model that is interfaced with the inner LES solution. The use of equilibrium-flow wall functions goes back to early proposals of Deardorff [3] and Schumann [4], and a number of versions have subsequently been investigated, which are either designed to satisfy the log-law in the time-averaged field or, more frequently, involve an explicit log-law or closely related power-law prescription of the instantaneous near-wall velocity (e.g., Werner and Wengle [5], Hoffman and Benocci [6], Temmerman et al. [2,7]). These can provide useful approximations in conditions not far from equilibrium, but cannot be expected to give a faithful representation of the near-wall layer in separated flow. The alternative of adopting a RANS-type turbulence-model solution for the inner near-wall layer is held to offer a more realistic representation of the near-wall flow in complex flow conditions at cell-aspect ratios much higher than those demanded by wall-resolved simulations.

The best-known realization of the combined RANS-LES concept is Spalart et al. [8] DES method. A key feature of the method is that a single turbulence model – the one-equation Spalart–Allmaras model – is used, both as a RANS model in the inner region and a subgrid-scale model in the outer LES region, with a switching of a length-scale in the model controlling the transition from the RANS to LES at the "interface" location $y_{\text{int}} = \min(y, C_{\text{DES}} \max(\Delta x, \Delta y, \Delta z))$. The scheme is designed to return a RANS solution in attached flow regions and revert to LES once separation is predicted. This is done by arranging the wall-parallel cell dimensions Δx or/and Δz to be much larger than Δy, the consequence being an outward shift of y_{int} away from the wall and a dominance of the RANS scheme. This concept of extensive steady patches co-existing, seamlessly, with unsteady resolved portions raises a whole host of questions and uncertainties, among them:

- How is the statistically-resolved turbulence that exists in an attached upstream flow and is convicted toward and around a separated region interact with the LES-resolved separated region next to it?
- What is the consequence of the spectral content of the upstream flow being ignored on the separation process and the subsequent flow evolution?
- Following reattachment, how does the transition from the resolved to statistical turbulence controlled in any physically meaningful way?

In general flows, the streamwise grid density often needs to be high to achieve adequate resolution of complex geometric and flow features, both close to the wall (e.g., separation and reattachment) and away from the wall. Thus, another problem with DES is that the interface can be forced to move close to the wall, often as near as $y^+ = O(50\text{--}100)$, in which case RANS and LES regions co-exist even in fully attached flow. In such circumstances, it has been repeatedly observed, especially at high Reynolds numbers, that the high turbulent viscosity generated by the turbulence model in the inner region extends, as subgrid-scale viscosity, deeply into the outer LES region, causing severe damping in the resolved motion and a misrepresentation of the resolved structure as well as the time-mean properties.

A hybrid method allowing the RANS near-wall layer to be predefined and to be interfaced with the LES field across a prescribed boundary has recently been proposed by Temmerman et al. [9]. With such a method, one important issue is compatibility of turbulence conditions across the interface; another (related one) is the avoidance of "double-counting" of turbulence effects – that is, the over-estimation of turbulence activity due to the combined effects of modeled and resolved turbulence. This approach, outlined, is one of those used later for simulating separation from curved surfaces. A general problem often observed with this type hybrid scheme is an insufficient level of turbulence activity just beyond the interface, as a consequence of the near-wall RANS model misrepresenting the near-wall (streaky) structure and the fact that the turbulence in the LES region close to the interface is not sufficiently vigorous, because this region is subjected to wrong or distorted structural information at the interface. Several attempts have thus been made to inject synthetic turbulence into the interface in an effort to at least partially recover the influence of the small-scale structures lost by the application of the RANS model. Alternative approaches have been proposed by Piomelli et al. [10], Davidson and Billson [11], Davidson and Dahlstrom [12]. Also, in what follows, a method of "spectral enrichment" is included in a scheme presented later. While these measures have some beneficial effects, in terms of reducing mean-velocity anomalies, they do not – and cannot – cure the most of the defects arising from the inevitable misrepresentation of the turbulence structure near the wall. They are also not practically usable in a general computational environment.

It is arguable that any near-wall approximation that circumvents a detailed resolution of the near-wall structure cannot be expected to return a physically correct spectral state and therefore cannot provide the correct "boundary conditions" for the LES portion above the approximated near-wall layer. It can further be claimed, with some justification, that the most that a near-wall model can be expected to provide is a realistic representation of the wall shear stress, and that this should be the only quantity that is fed into the LES procedure. This is the basis of a second group of approaches, collectively referred to as "zonal schemes." Like hybrid strategies, zonal schemes involve the application of a RANS model in the near-wall layer. However, they involve a more distinct division, both in terms of modeling and numerical treatment, between the near-wall layer and the outer LES region. Such schemes have been proposed and/or investigated by Balaras and Benocci [13], Balaras et al. [14], Cabot and Moin [15] and Wang and Moin [16]. In all these, unsteady forms of the boundary-layer (or thin shear-flow) equations are solved across an inner-layer of prescribed thickness, which is covered with a fine wall-normal mesh, with a mixing-length-type algebraic model providing the eddy viscosity. Computationally, this layer is partially decoupled from the LES region, in so far as the pressure field just outside the inner layer is imposed across the layer, i.e., the pressure is not computed in the layer. The wall-normal velocity is then determined from an explicit application of the mass-continuity constraint, one consequence being a discontinuity in this velocity at the interface. The principal information extracted from the RANS computation is the wall shear stress, which is fed into the LES solution as an unsteady boundary condition.

In this paper, the effectiveness of the hybrid LES-RANS scheme of Temmerman et al. [9] and the zonal scheme of Balaras et al. [14] is investigated in the context of simulating separation from curved surfaces at high Reynolds number. The emphasis is on two particular configurations: a statistically homogeneous flow over the rear portion of a hydrofoil, separating from the upper suction surface, and a three-dimensional flow, separating from the leeward side of a circular hill in a duct.

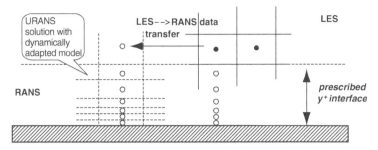

Fig. 1. Schematic of the hybrid LES-RANS scheme

2 Outline of Methods

2.1 Hybrid RANS-LES Scheme

The principles of the hybrid scheme are conveyed in Fig. 1, which also indicates the numerical coupling between the RANS and LES portions.

The thickness of the near-wall layer may be chosen freely, although in applications to follow, the layer is simply bounded by a particular wall-parallel grid surface. The LES and RANS regions are bridged at the interface by interchanging velocities, modelled turbulence energy and turbulent viscosity, the last being subject to the continuity constraint across the interface,

$$\nu_{\text{SGS}}^{\text{LES}} = \nu_t^{\text{RANS}}. \tag{1}$$

The turbulent viscosity can be determined, in principle, from any turbulence model. In the case of a two-equation model,

$$\nu_t = C_\mu f_\mu \frac{k^2}{\varepsilon}, \tag{2}$$

and matching the subgrid-scale viscosity to the RANS viscosity at the interface is effected by:

$$C_\mu = \frac{\langle f_\mu \left(k^2/\varepsilon\right) \nu_{\text{SGS}} \rangle}{\langle (f_\mu \left(k^2/\varepsilon\right))^2 \rangle}, \tag{3}$$

where $<...>$ denotes averaging across any homogeneous direction, or over a predefined patch, in case no such direction exists. An analogous approach may be taken with any other eddy-viscosity model. With the interface C_μ determined, the distribution across the RANS layer is needed. Temmerman et al. [9] investigate several sensible possibilities, and the one adopted here, based on arguments provided in the aforementioned study, is:

$$C_\mu(y) = 0.09 + (C_{\mu,\text{int}} - 0.09) \frac{(1 - e^{-y/\Delta})}{(1 - e^{-y_{\text{int}}/\Delta_{\text{int}}})}. \tag{4}$$

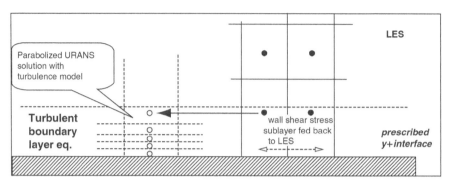

Fig. 2. Schematic of the (two-layer) zonal scheme

2.2 Zonal Scheme

As noted earlier, the objective of the zonal strategy is to provide the LES region with the wall-shear stress, extracted from a separate modeling process applied to the near-wall layer. The wall-shear stress can be determined from an algebraic law-of-the-wall model or from differential equations solved on a near-wall-layer grid refined in the wall-normal direction. The method is shown schematically in Fig. 2, was originally proposed by Balaras and Benocci [13] and tested by Balaras et al. [14] and Wang and Moin [16] to calculate the flow over the trailing edge of a hydrofoil with an immersed-boundary method.

At solid boundaries, the LES equations are solved up to a near-wall node which is located, typically, at $y^+ = 50$. From this node to the wall, a refined mesh is embedded into the main flow, and the following simplified turbulent boundary-layer equations are solved:

$$\frac{\partial \rho \tilde{U}_i}{\partial t} + \frac{\partial \rho \tilde{U}_i \tilde{U}_j}{\partial x_j} + \frac{d\tilde{P}}{dx_i} = \frac{\partial}{\partial y}\left[(\mu + \mu_t)\frac{\partial \tilde{U}_i}{\partial y}\right] \quad i = 1, 3, \qquad (5)$$

where y denotes the direction normal to the wall and i identify the wall-parallel directions (1 and 3). In the present study, only the pressure-gradient term has been included in the left-hand side of (5), this left-hand side being identified later by F_i. The effects of including the remaining terms is the subject of ongoing studies investigated and will be reported in a future paper. The eddy viscosity μ_t is here obtained from a mixing-length model with near-wall damping:

$$\frac{\mu_t}{\mu} = \kappa y_w^+ (1 - e^{-y_w^+/A})^2. \qquad (6)$$

The boundary conditions for (5) are given by the unsteady outer-layer solution at the first grid node outside the wall layer and the no-slip condition at $y = 0$. Since the friction velocity is required in (6) to determine y^+ (which depends, in turn, on the wall-shear stress given by (5)), an iterative procedure had to be implemented wherein μ_t is calculated from (6), followed by an integration of (5).

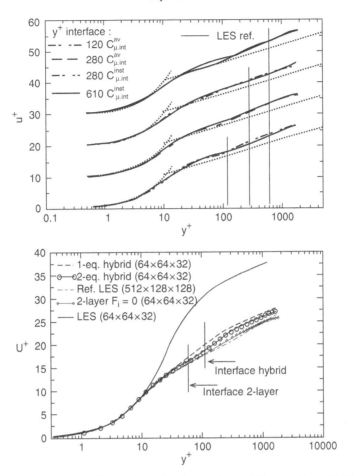

Fig. 3. Velocity profiles in fully-developed channel flow at $Re = 42,000$

3 Illustrative Applications

Extensive testing has been done for fully-developed channel flows, but much of this cannot be reported herein (see Temmerman et al. [9]). Some examples are shown in Fig. 3, all results relating to flow at $Re = 42,000$ ($Re_\tau = 2,000$).

The upper plot illustrates, for the hybrid RANS-LES scheme, a weak sensitivity to the location of the interface and the benefit arising from the use of the *instantaneous* C_μ in (3) (i.e., C_μ^{inst}), relative to a spanwise-averaged value (C_μ^{av}), the former constituting a form of "spectral enrichment." The lower plot contains results arising from the hybrid scheme, with the RANS model based on either a one-equation k–l or a two-equation k–ε model, the zonal two-layer model and a reference simulation obtained with $512 \times 128 \times 128$ grid, relative simulations with both near-wall approximations and a pure (under-resolved) LES with a $64 \times 32 \times 32$.

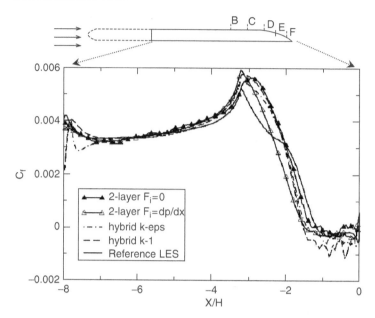

Fig. 4. Friction factor for a separated flow along the rear upper side of hydrofoil (F_i represents the terms on the left-hand side of (5) that have been accounted for in the zonal implementation)

Sample results for a flow separating from the upper side of an asymmetric trailing edge of a hydrofoil, at a Reynolds number 2.15×10^6, based on free stream velocity U_∞ and the hydrofoil chord, are shown in Fig. 4. Simulations were performed over the rear-most 38% of the hydrofoil chord, using inflow conditions taken from Wang and Moin [16], over a $512 \times 128 \times 24$ grid, with the RANS-LES interface on the horizontal upper side at $y^+ = 60$ and 40 for the hybrid and zonal schemes, respectively. These are compared to a wall-resolved LES performed by Wang and Moin with a mesh of $1{,}536 \times 48 \times 96$ nodes.

It needs to be appreciated that the skin friction is an especially sensitive parameter, being linked to the velocity derivative at the wall. All formulations return reasonable agreement with the reference LES data, although there are some not insignificant variations downstream of the start of the curved section, where wall curvature experiences a sudden change. Especially interesting is the observation that the inclusion of the pressure gradient in the zonal formulation (i.e., $F_i = \partial P/\partial x_i$) results in the effects of the sudden change of wall curvature being reproduced. Also, as might be expected, this formulation returns earlier separation relative to the implementation from which the pressure gradient has been omitted ($F_i = 0$).

The third illustrative application is a 3d circular hill, of height-to-base ratio of 4, is located on one wall of a duct, as shown in the inset of Fig. 5. This flow, at a Reynolds number of 130,000 (based on hill height and duct velocity) has been the subject of extensive experimental studies by Simpson

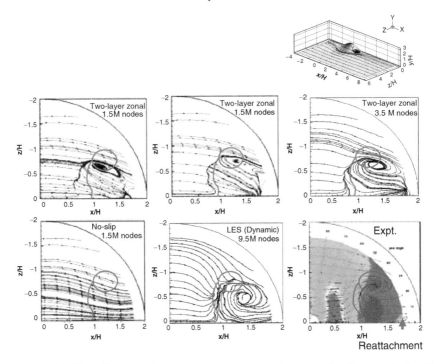

Fig. 5. 3d-Hill-surface topology maps (on rear quarter of surface) computed with different grids, the coarser two using the zonal two-layer model

et al. [17] and Byun and Simpson [18]. The size of the computational domain is $16H \times 3.205H \times 11.67H$, with H being the hill height. The hill crest is $4H$ downstream of the inlet plane. The inlet boundary layer, at $-4H$, was generated by a combination of a RANS and LES precursor calculations, the former matching the experimental mean-flow data and the latter providing the spectral content.

Simulations were undertaken with meshes containing between 1.5 and 9.5 million nodes, the finest-grid simulation approaching full wall resolution. Coarser grids were used in conjunction with the zonal scheme, wherein the interface was placed within $y^+ = 20$–40 and 40–60, using 3.5 and 1.5 million nodes, respectively.

Figure 5 shows predicted hill-surface topology maps, identifying the separation patterns that are returned by the various simulations. The designation "LES (Dynamic)" is to signify the use of the dynamic Smagorinsky model in the pure LES performed on the finest grid used. The designation "No-slip" indicates that no wall model was used on this coarsest-grid computation – i.e., this is again a pure LES computation, although one that is far from being wall-resolving. The inserted circles and connected curved lines indicate, respectively, the location of experimental focal point of the vortex and the

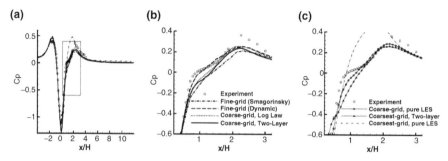

Fig. 6. Pressure-coefficient distributions on the center-line along the surface of the 3d hill

separation line on the surface. The most important feature shown by Fig. 5 is that no separation at all is predicted with the 1.5 million-node grid when a no-slip condition is applied, while a fair representation of the separation pattern is returned with the zonal scheme.

Figure 6 shows predicted surface-pressure-coefficient distributions along the hill center line. "Coarse-grid" and "Coarsest-grid" signify 3.5 and 1.5 million nodes, respectively, while "Log-Law" indicates that a log-law-based wall function has been used in the related simulation, included for comparison purposes. Most computed results agree fairly well with the experimental distribution, except the one arising from the 1.5-million-node solution with no-slip condition applied (i.e., a pure LES on a grid that does not resolve the near-wall region). As shown in Fig. 5, this computation fails to resolve the separation, thus also returning an entirely wrong pressure-recovery behavior.

Finally, Fig. 7 compares the experimentally recorded velocity field in the center plane bisecting the hill with a simulation performed on the coarse (3.5-million-node) grid with the zonal model. A characteristic feature of the flow is a thin recirculation region located close to the hill surface, followed by reattachment close to the foot of the hill. Wang et al. [1] show that RANS

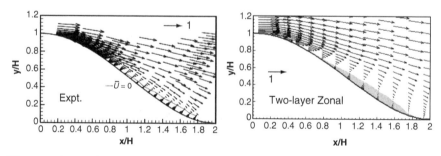

Fig. 7. Velocity field in center plane bisecting the hill, prediction on r.h.s. plot with 3.5 million nodes (note thin reverse-flow region at the leeward side of the hill)

models, at whatever level of sophistication, seriously over-estimate the size of and flow intensity in this region. As seen, the simulation with the zonal model returns a creditable representation of the reverse-flow region, and this is consistent with the favorable result for the pressure coefficient. Use of the no-slip condition on this grid results in a far too thin and short reverse-flow layer (not shown here). A more detailed exposition may be found in Tessicini et al. [19, 20].

4 Conclusions

Two approaches to approximating the near-wall region within a combined LES-RANS strategy have been introduced in the context of simulating high-Reynolds-number flows involving separation from curved surfaces. One of the two – the hybrid RANS-LES scheme – performs well in channel flow, but has been found, in separated flow, to be rather sensitive to the aspect ratio of the grid around the interface and to give rather poor result for the separation process when the aspect ratio is excessively high. This is not an unexpected outcome, and it suggests that the grid around the interface needs to adhere to known LES constraints, in terms of aspect ratio, in particular.

The two-layer zonal approach has been implemented herein in a simplified form, namely with advection ignored in the solution of the RANS equations in the near-wall layer. With pressure gradient also omitted, as done in some simulations, this constrains the zonal method to returning instantaneous representations of the log-law in the near-wall region. Thus, as expected, these simulations display a behavior close to one in which log-law-based wall functions are imposed explicitly in the course of the simulation. Both implementations have been found to return surprisingly good results for both separated flows examined. In particular, the separation behavior is broadly correctly predicted, and the consequence is a generally realistic representation of other flow features. These observations suggest that the detailed representation of the turbulence processes close to the wall may not be as important as is generally assumed, and that the most important issue is the balance between the wall shear stress and the pressure gradient that provokes separation.

References

1. Wang C, Jang YJ, Leschziner MA (2004) Int. J. Heat Fluid Flow 25:499–512
2. Temmerman L, Leschziner MA, Mellen CP, Froehlich J (2003) Int. J. Heat Fluid Flow 24:157–180
3. Deardorff JW (1970) J. Fluid Mech. 41:453–480
4. Schumann U (1975) J. Comput. Phys. 18:376–404
5. Werner H, Wengle H (1991) Large-eddy simulation of turbulent flow over and around a cube in a plate channel, 8th Symposium on Turbulent Shear Flows pp. 155–168

6. Hoffmann G, Benocci C (1995) Approximate wall boundary conditions for large-eddy simulations, in Advance in Turbulence V, Benzi, R. (ed.) pp. 222–228
7. Temmerman L, Leschziner MA, Hanjalić K (2002) A-priori studies of a near-wall RANS model within a hybrid LES/RANS scheme, in Engineering Turbulence Modelling and Experiments, Rodi, W. and Fueyo, N. (eds.), Elsevier, Amsterdam, pp. 317–326
8. Spalart PR, Jou W-H, Strelets M, Allmaras SR (1997) Comments on the feasibility of LES for wings and on the hybrid RANS/LES approach, in Advances in DNS/LES, 1st AFOSR International Conference on DNS/LES, Greyden Press pp. 137–148
9. Temmerman L, Hadziabdic M, Leschziner MA, Hanjalic K (2005) Int. J. Heat Fluid Flow 26:173–190
10. Piomelli U, Balaras E, Pasinato H, Squires K, Spalart P (2003) Int. J. Heat Fluid Flow 24:538–550
11. Davidson L, Billson M (2006) Int. J. Heat Fluid Flow 27(6):1028–1042
12. Davidson L, Dahlstrom S (2004) Hybrid LES-RANS: an approach to make LES applicable at high Reynolds number, in Procs CHT-04 Advances in Computational Heat Transfer III, G. de Vahl Davis and E. Leonardi (eds.), Norway, April 2004
13. Balaras E, Benocci C (1994) Subgrid-scale models in finite difference simulations of complex wall bounded flows, in Applications of Direct and Large Eddy Simulation, AGARD pp. 2.1–2.6
14. Balaras E, Benocci C, Piomelli U (1996) AIAA J. 34(6):1111–1119
15. Cabot W, Moin P (1999) Flow, Turb. Combust. 63:269–291
16. Wang M, Moin P (2002) Phys. Fluids 14:2043–2051
17. Simpson RL, Long CH, Byun G (2002) Int. J. Heat Fluid Flow 140:233–258
18. Byun G, Simpson RL (2005) Structure of three-dimensional separated flow on an axisymmetric bump, AIAA Paper 2005-0113
19. Tessicini F, Temmerman L, Leschziner MA (2006) Int. J. Heat Fluid Flow 27:789–799
20. Tessicini F, Li N, Leschziner MA (2007) Large eddy simulation of three-dimensional flow around a hill-shaped obstruction with a zonal near-wall approximation, Int. J Heat and Fluid Flow (to appear)

Numerical Experimentation: A Third Way to Study Nature

Marie Farge

Summary. We will outline the history of the numerical approach and trace back the origins of the use of computers to carry out simulations in mathematics and physics. We will then present the techniques used, by taking as example the finite-difference method to solve PDEs and discuss the nature and impact of numerical errors. Finally, we will argue that numerical simulation pertains more to experiment than to theory.

> *'Although this may seem a paradox,*
> *all exact science is dominated*
> *by the idea of approximation'*
>
> (Bertrand Russell, The Scientific Outlook, 1931)

1 Historical Sketch

The numerical approach goes back much further than the appearance of the first computers. In a paper submitted in 1822 [1], Charles Babbage already suggested using numerical machines to calculate astronomical tables. These machines were made up of an 'attic', where data were stored, and of a 'mill', where calculation took place. However, this consisted merely in numerically evaluating some solutions, already known analytically, and not, in fact, performing simulations in the modern sense. Numerical simulation is defined as solving the equations that describe the physical laws governing the system studied by using algorithms. One can trace its origin back to the year 1899, with the development of the finite difference method by Sheppard [2]. It was then developed by Richardson [3] who used it, from 1910 onwards to calculate the stress exerted upon a dam. Richardson later had the idea of numerically solving the equations of atmospheric dynamics in order to predict the weather. He designed for this a finite difference scheme, which now bears his name, and applied it to find out the atmospheric situation on May 20th 1910. After six weeks of hard work he had only managed to compute the state of the atmosphere for two vertical columns, and the values he was obtaining then, for the wind and pressure fields, were already quite different from those observed. He then realized that to perform this calculation by hand, at the same speed as the current atmospheric evolution (which is the minimum required to have prediction rather than 'postdiction'), it would require 64,000 people to get a new state of the atmosphere every 3 h using a computational grid of $200 \times 200\,\text{km}^2$ over the whole Earth. In a book published in 1922 [4] he went as far as to imagine a city with a huge theater at its center where 'a myriad computers are at work upon the weather of the part of the map where each sits, but each computer attends only the one equation or part of an equation'. Each one at a grid point, under the leadership of one person in charge of synchronizing the computation 'like the conductor of an orchestra in which the instruments are slide-rules and calculating machines!'. Such a project never came to pass, which is just as well since O'Brien et al. [5] showed in 1950 that Richardson's scheme is unconditionally unstable. The mathematical justification of the numerical approach was given by Courant, Friedrichs and Lewy in a paper published in 1928 [6]. They proved that discrete equations actually constitute an approximation of continuous partial differential equations (PDEs), as long as some stability conditions, known as 'CFL conditions', are satisfied.

Numerical simulation, in its modern sense, implies the use of computers to carry out calculations. Computers appeared at the beginning of the Forties. The first were the Z2 and the Z3, built in Germany by Konrad Zusse in 1939 and 1941, respectively, but they were destroyed during the war. Independently Turing at Cambridge, England, was developing calculators to decipher codes, but unfortunately most of the information about his work is still classified. The first computer built across the Atlantic was the Mark I, which was completed by Aiken in 1944 at Harvard University, with the help of IBM. Meanwhile, at the University of Pennsylvania, Mauchly and Eckert were designing the ENIAC, which was installed in 1946 at the Ballistic Research Laboratories of Aberdeen, Maryland. They then founded the company Univac, which in 1951 was the first to launch a computer on the market.

The first two people to have foreseen the impact that computers could have in mathematics and physics were von Neumann and Ulam. The latter recounts in his autobiography [7] that: 'Almost immediately after the war Johny and I also began to discuss the possibilities of using computers heuristically to try to obtain insights into questions of pure mathematics. By producing examples and by observing the properties of special mathematical objects one could hope to obtain clues as to the behavior of general statements which have been tested on examples'. He went on to show, for the sake of example, how numerical experiments could help study the regularity of some solutions to PDEs, and concludes by explaining that: 'In the following years in a number of published papers, I have suggested – and in some cases solved – a variety of problems in pure mathematics by such experimenting or even merely observing'. Thus, right from the beginning, two of the essential contributions of computers to research in mathematics and physics were already foreseen, namely the possibility of experimenting with equations and of 'seeing' their solutions. These two points will be illustrated in the third part.

During the Second World War, part of von Neumann's research in Los Alamos was about the development of numerical schemes and the definition of a stability criterion, which bears his name, allowing one to choose optimal space and time steps when discretising parabolic equations with a finite difference scheme. Von Neumann never published this work (which can, nevertheless, be consulted in his complete works [8]); indeed, he deemed it too approximate, since it did not take into account, either the nonlinear terms, or the boundary conditions. Nonetheless, such a method is still widely used and is being developed, yielding very general extensions where the discretization of the boundary conditions plays a role. In 1949 von Neumann and Richtmyer conceived a numerical technique to calculate the shocks which appear in compressible flows [9]; at the time, the problem was still out of reach analytically due to the presence of singularities. The method which they adopted consisted in introducing some dissipative terms in the equations to spread the shocks and creates boundary layers that were at least of the size of the computation in order to smooth out any singularities. The technique is still very much in use today. Von Neumann then set out, along with Charney and Fjörtoft, to integrate numerically an atmospheric barotropic circulation model, i.e. which neglects the temperature variations along the isobar surfaces. This model used the vorticity equation instead of the primitive equations with pressure–velocity form, and suppressed the flow to be quasi-geostrophic – i.e. with a stable horizontal stratification – and nondivergent, i.e. supposing the fluid to be incompressible. The computational grid was of 15 times 18 grid points and only covered the US, which corresponded to a space resolution of 736 km. The numerical resolution, carried out in 1950 on the ENIAC of Aberdeen, made it possible to calculate a 24 h meteorological evolution on the 5th, the 30th and the 31st of January, and of the 13th of February 1949. In their article [10], the authors made the following remark: 'It may be of interest to remark that the computation time for a 24-h forecast was about 24 h, that is, we were just able to keep pace with the weather. However, much of this time was consumed by manual and I.B.M. operations, namely by reading, printing, reproducing, sorting, and interfiling of punch cards. In the course of the four 24-h forecasts about 100,000 standard I.B.M. punch cards were produced and 1,000,000 multiplications and divisions were performed'. The authors then compared the predicted fields with those obtained and endeavoured to account for the prevision errors that were observed. According to them, they were due to too coarse a computational grid and to the fact that the baroclinic effects (i.e. the pressure variations) had been disregarded. This first numerical simulation, performed in collaboration with Smagorinsky from the US Weather Bureau and Mrs. von Neumann who was programming the ENIAC, is at the origin of the present dynamical models used to study meteorology and climatology, e.g. general circulation model (GCM) which take into account the whole planet.

In the beginning of the 1950s John von Neumann completed his computer, called EDVAC, at the Institute for Advanced Studies in Princeton and then built a second one, which was installed by Metropolis in Los Alamos where it remained in use until 1971. On this latter machine, von Neumann and Fermi undertook the simulations of the first hydrogen bomb, and particularly the study of hydrodynamic instabilities such as the Rayleigh–Taylor instability [11], as well as the calculation of neutron cross-sections. On this occasion Ulam proposed the Monte-Carlo method, for which, in order to solve problems involving a large number of particles, one chooses a subset of statistical samples instead of studying all possible configurations. This technique, which is widely used nowadays, does not give an exact solution, but merely an estimation for a given error, and it allows the resolution of problems that would otherwise remain unsolved.

From very early on, Fermi has foreseen the use of computers and he undertook, along with Pasta and Ulam, to study numerically the evolution of a system of interacting particles that was marginally nonlinear. To their great surprise, they discovered that, instead of leading to energy equipartition, on the contrary, the system presented quasi-periodical solutions, which contradicted the ergodic hypothesis of statistic mechanics [12]. At the beginning of the 1960s, Kruskal and Zabusky took over the same research program but considered a quadratic rather than a cubic nonlinear interaction; they showed that the system was described by the Korteveg-de Vries equation. They integrated it numerically and found wavelike solutions, whose behaviour recalled that of the particles (since they preserve their shape and their velocity after interaction). They called them 'solitons' [13–15]. These numerical experiments paved the way for a whole row of new problems concerning nonergodic dynamical systems where are still at the frontier of statistical physics.

To conclude this first part, in which we have tried to cast light upon some of the works grounding the numerical approach, we would like to quote Ulam once more. His vision of mathematics is after all couched in a language that does somehow recall that of Thom [7]: 'The recent study of the mathematics of morphogenesis and the possibility of studying experimentally on the computer the dynamics of competitions and conflicts between different geometrical configurations, on the model of life's struggle, could lead to new mathematical concepts. [...] The use of the computer seems, not only practical, but absolutely essential for such experiments which require to follow those games and fights over a large number of stages and steps. I believe that the experience gained by following the evolution of such processes will have a fundamental influence that may, one day, finally generalise, or even replace, in mathematics the exclusive immersion in formal axiomatics we presently have'.

2 Simulation or Simulacrum?

In a footnote from his book 'Stabilité Structurelle et Morphogénèse', René Thom remarks: 'classical Euclidian geometry can be considered as magic. At the price of a minimal distortion of appearances (the point with no extension, the straight line without any thickness), the formal language of geometry adequately describes spatial reality. In this sense one could say that geometry is successful magic'. We believe that numerical simulation can also be considered as 'successful magic', at the cost, this time, of a minute distortion of the equations, this distortion being exerted the other way around. Indeed, this method consists in replacing the differential equations by discretized algebraic equations; and one notices – here comes the magic – that their numerical processing allows to reach solutions which adequately describe the physical reality. However, such an astonishment is not so much the lot of the sole numerical approach as that of the entire physics: 'The most incomprehensible thing of the world is that it is comprehensible'(Einstein).

A numerical simulation unfolds in five stages:

1. *Defining the problem.* The phenomenon (e.g. physical, biological, economical) to be studied is defined and the questions the numerical simulation aims to answer are stipulated.
2. *Choosing the mathematical model.* The problem is typically, but not always, described using a system of integro-differential equations defined in space and time, with suitable initial and/or boundary conditions chosen to ensure that the problem is well posed, i.e. such that solution depends continuously on them. In practice, many problems are actually ill-posed, e.g. inverse problems such as deconvolution or tomography using Radon transform, and require adhoc procedure to deal with.
3. *Discretizing of the continuous equations.* The continuous variables are replaced by a set of discrete and finite values, given only for a subset of points, evenly distributed in space and time, which defines the computational grid. Moreover, each integro-differential equation is replaced by a finite set of algebraic equations which are verified at each point of the computational grid. This procedure corresponds to the finite-difference method, although there exist other methods based on a similar principle. For instance, spectral methods decompose each continuous variables into an infinite series of orthogonal functions, chosen according to the problem and its geometry; the series is then truncated to a finite number of terms, and the norm of the discarded terms evaluates the truncation error.
4. *Choosing of the numerical algorithm.* The algebraic equations are solved using a numerical algorithm chosen according to its computational efficiency, memory requirement and numerical precision on the computer one uses, which could be of different types (scalar, vectorial, parallel, or a combination of these).

5. *Programming the computational code.* One selects the programming language, according to his preference and practice, and then the most efficient compiler available on his computer. These choices should be made in order to maximize computational performances, but also to guarantee readability and portability of the code, keeping in mind that several people will have to use it, modify it and implement it on different computers having different compilers, which both evolve on very short time scales.
6. *Testing the code.* After having written down the computer program, one checks the convergence of the numerical scheme by reducing the space and time steps until the solution does not change anymore. The code is then used to solve a known case or, should such 'test case' be missing, at least verify that the structural consistency (the conservation of symmetries and invariants) of the numerical scheme is correct. These tests are also important to estimate the observed numerical errors and compare them to their theoretical values predicted by numerical analysis theorems.
7. *Analysing and visualizing the results.* Present computers perform number crunching at a tremendous rate (up to several teraflops, i.e. trillions floating point operations per second) and therefore generate huge datasets which have then to be analyzed and represented. A numerical simulation then requires signal processing (e.g. statistical analysis, filtering, denoising) and visualization as a laboratory experiment does.

By describing the nature of the numerical errors, we would like to evaluate the risk of simulation being a *simulacrum*, i.e. the risk that one might be misled by a fake representation of the *phenomena*. For the sake of example, we will confine ourselves to studying numerical errors encountered in the finite-difference method, and will briefly tackle the case of the Monte-Carlo method. This discussion boils down to the following problems: existence and uniqueness of the solution (are the equations well-posed?), consistency (is the phenomenon under study akin to that predicted by the simulation), stability (does the numerical code amplify errors and diverge?) and convergence (does the approximate solution remain sufficiently close to the exact solution?).

2.1 Existence and Uniqueness

Regarding the mathematical formulation of the phenomenon one wishes to simulate, the first question to arise deals with the existence and, for evolution problems, the uniqueness of the solutions of the equations, for the chosen boundary conditions. Typically, the problems solved by using numerical methods are two-fold:

- *Initial value problems* (Cauchy problems), where the computation is performed from an initial state, considering appropriate boundary conditions, by solving PDEs, which could be, either elliptic (e.g. diffusion equation), parabolic (e.g. transport equations), or hyperbolic (e.g. wave equation),

– *Boundary value problems*, where there is no time evolution and the values inside the computational domain should only satisfy the boundary conditions; they are also called, figuratively, 'jury problems', since the values inside the computational domain should 'agree' with those at the boundary. Elliptic equations (e.g. Poisson's or Laplace's equations) in general belong to this category.

If the equations are linear, one can generally rest assured that one solution does exist, insofar as one has a sufficient number of boundary conditions to close the system and that one only studies the evolution forward in time. Nonetheless, one of the most useful applications of the numerical approach is the treatment of nonlinear problems, which are described by nonlinear equations for which there seldom exist theorems proving existence and uniqueness of their solutions, neither analytical methods to solve them. Let us take, for the sake of the example, the case of fluid dynamics, whose fundamental equation are the so-called 'Navier–Stokes equations'. One can guarantee the existence and uniqueness of their solutions, but only for Reynolds numbers (defined as the ratio between the nonlinear and the linear terms of Navier–Stokes equations) below order one [17]. Conversely, for large Reynolds numbers, if existence of the solutions (in the weak sense) is well established [18, 19], it is not the case for their uniqueness, except in dimension two [20, 21]. Indeed, beyond a critical value, which corresponds to the transition laminar and turbulent regimes, uniqueness has not yet been proven in dimension three, since one then does not know any longer how to control the nonlinear terms [22] without increasing dissipativity (e.g. the power of the Laplacian which takes care of viscous damping) [23, 24]. As for Euler equation (which corresponds to zero viscosity), the problem of existence and uniqueness for all times has been solved in dimension two, but not yet in dimension three [25].

When there is no uniqueness of the solutions, then, the problem is ill posed (in Hadamard's sense), i.e. there is no continuity between the solution and the data. In most cases uniqueness is intimately linked to the regularity for all times. Nowadays, computers play an important heuristic role to study such properties, either by studying Taylor expansions of the solution, or, thanks to direct numerical simulations (DNS), by estimating when a singularity may intersect the real axis [26] which would result in finite-time blow-up of the solution.

2.2 Consistency and Precision

When one discretizes a differential equation one must wonder whether the algebraic approximation is indeed consistent, i.e. is the discretized equation equivalent to the differential equation which one is trying to integrate? The idea of consistency spans the two following properties:

 – The discretized equation must tend towards the continuous equation when the space and time steps tend towards zero.

- The discretized equation must preserve the structure (e.g. Hamiltonian, sympletic) structure and the symmetries of the differential equations, or, more precisely, its invariance group must be a subgroup (as large as possible) of that of the original differential equation.

If we consider as example finite-difference schemes, it is of much importance that the first property should be verified, and it is, indeed, the case of the most known discretization schemes. As for the second property, on the other hand, it is only verified for centred finite differences, there, unknown values at grid points are expressed in terms of an even number of values at neighbouring points.

In practice, of course, one uses time and space steps that are not equal to zero. Therefore, even if a discretization scheme is consistent, it will nonetheless introduce errors, which are called truncation errors. The precision of the numerical results depends on both:

 - The *round-off error*, which stems from the limited size of the memory words (usually 32 or 64 bits) of the computer which is used
 - The *truncation error*, which is directly linked to the order of the discretization scheme (i.e. to the order of the neglected terms in the Taylor's series expansion of the differential operators).

Typically, for first-order schemes the truncation error has a diffusive effect on the solutions, which thus smooth the gradients. For second-order schemes the truncation error does have, conversely, a dispersive effect that is characterized by the appearance and spreading of spurious oscillations in high gradient regions (e.g. shocks, fronts) [27]. The behaviour of the truncation error varies according to the order of the discretization scheme, but its amplitude, on the other hand, does only depend on the space step. Therefore, it is always possible to limit the truncation error by reducing the size of the computational grid. As for the round-off error, however, it does not depend on the discretization scheme but on the precision of the computer one uses. In practice, this error is always negligible in comparison to the truncation error.

2.3 Stability and Convergence

Once it is verified that the chosen discretization scheme really is an approximation of the PDEs problem, one should raise two questions:

 - Is the calculation process stable, that is to say, does not it amplify the round-off errors at the risk of diverging?
 - If it is stable, does the numerical solution converges towards that of the PDEs problem when space and time steps are tending towards zero?

Lax's [28] equivalence theorem does state that: 'Given a properly posed initial-value problem and a finite-difference approximation to it that satisfies the consistency condition, stability is the necessary and sufficient condition for convergence'. Thus, insofar as one has already answered affirmatively to the questions of paragraphs II.1. and II.2., the problem of stability and of convergence does come to one sole and only question: is the numerical scheme converges?

Generally one studies the stability of a scheme by carrying out the harmonic analysis of a perturbed solution. One checks that the frequencies' amplitudes in the power spectrum remain bounded when the number of time steps increases, so as to prevent the round-off errors from increasing and the computation from diverging. This is the principle of von Neumann's analysis. It shows that the explicit schemes (i.e. those for which the unknowns are explicitly expressed in terms of the values computed at the previous time step) are unstable or conditionally stable, whereas the implicit schemes (i.e. those for which the unknowns are implicitly expressed in terms of both previous values and unknown values at the neighbouring points) are unconditionally stable [27]. Von Neumann's analysis thus allows us to estimate, depending on each case, the maximal time step able to preserve the stability of the solutions, i.e. a time step that is sufficiently small for the computation to follow the flow evolution [27]. (To draw an analogy, one can think of the focusing of the obturation time of a camera lense in function of the speed at which the photographed object moves). If one adopts the viewpoint of stability, the analogy between the numerical experience and the laboratory experience allows us, for a given price, to reach a certain spatial and temporal resolution, just as the resolution and the dynamical behaviour of a numerical simulation depend on the chosen grid size and consequently on the available computation resources.

Nevertheless, von Neumann's criterion does merely apply to the simplest linear problems, since it does take into account neither nonlinear effects, nor incidence on stability of the discretized boundary conditions. In view of the impossibility in which the numerical analysts are to establish stability criteria that are adapted to the numerical treatment of nonlinear equations, we are suggesting to resort to a more physical analysis, that some may condemn as a rather too intuitive one. This approach draws upon the analogy there is between the mechanisms which govern the physical problem (e.g. physical instability) and those which ensure numerical stability. It proceeds from the following principle: 'To obtain numerical stability there must be hydrodynamic stability at the scale of the computational mesh' [27]. The hydrodynamic stability is conditioned by the balance which exists, at a certain scale, called the dissipation scale (Kolmogorov's scale built upon the molecular viscosity), between the advective forces governed by the nonlinear terms and the diffusive forces described by the linear terms of the Navier–Stokes equation. One incurs the risk that instabilities may develop if the nonlinear terms take over the linear terms, i.e. when, the advective forces dominate the diffusive forces. Similarly, numerical instabilities appear in a simulation when, on the scale of the computational mesh, the nonlinear advective terms take over the linear diffusive terms; in this case the smallest perturbation, introduced for instance by the discretization of the boundary conditions or by truncation errors, is amplified by the nonlinear terms before having had time to be damped by diffusion, and the calculation diverges. Thus, to guarantee numerical stability, one must choose a computational mesh which is sufficiently fine for diffusion to dominate at the smallest scale. This is further expressed by saying that the grid Reynolds number (ratio of the advective terms upon the diffusive terms at the scale of the grid) must always remain inferior to 1. In practice, the destabilizing effect of the nonlinear terms cannot really be felt below a Reynolds of 4 [29].

Until now, we had only been dealing with models built according to the equations that govern the physics of the *phenomena* which we are studying: the computer is used to integrate these equations and, given the numerical errors, the link between the underlying theory and reality solely hinges upon the adequation of underlying theory and reality. There is a second category of numerical models for which this link is much more questionable, namely Monte-Carlo methods, that we have already mentioned in the first part. Indeed, these models are based on stochastic processes, which seem to have no relation a priori to the phenomenon under study but whose statistical behaviour proves to be similar as long as one chooses adequate statistical samples. In this case, the computer is used as a random number generator and, among all realizations thus obtained, one only keeps those having enough physical 'realizability', e.g. which verify the conservation and symmetry properties of the phenomenon one wishes to simulate. The relevance of such models to reality rests upon the partial isomorphism which exists between stochastic processes and differential equations. This approach turns out to be highly heuristic, and in practice there are only two ways in which to test these models: either by comparing them to laboratory experiences, or by performing a large number of simulations and check that the variance of the set of solutions is sufficiently small value; unfortunately this is usually is too costly in terms of computation time to be done thoroughly.

To conclude, we think that the question 'simulation or simulacrum of the phenomena?' is not specific to the numerical approach, but to the entire field of physics. The study of the phenomena with the help of numerical simulation does not seem to us to be more 'artificial' than this very study carried out thanks to experiments as complex as those used in physics nowadays: the risk of *simulacrum* is not greater when one studies the behaviour of dynamical systems thanks to computers than when one observes that of the particles thanks to accelerators or that of galaxies thanks to telescopes. Indeed, the distance between the phenomena and their observation is such that one always needs theoretical principles to discern what one holds as true from the rest of the artifacts.

The terms '*simulacrum*' and 'simulation' have a pejorative connotation and mean, for the first, 'image, idol, representation of false gods' but also 'action faked to mislead', and for the latter, 'disguise, fiction' [30]. Such a terminology may well hide some symbolic meaning, where one finds again the opposition between ideas and idols [31], laws and observables, where one encounters both the risk of an illusory fascination exerted by the computer and the original dependency which links that same computer to the military-industrial *nomenclature*. But, to answer the question, we had rather avoid using both terms, 'simulation' or 'simulacrum', and use instead the word 'experimentation', which corresponds a lot better to what simulation is used for in mathematics and physics, that is to say, the possibility to experiment with the equations or with the principles governing the phenomena under study.

3 Numerical Experiment

We would like to illustrate the role played by the computer in physics by choosing, for the sake of the example, the study of turbulence, that is to say, the study of dynamical systems having a chaotic behaviour. The turbulence is a rather beautiful source of 'formless forms', which is the first type of unstable morphologies according to Thom's classification, who gives the following description: 'Certain forms are shapeless because they present an extremely complicated structure; being chaotic, they offer to analysis only little or no element to identify' [16]. It is precisely that type of morphologies which is encountered in the two-dimensional turbulent flows (see the figure earlier). The study of turbulent flows has a distinctive status from the epistemological viewpoint, insofar as the Navier–Stokes equations governing the fluid dynamics have always been unanimously admitted and only their integration poses problem. In an unpublished (but available in his complete works) article from 1946, written in collaboration with Goldstine [32], von Neumann emphasized the originality of this field in comparison with the rest of physics: 'The phenomenon of turbulence was discovered physically and is still largely unexplored by mathematical techniques. At the same time, it is noteworthy that physical experimentation which leads to these and similar discoveries is a quite peculiar form of experimentation; it is very different from what is characteristic in other parts of physics. Indeed, to a great extent, experimentation in fluid dynamics id carried out under conditions where the underlying physical principles are not in doubt, where the quantities to be observed are completely determined by known equations. The purpose of the experiment is not to verify a proposed theory but to replace a computation from an unquestioned theory by direct measurements. Thus wind tunnels are, for example, used at present, at least in large part, as computing devices of the so-called analogy type to integrate the nonlinear partial differential equations of fluid dynamics. Thus it was to a considerable extent a somewhat recondite form of computation which provided, and is still providing, the decisive mathematical ideas in the field of fluid dynamics. It is an analogy method, to be sure. It seems clear, however, that digital devices have more flexibility and more accuracy, and could be made much faster under present conditions. We believe, therefore, that it is now time to concentrate on effecting the transition to such devices, and that this will increaser the power of the approach in question to an unprecedented extent'. Thus, with a stunning intuition given the possibilities of computers in 1946, von Neumann was suggesting to replace laboratory experiments by numerical experiments in order to study fluid mechanics problems, and especially turbulence. Such foreknowledge/insight was even rather optimistic when one thinks that the idea to replace wind tunnel experiments by numerical experiments was revived by NASA in 1978 ('Numerical Wind Tunnel' project of NASF) but had to be postponed sine die since the performances of the most powerful computers are still insufficient to numerically compute, without any adhoc turbulence model, flows with large Reynolds around complex three-dimensional geometries [33]. However, the progresses that have been made over the last thirty years, regarding as much the computers as the algorithms, are hardly imaginable. One example will be sufficient to illustrate this: to compute the flow past an airplane wing using Reynolds equations (averaged Navier–Stokes equations with first-order closure), one needs less than half an hour nowadays, whereas the same simulation tried forty years ago with the algorithms and computers then available would have been 10,000 times more expensive and the calculation of only one realization of the flow would have taken about 30 years [34]. But the advantage of numerical experimentation is not to replace laboratory experimentation, which would be, really, a dangerously illusory program insofar as simulation, just as theory, needs laboratory experiment to confirm or infirm its predictions, at least on a few test cases. No, the advantage of numerical approach is to open up new fields of experimentation that are out of reach in a laboratory: numerical experimentation should not replace laboratory experimentation but complement it.

The study of the two-dimensional turbulence is a particularly interesting case to illustrate the originality of the numerical approach in physics (see Fig. 1). Indeed, two-dimensional turbulence is practically out of reach in laboratory since it is encountered in large scale atmospheric and oceanic flows (and even then, this is a mere approximation). To study it, one needs to consider planetary scale, which is why simulation is the only method left in this case to actually experiment rather than observe. The numerical approach represents a break from the traditional scheme Thom had in mind when he wrote: 'From the epistemologist's point of view two types of morphological disciplines can be distinguished. Some disciplines are experimental: man can create the morphology under study (in physics, chemistry), or intervene, more or less brutally to say the least, in its development (as in biology). Other disciplines on the other hand are solely observational: no experimentation is possible either because of spatial distance (astronomy) or temporal distance (geology, paleontology, ethnography, history) or finally for ethical reasons (psychological and social phenomena)' [35]. This classification may indeed be widened since, some purely observational sciences could now be considered as numerically experimental sciences; for instance it is the case of paleoclimatology, the study

Fig. 1. Vorticity field computed by direct numerical simulation of a two-dimensional turbulent flow using a pseudo-spectral method [39]

of planetary atmospheres or that of galaxies' evolution. Numerical simulation thus allows us to turn purely observational sciences into experimental ones.

Numerical simulation is half-way between theory and experiment without replacing either, since theory, simulation and experiment are all interrelated. Just as an experiment requires a theory to be interpreted, and a theory requires an experiment to be refuted (or, as some would say, 'falsified'), a numerical simulation equally requires a theoretical model for its formulation, and also experimental results for its validation. In return, it can enrich the intuition of theoreticians and allow them to solve problems out of reach analytically, to imagine a larger number of cases, and to present the results in a graphical form which represents information in a more concise way. A numerical simulation allows the experimentalist to study the behaviour of a system by varying one parameter independently of the others, to know the value of the fields at all points in a given domain and to visualize phenomena which are too fine, too fast or too remote to be observed through conventional means. We think that numerical simulation reintroduces in physics the traditional notion of 'thought experiment' ('Gedankenexperiment'), now on a new scale, just as the thought experiments of Lucretius, Bruno, Galileo, Einstein or Bohr once did: by reasoning on the basis of an 'imaginary' experiment, one tries to predict a behaviour that might challenge theory at its very foundations. For instance, the numerical experiment of Fermi-Pasta-Ulam has questioned the ergodic hypothesis on which all statistical mechanics is based. The numerical approach could also reveal the phenomenology hidden in fundamental equations, and hence improve the validation of simplified models: for example the numerical solution of Newton's equation to go beyond the three-body problem and study the formation of galaxies, the solar system's stability or any other n-body problems (for n larger than 3). Likewise, numerical experiments allow us to solve the Navier–Stokes equation for a large number of degrees of freedom in order to better understand turbulence in fluids and plasmas.

Numerical experiment is thus a third way whose specificity is the heuristic use of computers. This new approach is quite distinct, by its methods and its requirements (in particular in terms of academic curriculum), from both theory and experiment. Kenneth Wilson [36] recounts that: 'In the Sixties most scientists believed that a good theorist seated in front of a computer could instantly produce good science. It was overlooked that before gaining any decent results a long and difficult training period was necessary to overcome the very strict constraints dictated by computers (whatever their power). A large scale numerical experiment is as difficult to succeed as a proper experiment or an analytical calculation leading to good results [...] For three centuries students have been trained to perform experiment and theory; the best are selected and only those can continue; a similar system should now be implemented for numerical simulation. This would not be easy, students and teachers alike will have to carry out many unsuccessful simulations before the training effort bears fruit'. The numerical approach has its own practice, one which is indeed difficult to transmit. Zabusky [37] alludes to 'synergetic computational style' and he goes on to write: 'I have found it difficult to relate this mode of working via lectures. Perhaps this mode is still an art form understood by committed practitioners in benign computer environments and learned only by apprenticeship'. This impression is confirmed by Roache who, in the introduction to his classic book on numerical simulation, 'Computational Fluid Dynamics' [29], stipulates that: 'the newcomer to computational fluid dynamics is forecast in this field there is at least as much artistiry as science'. Similarly, Turkel, in his review published in 1983 in 'Progresses in Numerical Physics' [38], states that: 'the coding of large scale problems is as much art as it is science with a large reliance on intuition and folklore'.

Should we regard this as an indiscretion of youth or is it intrinsic to the numerical approach? Is the fact that numerical simulation is still perceived as 'amateur practice' due to its youth or is intrinsic to its methodology? Is this not always the case for any research domain too young to have become ossified? We believe that theoreticians, experimentalists and numericians will better contribute to their common enterpriser, namely the evolution of concepts and the explanation of phenomena, by keeping and affirming their singular identity, while yet engaging in vigorous core interaction on the foundations. As a matter of fact, this 'differentiation of the species' should not surprise us, since the distinction between theoreticians and experimentalists, just as that drawn between mathematicians and physicists, is quite recent in our history. Should the choice of the numerical approach not ultimately be, as elsewhere, a matter of taste, of personal sensitivity, and, indeed, why not, of vocation?

> *'One cannot escape the feeling that these mathematical formulae have an independent existence and an intelligence of their own, that they are wiser even than their discoverers, that we get more out of them than was originally put into them'.*
>
> (Heinrich Hertz)

Acknowledgements

Most material of this text has been translated from a conference given on September 1983 at Cerisy for the 70th birthday of the French mathematician René Thom. I thank Claude Bardos, Julie Barrau, Dominique Escande, Susan Friedlander, Edriss Titi, Pierre Mounier, Kai Schneider, Emma Wagstaft and Kerry Wallart for their kind help. I am also very grateful to the fellows of Trinity College, Cambridge, UK, in particular to Keith Moffatt, for their kind hospitality while writing this paper.

References

1. Babbage C. (1822), Mem. Astron. Soc., 1:309
2. Sheppard W. F. (1899), Proc. Math. Soc., 21
3. Richardson L. F. (1910), Philos. Trans. A, 210:307–357
4. Richardson L. F. (1922), *Weather prediction by numerical process*, Cambridge University Press, Cambridge, 219
5. O'Brien G. G., Hyman M. A., Kaplan S. (1950), J. Math. Phys., 29:223–251
6. Courant R., Friedrichs K., Lewy H. (1928), Math. Annalen, 100:32–74, English translation published in I.B.M. J. Res. Dev. (1967), 11:215–234
7. Ulam S. M. (1976), *The adventures of a mathematician*, Scribners, New York
8. von Neuman J. (1961), *First report on the numerical calculation of flow problems*, unpublished report for Office of Naval research, published in von Neumann J., *Collected Works*, 5:664–750, Pergamon, New York
9. von Neumann J., Richtmyer R. D. (1950), J. Appl. Phys., 21:232–237. Republished in von Neumann J. (1961), *Collected Works*, 6:380–385, Pergamon, New York
10. Charney J. G., Fjörtoft R., von Neumann J. (1961), republished in von Neumann J., *Collected Works*, 6:413–430, Pergamon, New York
11. Fermi E., von Neumann J. (1953), Los Alamos Report AECU-2979, unpublished, published in von Neumann J. (1961), *Collected Works*, 6:431-434. Pergamon, New York

12. Fermi E., Pasta J. R., Ulam S. M. (1955), Los Alamos Report La-1940, unpublished, published in Fermi E. (1965), *Complete Works*, 2:978–988, University of Chicago Press, IL, USA
13. Kruskal M. D. (1960), Lecture given at the Plasma Physics Laboratory, Princeton, Princeton, NJ
14. Zabusky N. J. (1962), J. Math. Phys., 3:1028–1039
15. Kruskal M. D. (1978), *The birth of soliton. Nonlinear evolution equations solvable by the spectral transform*, 1–8, Pitman, London
16. Thom R. (1972), *Stabilité structurelle et morphogénèse*, Benjamin, New York
17. Leray J. (1933), J. Math. Pures Appliquées, 12:1–82
18. Leray J. (1934), J. Math. Pures Appliquées, 13:131–418
19. Leray J. (1934), Acta Math., 63:193–248
20. Rose H. A., Sulem P. L. (1978), J. Phys., 39:441–484
21. Frisch U. (1981), Les Houches Summer School 1981 on *Chaotic behaviour in deterministic system*, North Holland
22. Teman R. (1977), *Navier-Stokes equation*, North Holland
23. Ladyzhenskaya O. A. (1963), *The mathematical theory of viscous incompressible flow*, Gordon and Breach, London
24. Lions J. L. (1969), *Quelques méthodes de résolution des problèmes aux limites non linéaires*, Dunod–Gauthier–Villars
25. Sulem C., Sulem P. L. (1983), J. Mécanique Appliquée, special issue on *Two-dimensional turbulence*: 217–242
26. Sulem C., Sulem P. L., Frisch U. (1983), J. Comput. Phys., 50:138–161
27. Farge M. (1983), Goutelas Spring School on *Hydrodynamical instabilities in astrophysics*, Editions de l'Observatoire de Paris
28. Richtmyer R. D., Morton K. W. (1957), *Difference methods for initial-value problems*, Interscience, New York
29. Roache P. J. (1972), *Computational Fluid Dynamics*, Hermosa, 1
30. Didot F. (1879), *Dictionnaire de l'Académie Française*, 7th edition
31. Serres M. (1977), *La naissance de la physique dans le texte de Lucrèce, fleuves et turbulences*, Editions de Minuit
32. Goldstine H. M., von Neumann J. (1946), unpublished, published in von Neumann J. (1961), *Collected works*, 5:1–32, Pergamon, New york
33. Lax P. D. (1982), *Large scale computing in science and engineering*, DOD-NSF report
34. Chapman D. R. (1979), Amer. Inst. Aero. Astro. J., 17:1293–1313
35. Thom R. (1974), *Modèles mathématiques de la Morphogenèse*, Collection 10–18, Union Générale d'Editions
36. Wilson K. (1983), La Recherche, 14:1004–1007
37. Zabusky N. J. (1981), J. Comput. Phys., 43:195–249
38. Turkel E. (1983), Comput. Fluids, 11:121–144
39. Farge M. (1983), Phys. Fluids A, 3:2029

Recent Computational Developments in Quantum Conductance and Field Emission of Carbon Nanotubes

Jisoon Ihm

Summary. As the computational power has explosively increased for the last decade, the material science taking advantage of computational methods to study the electronic and atomic properties of solids has developed enormously. Especially, it becomes possible to investigate the electronic transport in the nanostructure with the accuracy on the level of the precise quantum mechanics. The computational achievements in quantum transport of carbon nanotubes are reviewed here mainly based on the recently published paper. The electrical resistance of the carbon nanotube with defects on it is shown to be controllable by the transverse applied electric fields. We also briefly mention that accurate calculations of the field emission from carbon nanotubes are possible by solving the time-dependent Schrödinger equation.

The last decade has witnessed an astonishing progress in the computational capacity and techniques for scientific researches and developments. In the field of nanoscience and nanotechnology, a lot of efforts have been devoted to the computational simulation for synthesis and fabrication of nanostructured materials of specific properties. For example, in order to be used for nanoelectronic devices [1], nanostructured materials are required to have tunable electric conductance. Semiconducting carbon nanotubes are well suited for these applications [2–4]. But metallic tubes reportedly have electrical resistances which are not sensitive to transverse electric fields [5,6], and this insensitivity has discouraged device applications of metallic tubes. Here, we present first-principles calculations of quantum transport [7] which show that the resistance of metallic carbon nanotubes can be changed dramatically with transverse electric fields if the nanotubes have impurities or defects. Furthermore, a particular complex of impurities are found to exhibit total backscattering of electrons, resulting in zero conductance at a certain energy. These findings are made possible by the state-of-the-art computational methods using quantum mechanics. Based on such scattering behaviors of defects, we can now "design" material structures with desired properties for application. We also mention very briefly that the field emission current from carbon nanostructures, which are being studied in the industry for the application to the field emission display, may be calculated very accurately employing quantum mechanical computational method.

A clean metallic carbon nanotube should have an electrical resistance of 6.5 kΩ in two-probe measurements with perfect electrical contacts [8,9]. The value of 6.5 kΩ arises from 12.9 kΩ [10,11] (which is the quantum of conductance $h/2e^2$ with h = Planck's constant and e = charge of an electron) divided by the number of bands at the Fermi energy (which is two for the metallic carbon nanotubes). The resistance of a clean metallic carbon nanotube is insensitive to the electric field when a uniform electric field is applied perpendicularly to the tubular axis. Although the applied electric field polarizes the nanotube along the field direction [12] and the band dispersion is modified near the Fermi energy, electric fields of moderate strength do not change the number of bands at the Fermi energy [5,6,13], which is the only material parameter that determines the resistance of a clean nanoscale sample. However, as we will show, contrary to the clean tube case, a defective metallic carbon nanotube can have electrical resistance which is very sensitive to electric fields perpendicular to the tubular axis.

We study the resistance of defective metallic carbon nanotubes in transverse electric fields using first-principles calculations [7]. We introduce either impurities or structural defects into metallic nanotubes [14] and describe their atomic and electronic structures using the ab initio pseudopotential [15, 16] density functional method [17]. By calculating the scattering wavefunctions of electrons around the defects, we obtain the quantum-mechanical probability for an electron at the Fermi energy to transmit through the defects. The two-probe resistance of the sample is inversely proportional to the obtained transmission probability as well as the number of bands [18]. Our calculational results show that the resistance of defective metallic carbon nanotubes is very sensitive to the strength and/or direction of the applied transverse electric field (Fig. 1).

With impurities, a metallic carbon nanotube shows an on-to-off switching behavior with an applied electric field (Fig. 1a,b). When two carbon atoms in a (10, 10) nanotube are replaced with a nitrogen and a boron atom (Fig. 1a), the resistance of the three nanotube is slightly increased when no transverse electric field is present. It remains in the low-resistive "on-state" (Fig. 1b). When an electric field of 0.4 V Å^{-1} is applied from the nitrogen-doped side to boron-doped side (Fig. 1a), the resistance of the nanotube increases to 607 kΩ, switching to a high-resistive "off-state." With further increase in the strength of the electric field, the resistance decreases back to 20 kΩ. In this case, the direction of the electric field is important; the nanotube remains in the low-resistive "on-state" with the reversed direction of the electric field (Fig. 1b).

The opposite behavior occurs when considering a carbon nanotube with a vacancy [19]. There is an off-to-on switching with a transverse applied electric field (Fig. 1c,d). When four carbon atoms are removed in a (10, 10) carbon

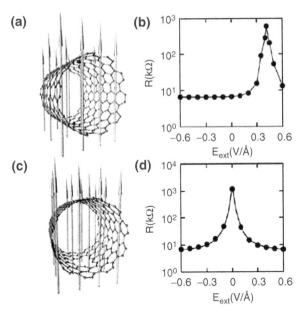

Fig. 1. Electrical resistance of defective carbon nanotubes in transverse electric fields (E_{ext}). (**a**) A schematic diagram of a nitrogen (*a dot at the top*) and boron (*a dot at the bottom*) co-doped carbon nanotube in a transverse electric field. The externally applied electric field is static and uniform, and it is perpendicular to the tubular axis. (**b**) Calculated two-probe electrical resistance of the nitrogen and boron co-doped $(10, 10)$ carbon nanotube. Without the transverse electric field, the resistance is $6.7\,\text{k}\Omega$. It increases to $607\,\text{k}\Omega$ with an electric field of $0.4\,\text{V}\,\text{Å}^{-1}$ when the direction of the field is from the nitrogen-doped side to the boron-doped side as shown in (a) (defined as positive in the graph, $E_{\text{ext}} > 0$). When the direction of the electric field is reversed ($E_{\text{ext}} < 0$), no increase of the resistance occurs. (**c**) A schematic diagram of a carbon nanotube with an extended vacancy in a transverse electric field. The vacancy is made by eliminating four adjacent carbon atoms. (**d**) Calculated two-probe electrical resistance of the nanotube with the vacancy. The direction of electric field drawn in (c) is defined as positive ($E_{\text{ext}} > 0$) and the opposite as negative ($E_{\text{ext}} < 0$). Without the electric field, the resistance is over $1\,\text{M}\Omega$, but it decreases to less than $10\,\text{k}\Omega$ when an electric field of $0.4\,\text{V}\,\text{Å}^{-1}$ is applied. The resistance is symmetric with respect to the direction of the electric field. See Ref. [14] for more details.

nanotube (Fig. 1c), the nanotube is in a high-resistive "off-state" with no electric field and its resistance is $1.2\,\text{M}\Omega$ (Fig. 1d). The nanotube switches to a low-resistive "on-state" when a transverse electric field is applied (Fig. 1d). In this case, the switching behavior occurs for either direction of the electric field (Fig. 1d).

The high-resistive off-state of the metallic carbon nanotubes originates from the impurity- or defect-induced resonant back-scattering of electrons [20].

When impurities or structural defects are introduced into metallic nanotubes, they may produce quasibound states which give rise to resonant backscattering of conducting electrons with energy close to these states. A metallic nanotube will be in the high-resistive "off-state" if the resonant scattering reflects all the conducting electrons at the Fermi energy; otherwise, it will be in the low-resistive "on-state." In the cases of the nitrogen and boron co-doped nanotube, resonant scattering due to the quasibound states occur away from the Fermi energy with no electric field, but they move to the Fermi energy at specific applied electric field. The resonant levels cross the Fermi energy with further increase of the applied electric field and then the resistance decreases. In this case, the relative position between boron and nitrogen can change the maximum resistance at "off-state" but overall switching behavior does not change. On the other hand, in case of the nanotube with the vacancy, the resonant level which appears at the Fermi level moves away when the electric field is applied. These switching behaviors are made possible because the energies of the resonant states are changeable with the applied transverse electric fields.

In a larger-diameter nanotube, the electrical switching requires a weaker electric field. First-principles calculations are preformed for defective $(20, 20)$ carbon nanotubes, and tight-binding calculations, for even larger-diameter nanotubes. The required field strength is found to be approximately inversely proportional to the square of the diameter of the nanotube. For example, the "off-state" for a nitrogen–boron co-doped $(30, 30)$ nanotube can be reached by a field of $0.07\,\mathrm{V\,\text{Å}^{-1}}$. Thus, the switching behavior will be more easily achieved in experiments with large-diameter nanotubes.

We have performed quantum conductance calculations for a $(5, 5)$ carbon nanotube with three nitrogen impurity atoms. Three nearest-neighbor nitrogen substantial impurity atoms form an equilateral triangle. Doubly-degenerate localized states occur above the Fermi level because of the symmetry of the equilateral triangle. Even without applied electric fields, the conductance practically vanishes at the energy level due to the backscattering of all incoming electronic states by the degenerate localized states.

We also want to mention that the electron field emission from carbon nanostructures [21] can be accurately calculated based on the first-principles computational method. The carbon-nanostructure-based field emission display is being developed in the industry. We can optimize the structure of the emitting tip (such as carbon nanotubes) by extensive computer simulation of the field emission. By solving the time-dependent Schrödinger equation, we obtain the field emission current from various carbon nanotubes. The tunneling current is found to depend critically on the character of the electronic wavefunction. Typically, the zero-angular-momentum state contributes most to the tunneling current. Furthermore, we find that the tunneling beam is best focused for the zero-angular-momentum state. Our prediction can lead to the optimized design of the electron emission display.

References

1. McEuen, P. L., Fuhrer, M. S. & Park, H. (2002) Single-walled carbon nanotube electronics. IEEE transactions on nanotechnology 1(1), 78–85
2. Tans, S. J., Verschueren, A. R. M. & Dekker, C. (1998) Room-temperature transistor based on a single carbon nanotube. Nature 393, 49–52
3. Javey, A., Guo, J., Wang, Q., Lundstrom, M. & Dai, H. (2003) Ballistic carbon nanotube field-effect transistors. Nature 424, 654–657
4. Martel, R., Derycke, V., Lavoie, C., Appenzeller, J., Chan, K. K., Tersoff, J. & Avouris, Ph. (2001) Ambipolar electrical transport in semiconducting single-wall carbon nanotubes. Phys. Rev. Lett. 87, 256805
5. Li, Y., Rotkin, S. V. & Ravaioli, U. (2003) Electronic response and bandstructure modulation of carbon nanotubes in a transverse electrical field. Nano Lett. 3, 183–187
6. O'Keeffe, J., Wei, C. & Cho, K. (2002) Bandstructure modulation for carbon nanotubes in a uniform electric field. Appl. Phys. Lett. 80, 676–678
7. Choi, H. J., Cohen, M. L. & Louie, S. G. First-principles scattering-state approach to nonlinear electrical transport in nanostructures. to be published.
8. Saito, R., Dresselhaus, G. & Dresselhaus, M. S. (1998) Physical Properties of Carbon Nanotubes. Imperial College Press
9. Frank, S., Poncharal, P., Wang, Z. L. & de Heer, W. A. (1998) Carbon nanotube quantum resistors. Science 280, 1744–1746
10. Landauer, R. (1970) Electrical resistance of disordered one-dimensional lattices. Philos. Mag. 21, 863–867
11. Fisher, D. S. & Lee, P. A. (1981) Relation between conductivity and transmission matrix. Phys. Rev. B 23, 6851–6854
12. Benedict, L. X., Louie, S. G. & Cohen, M. L. (1995) Static polarizabilities of single-wall carbon nanotubes. Phys. Rev. B 52, 8541–8549
13. Novikov, D. S. & Levitov, L. S. Electron properties of carbon nanotubes in the field effect regime. cond-mat/0204499
14. Son, Y.-W., Ihm, J., Cohen, M. L., Louie, S. G. & Choi, H. J. (2005) Electrical switching in metallic carbon nanotubes. Phys. Rev. Lett. 95, 216602. 1–4
15. Troullier, N. & Martins, J. L. (1991) Efficient pseudopotentials for plane-wave calculations. Phys. Rev. B 43, 1993–2006
16. Kleinman, L. & Bylander, D. M. (1982) Efficacious form for model pseudopotentials. Phys. Rev. Lett. 48, 1425–1428
17. Soler, J. M., Artacho, E., Gale, J. D., Garciá, A., Junquera, J., Ordejón, P. & Sánchez-Portal, D. The SIESTA method for ab initio order-N materials simulation. J. Phys.: Cond. Mater. 14, 2745–2779 (2002).
18. Datta, S. (1995) Electronic Transport in Mesoscopic Systems. Cambridge University Press, Cambridge.
19. Hashimoto, A., Suenaga, K., Gloter, A., Urita, K. & Iijima, S. (2004), Direct evidence for atomic defects in grapheme layers. Nature 430, 870–873
20. Choi, H. J., Ihm, J., Louie, S. G. & Cohen, M. L. (2000), Defects, quasi-bound states, and quantum conductance in metallic carbon nanotubes. Phys. Rev. Lett. 84, 2917–2920
21. Son, Y.-W., Oh, S., Ihm, J. & Han. S. (2005) Field emission properties of double-wall carbon nanotubes. Nanotechnology 16, 125–128

Energy Landscapes of Protein Self-Assembly: Lessons from Native Topology-Based Models

Yaakov Levy and José N. Onuchic

Many cellular functions rely on interactions among proteins and between proteins and nucleic acids. Our understanding of the principles that govern protein folding has been advanced in the recent years using the energy landscape theory and thanks to tight collaborations between experimentalists and theoreticians. It is likely that our current understanding of protein folding can be applied to understand more complex cellular self-organization processes. The limited success of binding predictions may suggest that the physical and chemical principles of protein binding have to be revisited to correctly capture the essence of protein recognition. In this review, we discuss the power of reduced models to study the physics of protein assembly. Since energetic frustration is sufficiently small, native topology-based models, which correspond to perfectly unfrustrated energy landscapes, have shown that binding mechanisms are robust and governed primarily by the protein's native topology. These models impressively capture many of the binding characteristics found in experiments and highlights the fundamental role of flexibility in binding. The essential role of solvent molecules and electrostatic interactions in binding is also discussed. Despite the success of the minimally frustrated models to describe the dynamics and mechanisms of binding, the actual degree of frustration has to be explored to quantify the capacity of a protein to bind specifically to other proteins. We have found that introducing mutations can significantly reduce specificity by introducing an additional binding mode. Deciphering and quantifying the key ingredients for biological self-assembly is invaluable to reading out genomic sequences and understanding cellular interaction networks.

1 Introduction

The life of cells is orchestrated by a network of chemical reactions involving numerous proteins, nucleic acids, and the transport of those molecules between cellular compartments. The remarkable efficiency of organizing these processes to yield a cellular function presents a major theoretical puzzle given the large number of molecular species and the crowded environment they inhabit. In the recent years, we have come to understanding the assembly of the individual actors in this drama thanks to many cooperative efforts between experiments and theory. We now understand the main principles of folding kinetics [1], can often predict monomeric protein structure, and can even design novel protein structures. However, knowing everything about isolated monomeric proteins does not give a complete understanding of function. Function requires change of structure and specific recognition to form large assemblies. These processes must be governed by the information stored in their sequences and structures. Furthermore, bio-macromolecules are flexible with a rich repertoire of movements on various length and time scales. These motions are essential to determine the ability of a protein to bind different ligands at the same or different binding sites [2]. Deciphering the molecular and structural origins of high-specificity as well as the catalytic promiscuity and multitasking of proteins is prerequisite for a quantitative understanding of the complexity and multidimensionality in genomes. This co-operation of many proteins and nucleic acids, which is largely "wire-less," is quite intricate. Understanding the principles of biomolecular assembly in quantitative detail constitutes the basis for the molecular theory of biological networks.

Theoretical and computational studies of protein binding have concentrated on analyzing the structural and chemical properties of interfaces [3, 4] as well as predicting the structure of the formed complexes and their binding affinity [5]. Understanding the organization of proteins into large complexes is required to understand their function and irreversible aggregation. The challenge of predicting the complex formed between pairs of proteins has been addressed for several years by docking two proteins using various models, which range from reduced models to atomistic ones [5] and include different flavors. Approaches to predict the structures of higher complexes, which are often defined as cellular machines, have been recently developed too. These approaches include, for example, combinatorial docking schemes or fitting to cryoEM density maps at low resolution [6]. Some progress has been made in recent years in the performance of docking algorithms, yet, their successes in predicting the structure of the protein complex is limited mainly to docking of the bound conformations of the complex subunits.

The inferiority of binding prediction to folding prediction is surprising because the conformational search required in binding processes of two-folded proteins is smaller than that involved in protein folding. This shortcoming suggests that the physical and chemical principles of protein binding have to be revisited. The poor predictions of docking when using the conformations of the free subunits, obviously indicates that protein flexibility is an important component in binding. Neglecting conformational changes upon binding is common to other computational models of protein binding. As examples, one may mention the Brownian dynamics simulations that are used to study the kinetics of protein–protein binding and the continuum electrostatic

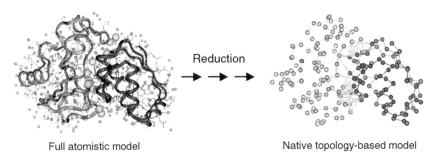

Fig. 1. An illustration of the reduction procedure from full atomistic presentation to topological presentation at the residue level. The illustration is done on the complex between barnase and barstar (pdb entry 1BRS). The atomistic presentation includes 204 water molecules at the protein surface and 16 water molecules at the interface that mediate contacts between the two monomers. In the reduced models each residue is represented by a single bead. In the native topology only model all the beads are represented the same way and differ in their connectivity to other residues

calculations to estimate free energies, both assuming rigid body association. Several docking approaches have introduced side-chain flexibility by using a rotamers library, however, it seems that backbone flexibility cannot be ignored [7]. It is likely, thus, that flexibility effects are still grossly underestimated as suggested from our recent association studies [8–10]. Solvent is also a critical component in protein association. The solvent molecules are well known to be essential in folding reactions by introducing the hydrophobic effect. While the protein cores are usually dry and contain a few water molecules, the interfaces in protein complexes are often very wet [11] (see Fig. 1). Recently, it was found that a funneled potential for binding between proteins was obtained only upon solvation of assembly interfaces [12]. These observations provide a strong indication that water can be indispensable in protein assembly and undoubtedly in protein binding to DNA due to its highly charged surface.

Beyond structure prediction, we also have to understand the basic mechanisms of association processes in biology to understand protein function, binding specificity, and cross-talk in the cell. Quantifying the Beyond structure prediction, we also have to understand the basic mechanisms of capability of a single protein to bind multiple, unrelated ligands at the same binding site or at different sites.

In this review, we will discuss the power of reduced models to address fundamental aspects of self-assembly of biological macromolecules. Reduced models are useful because of their computational efficiency in comparison to the heroic efforts required to study protein binding using all-atom simulations on a microsecond to second time scale. Moreover, reduced models with simplified presentations of biological macromolecules are especially attractive as they can be tailored to address a unique question. As such they allow one to model specific side chain information, electrostatic interactions, solvent degrees of freedom, as well as nonadditive effects of a variety of levels of accuracy. In that respect one may mention that the most significant progress in the theory of protein folding has been achieved by the development of simplified model that aimed to capture the physics of protein folding [13–15]. In particular, we discuss models that are based on the complex native structure to study the mechanism of protein binding. We also discuss the need to develop other models to further investigate other aspects of biological self-organization processes.

2 Native Topology-Based Models of Protein Association

Native topology-based (Go) models for protein complexes include only the interactions that stabilize the native structure as determined by NMR or X-ray measurements. Since these models do not include non-native interactions, they correspond to a perfectly funneled energy landscape. Moreover, all the native interactions are modeled by Lennard-Jones potential, which is short-ranged and very specific, and without discrimination based on the chemical property of the interactions. Figure 1 illustrates the reduction procedure from a full atomistic presentation to a topological presentation at the residue level, where each residue is identically represented or characterized by its charge. It has to be emphasized that reduced models can be built in various ways with the aim to serve the question at hand.

The native topology-based model has been recently applied in several studies to study the mechanism of protein association [8–10] and successfully reproduces the experimental classification of homodimers on whether monomer folding is prerequisite to monomer association (Fig. 2). Obligatory homodimers that exhibit two-state thermodynamics are found to be formed by a coupled folding and binding reaction. Transient homodimers, which binds via a thermodynamic intermediate, are formed by association of already folded monomers. In general, we found that most of the gross and, surprisingly, many of the finer features of binding mechanisms can be obtained by Go simulations. The native topology-based models agree with the experimental binding mechanism regarding the existence of a monomeric intermediate (Fig. 3). The validity of the model to study protein binding is reflected by the good correlation obtained between the computational and experimental Φ values, which measure the degree of structure at the transition state ensemble (TSE) at the residue level [16]. For Arc repressor and the tetramerization domain of 53 (p53tet) a direct comparison between the simulated and experimental Φ values is available and indicates that the simple Go model captures the nature of the TSE reasonably well. For Arc-repressor there are detailed deviations between the simulated and experimental Φ values of particular residues (reflected by correlation coefficient of 0.31), but there is an agreement about the overall structure of the TSE. For p53tet, which was experimentally classified as a

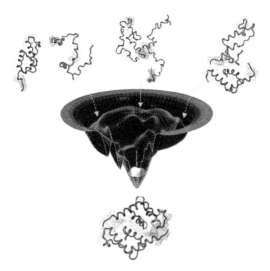

Fig. 2. Funneled energy landscape governs folding and binding of natural proteins. Self-organization processes in the cell ranges from folding of a monomeric protein up to formation of large molecular machines composed of large number of protein and nucleic acids molecules. We found that all these processes are governed by a funnel-shaped energy landscape. This means that natural protein sequences are evolutionarily selected to minimize interactions that are in conflict (frustrated interactions). Natural proteins, thus, have sequences that are minimally frustrated and recognition processes, such as folding and binding, can be described as a progressive organization toward the low-energy structure

dimer of dimers, the native-centric model not only reproduced the association mechanism (Fig. 4) but also the computational Φ values for the dimerization and tetramerization reactions are in agreement with the experimental ones. We have to point out that recently an all-atom molecular dynamics study was done on the dimerization reaction of p53tet [17]. The Φ values for the binding TSE from that study, which includes non-native interactions, displays qualitatively similar results to those obtained from the native topology-based model.

The ability of native topology-based models to reproduce the features of binding mechanisms is significant and suggests that the binding TSE and binding mechanism can be obtained by the knowledge of the final complex's structure alone. A support for the role of topology in protein assembly can be found in a recent study that has shown that the folding rate of a two-state homoheptamer can be predicted based on the topology of the native monomer [18]. Several studies have indirectly pointed to the central role of protein topology in determining the binding mechanism [19, 20]. We have recently found that protein complexes that are formed by the association of already folded subunits have structural and topological properties that are different from those with intrinsically unfolded subunits. More specifically, these two classes

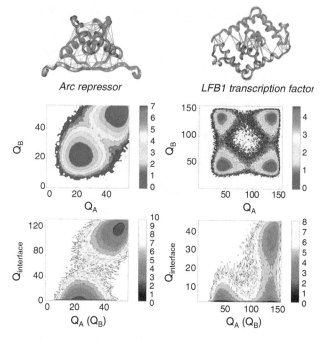

Fig. 3. The association mechanism of the dimerization of Arc-repressor (an obligatory homodimer) and LFB1 transcription factor (a transient homodimer). The binding free-energy landscapes are plotted against Q_A and Q_B (the native contacts in monomers A and B, respectively) and $Q_{Interface}$ (the interfacial native contacts). Q_A and Q_B correspond to folding/unfolding events, and $Q_{Interface}$ corresponds to binding/unbinding events. The free-energy surfaces indicate that the native topology-based model reproduced their experimentally determined binding mechanism

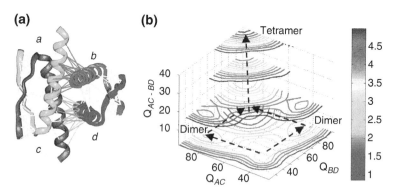

Fig. 4. Association mechanism of the tetramerization of p53 (p53tet). (**a**). The p53tet is a tetramer composed of four identical monomers (a–d). The interfacial contacts between the dimers composed of a and c as well as between b and d are shown by yellow lines. The interfacial contacts between these two dimers are shown by the red lines. (**b**). The free energy landscape for the assembly along the reaction coordinates of formation of the dimers ac and bd and the tetramer. The free energy surface produced based on the native topology-based model indicate that the tetramer is formed by dimerization of dimers, as found experimentally

of complexes differ in the topological properties (i.e., connectivity of residues, average clustering coefficient, and mean shortest path length) of the monomers and the interfaces [10]. Moreover, the structure of some protein complexes can be predicted by docking algorithms based on shape complementarity alone.

3 Frustration in Protein Binding

The native topology-based model, which is an energetically unfrustrated model, impressively reproduces many features of protein associations, yet, one may expect that adding non-native contacts will improve the agreement between simulations and experiments. In folding studies, perturbing a native topology-based model toward a more realistic protein model by introducing non-native interactions results in an enhanced folding rate [21]. In binding reactions, non-native interfacial interactions can assist specificity and increase binding rate. Alternatively, non-native contacts can lead to trap states. Frustration in binding has to be quantified to understand the role of non-native contacts and the degree of specificity in protein recognition by quantifying the effect of mutation at the binding site on the binding affinity. Moreover, frustration can result in several binding modes for a given pair of proteins where only one of them corresponds to the biological function. Frustration in binding is often observed in docking calculations that include many binding modes in addition to the correct one. Some of these binding modes can be physical other can be false positive as a results of errors in the potential. To better understand how proteins are evolutionary designed for specific binding, we have to be able to quantify the properties of all other binding mode with the aim to decipher the code for correct binding.

Frustration in the cellular level may lead to cross-talk and promiscuous binding. A recent study has shown that a yeast peptide selectively binds to a single SH3 domain among the 27 SH3 domains in yeast, however, it promiscuously binds to 12 nonyeast SH3 domains [22]. Binding specificity in protein–protein and protein–DNA binding is fundamental in cellular regulation and revealing its molecular origin is invaluable for understating the cellular machinery and to design molecules that will bind on a network level.

Nature avoids frustration in protein interactions by evolutionary design sequences to fold efficiently and robustly to a unique structure and to bind specifically and selectively. Models that rely on the principle of minimal frustration by focusing on native contacts and avoiding non-native contacts that are in conflict with the native structure, remarkably reproduce the experimental thermodynamic and kinetic features of both folding and binding processes. Yet, some degree of frustration in biological systems is unavoidable and it has to be quantified.

The effect of frustration on binding specificity has been recently illustrated in the association modes of repressor of primer (Rop) dimmer [23]. The kinetics of Rop dimer was studied experimentally and exhibits an anomalous behavior where the folding rates do not correlate with native stability and thus

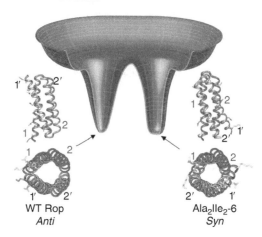

Fig. 5. A schematic presentation of a double-funneled energy landscape of Rop dimer. The existence of two competing topologies, each being preferred by different mutants or simultaneously populated in a mixture, resolves the non-Hammond kinetics of Rop mutants. Mutations can cause a conformational switch or lead to a mixture of both topologies. The Rop dimer can serve to quantify the interplay between the sequence and binding specificity

described as non-Hammond kinetics. This protein is an intriguing case because its folding/association kinetics cannot be explained by the simple funneled energy landscape. We proposed that its folding behavior might be resolved by the existence of a multifunneled energy landscape (Fig. 5) where the different basins correspond to distinct but similar topological structures. The slow folding kinetics can be the result of populating these alternative basins that act as traps. The existence of multiple structures for a given sequence is a manifestation of frustration. This frustration may be a consequence of the nonglobular elongated shape of these proteins or alternatively it may arise since these are not their functional forms (the sequence of wild-type Rop, which is much less symmetric than its mutants, shows a distinct conformational specificity).

We have shown that introducing mutations at the interface formed between the two identical monomers that composed the dimer can cause to a large conformational change from the Anti conformation to the Syn conformation of the dimer (Fig. 5), providing a plausible explanation for the non-Hammond effects [23]. The native topology-based model has shown that the Anti conformation has a higher barrier for folding than the Syn conformation. Accordingly, mutants that have similar stabilities as the wild-type Rop can fold and unfold much faster than the wild-type if they adopt the Syn conformation. Mutating the interface in a symmetric fashion results in shifting the binding pathway of the two monomers. The variation in the structure of Rop mutants illustrates the capacity of frustration in affecting binding specificity.

The role of minimally frustrated sequences of natural proteins in binding is also manifested by the prevalence of proteins to form domain-swapped

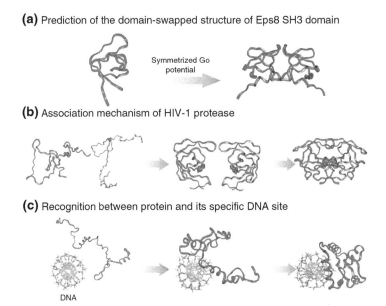

Fig. 6. Examples of biomolecular self-assembly processes studied with native topology-based models. (**a**) Prediction of the domain-swapped dimers of Eps8 SH3 protein with the symmetrized Go potential. The unique domain-swapped dimer obtained for Eps8 protein form the simulations is in agreement with the crystal structure. (**b**) The association mechanism of HIV-1 protease takes place through a monomeric HIV-1 protease. This accumulation of this intermediate offers a new strategy to design an inhibitor that will bind to residue (the orange regions) and will prevent dimerization rather than the traditional approach to design inhibitors that compete with the natural substrate on the binding to the active site (Asp25, red sphere). (**c**) Protein–DNA binding is guided not only by long-range electrostatic forces but is also enhanced by the protein flexibility

oligomers. In this binding mode, a domain or a secondary structure element of a monomeric protein is exchanged with an identical region of another monomer, using the interactions that stabilize the monomeric state [24]. It was shown that from the topology of the monomer alone, the domain-swapped dimer of the protein can be correctly reproduced [25]. Some proteins, however, exhibit a larger degree of frustration as the interactions that define the monomeric state result in several domain-swapped structures [26]. Figure 6 shows the unique domain-swapped structure resulted when simulating two monomers of Eps8 protein using the symmetrized Go potential that allows each intramonomeric contact to be formed intermolecularly as well [25]. The unique swapped structure of Eps8 obtained with the symmetrized Go potential is the experimentally observed structure. We found that swapping of the monomers of 434 repressor does not result in a single structure but with at least two structures. This suggests that the monomer of 434 repressor is frustrated for domain-swapping. A similar scenario was observed for the human prion protein [26]. However, introducing an intermolecular disulfide bond yielded the unique domain-swapped structure that was observed by X-ray crystallography. In other cases, where multiple swapped structures have been obtained, we conjecture that frustration in the monomer for binding via domain-swapping is needed to remain with a monomeric protein as the functional form of the protein.

4 The Need of Protein Flexibility in Protein Binding

Flexibility is pivotal in biological self-association processes. The role of flexibility in protein binding is acknowledged by the Koshland's venerable "induced fit" mechanism that expects conformational optimization upon binding. Structural rearrangement in binding is practically reflected by the very limited success of docking algorithms in predicting the complex structure when protein flexibility is not permitted or included side-chain motion only [7]. Protein binding to DNA often involves, additionally to conformational change of the protein, bending of the DNA [27]. Flexibility and plasticity are central to protein folding when coupled to protein association or binding via the domain-swapping mechanism. Domain swapping requires at least partial unfolding of the monomers and the mutual interchange of symmetrically identical swapped regions. Large conformational changes upon binding should not be treated as rare exceptions. For example, intrinsically disordered proteins that form a perfectly ordered structure in the presence of the appropriate ligand constitute more than 30% of the genomes of several organisms and the number of proteins observed in domain-swapped conformations constantly increases.

Our simulation studies indicate that protein flexibility is important also when the association is between folded proteins [8–10, 28]. Formation of a symmetric protein complex can take place via an asymmetric pathway where one monomer is more folded than the other monomer. For λ Cro repressor a more extreme case was observed where although its monomer is relatively stable on its own the association is between a folded monomer and an unfolded monomer that becomes folded in a later stage [8, 10]. The association mechanism of λ Cro repressor suggests that protein folding can be catalyzed in the presence of other monomers. The asymmetric binding pathway illustrates the demand for flexibility for an efficient binding that yield gradual recognition that is optimized upon complete binding. For λ Cro repressor the flexibility allows an unfolded monomer to bind to a folded monomer, which serves as a template, due to the simple geometrical and topological properties of the binding site. The minimalist model allows examining the contribution of flexibility to binding by restriction of the protein dynamics. To assess the role of flexibility, the formation of p53tet was studied by simulating a dimerization reaction with partially flexible monomers and by simulating a tetramerization with relatively rigid dimers. A dimerization of less flexible monomers results in less efficient binding, emphasizing the crucial role of flexibility in the association of the monomers. The association of rigid folded dimers to p53tet does not show, however, a significant effect on binding [10]. In a recent study on the binding mechanism of monomeric proteins to their binding sites on the DNA, we have found that the protein flexibility assists binding to DNA [29] (Fig. 6).

The need to understand the mechanism of binding and its main determinants is well illustrated by our study on the association pathway of dimeric HIV-1 protease [28, 30]. These studies have indicated that the monomeric HIV-1 protease is relatively folded in its free form (Fig. 6). The binding by association of prefolded monomers suggests a new way to inhibit the protease by designing an inhibitor that will bind to the monomer and thus prevent dimerization rather than designing an inhibitor that blocks the active site but will eventually becomes ineffective due to drug resistance.

5 The Role of Water in Binding

The solvent is widely appreciated to be important in governing protein folding and binding [31] (see Fig. 7). Yet, its exact effects and roles are not completely understood. Recently, it was reported that adding water to a Hamiltonian for structure prediction improved the predicted structures [32]. Accordingly, it was proposed that the water has more fundamental role in protein dynamics and stability and that the first layer of water is an integral part of the structure. A dominant role of water in binding is obviously expected in protein binding, especially when the binding is between two relatively folded proteins with hydrophilic surfaces [11, 33]. The abundance of water molecules in protein interfaces as detected by X-ray crystallography, although are underestimated, serve as a simple evidence for their structural role and that they are indispensable for protein recognition (see Fig. 1). For some complexes, it was proposed that water can contribute to exquisite specificity [34], whereas for others, water was found to allow promiscuous binding by acting as a buffer that weakens unfavorable polar interactions [35]. Thus, the enthalpy gain from water-mediated contacts is greater than the entropic cost that must be paid for immobilizing interfacial water. In a recent simulation study, it was proposed that a binding reaction can be driven by entropy increase as a result of an entropy increase of bound water in a relatively large pocket [36]. A structural or functional role of water is even more obvious in binding of proteins to DNA interfaces, which are highly solvated. Studies of DNA structures suggest that water molecules may increase specificity toward a given sequence acting as an integral part of the nucleic acid.

Our understanding of the role of water solvation in protein folding is improved, however, the limited successes of implicit solvent models to accurately represent protein stability and dynamics suggest that the physics of the interaction between biomolecule and the solvent is not completely captured

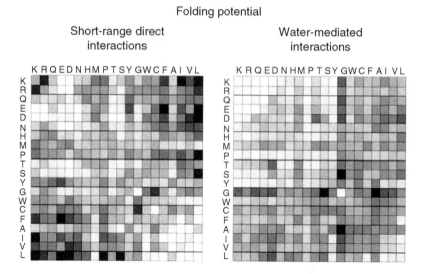

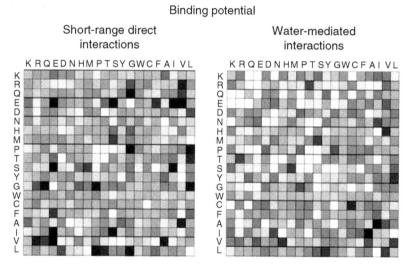

Fig. 7. Optimized knowledge-based potentials for folding and binding. Each pair of residue can interact directly or indirectly, mediated by a water molecule. Lighter color indicates a more favorable interaction. The potentials illustrate that water mediated interactions are dominant for both folding and binding, yet, with a stronger signal for binding processes

[37, 38]. Reduced models with simple representation of protein solvation has already started to shed light on the physics of protein–water interactions [32, 39] and are essential to understand fundamental questions in binding and recognition. The solvent molecules are likely to govern binding kinetics and stability. Kinetically, water molecules can guide a fully solvated protein to

recognize another fully solvated protein (or nucleic acids) by a gradual expulsion of water layers. Although the binding mechanisms and transition states are correctly characterized by topological-based models that can effectively take into account structural water molecules but not dynamic water molecules [10, 11], introducing solvent effects to binding models is required to study desolvation effect which can dominate the first stage of binding. Reduced models of water–biomolecules interactions should represent the discrete properties of water molecules. We had conjectured that our simulations of antibody–antigen complex using the topology-based model poorly reproduced the binding TSE due to lack of water molecules in our model [10]. The abundance of water in mediating contacts in other forms of the complex managed to explain the discrepancy between the experimental and computational characterization of the binding transition state [40]. Solvent molecules, thus, can assist the initial association to form the encounter complex, or, alternatively, the main binding transition state, which will be squeezed out at a later stage and result in a dry interface, which is stabilized by shape complementarity.

6 Conclusions

To understand the cellular network organization and how the genome is readout, we have to understand binding and recognition processes at the molecular level. Since all-atom molecular dynamics simulations of protein binding and protein–DNA recognition at physiological conditions will remain impractical in the foreseeable future, reduced models have to be developed to decipher the principles of self-organization in cell from a physical viewpoint. Several aspects of binding have been discussed in this paper and their investigation is needed to capture the wireless communication in the cell. Beyond improving our ability to predict the complex formed between interacting proteins, one has to be able to predict their binding affinity and association rate. While the structural aspects of protein complexes have been addressed and there are some successes in predicting complex structures, our understanding of the physics of macromolecular assembly is limited. We discussed in the paper the power of native topology-based models to study the mechanisms of biological binding. The models propose the importance of flexibility, as well as electrostatic, and water-mediated interactions for the assembly of some proteins, yet, the magnitude of these effects has to be quantified in the future.

Above that, in the genomic era, we have to study protein-protein interactions in the context of the cell. The crowded cellular environment may affect the complex stability and its formation pathway. Moreover, the coexistence of many proteins and nucleic acids requires the understanding of specificity in recognition and how proteins are evolutionary designed to avoid cross-reactivity. The interplays between binding specificity, affinity, promiscuity, and the protein plasticity in both protein–protein and protein–DNA interactions will ultimately give meaning to raw genomics.

Acknowledgments

This work has been funded by the NSF sponsored Center for Theoretical Biological Physics (grants PHY-0216576 and 0225630) with additional support from MCB-0543906.

References

1. Onuchic JN, Wolynes PG (2004) Curr Opin Struct Biol 14:70–75
2. Jeffery CJ (1999) Trends Biochem Sci 24:8–11
3. Jones S, Thornton JM (1996) Proc Natl Acad Sci USA 93:13–20
4. Conte LL, Chotia C, Janin J (1999) J. Mol Biol 285:2177–2198
5. Halperin I, Ma B, Wolfson H, Nussinov R (2002) Proteins 47:409–443
6. Russell RB, Alber F, Aloy P, Davis FP, Korkin D, Pichaud M, Topf M, Sali A (2004) Curr Opin Struct Biol 14:313–324
7. Ehrlich LP, Nilges M, Wade RC (2005) Proteins 58:126–133
8. Levy Y, Wolynes PG, Onuchic JN (2004) Proc Natl Acad Sci USA 101:511–516
9. Levy Y, Papoian GA, Onuchic J, Wolynes, PG (2004) Isr J Chem 44:281–297
10. Levy Y, Cho SS, Onuchic JN, Wolynes PG (2005) J Mol Biol 346:1121–1145
11. Levy Y, Onuchic JN (2004) Proc Natl Acad Sci USA 101:3325–3326
12. Papoian GA, Ulander J, Wolynes PG (2003) J Am Chem Soc 125:9170–9178
13. Bryngelson JD, Wolynes PG (1987) Proc Natl Acad Sci USA 84:7524–7528
14. Leopold PE, Montal M, Onuchic JN (1992) Proc Natl Acad Sci USA 89: 8721–8725
15. Plotkin SS, Onuchic JN (2002) Q Rev Biophys 35:111–167
16. Fersht A (1999) Structure and Mechanism in Protein Science: A guide to Enzyme Catalysis and Protein Folding, W. H. Freeman, New York
17. Chong LT, Snow CD, Rhee YM, Pande VS (2005) J Mol Biol 345:869–878
18. Bascos N, Guidry J, Wittung-Stafshede P (2004) Protein Sci 13:1317–1321
19. Rajamani D, Thiel S, Vajda S, Camacho CJ (2004) Proc Natl Acad Sci USA 101:11287–11292
20. Reichmann D, Rahat O, Albeck S, Meged R, Dym O, Schreiber G (2005) Proc Natl Acad Sci USA 102:57–62
21. Clementi C, Plotkin SS (2004) Protein Sci 13:1750–1766
22. Zarrinpar A, Park SH, Li, WA (2003) Nature 426:676–680
23. Levy Y, Cho SS, Shen T, Onuchic JN, Wolynes PG (2005) Proc Natl Acad Sci USA 102:2373–2378
24. Liu Y, Eisenberg D (2002) Protein Sci 11:1285–1299
25. Yang S, Cho SS, Levy Y, Cheung MS, Levine H, Wolynes PG, Onuchic JN (2004) Proc Natl Acad Sci USA 101:13786–13791
26. Cho SS, Levy Y, Onuchic JN, Wolynes PG. Phys Biol 2:S44–S55
27. Zhang Y, Xi Z, Hedge RS, Shakked Z, Crothers DM (2004) Proc Natl Acad Sci USA 101:8337–8341
28. Levy Y, Caflisch A, Onuchic JN, Wolynes PG (2004) J Mol Biol 340:67–79
29. Levy Y, Onuchic JN, Wolynes PG (2007, in press).
30. Levy Y, Caflisch A (2003) J Phys Chem B 107:3068–3079
31. Levy Y, Onuchic, JN (2006). Annu Rev Biophys Struct 35:389–415

32. Papoian GA, Ulander J, Eastwood ME, Wolynes PG (2004) Proc Natl Acad Sci USA 101:3352–3357
33. Papoian GA, Wolynes PG (2003) Biopolymers 68:333–349
34. Bhat TN, Bentley GA, Boulot G, Greene MI, Tello D, Dall'Acqua W, Souchon H, Schwartz FP, Mariuzza RA, Poljak, RJ (1994) Proc Natl Acad Sci USA 91:1089–1093
35. Sleigh SH, Seavers PR, Wilkinson AJ, Ladbury JE, Tame JR (1999) J Mol Biol 291:393–415
36. Petrone PM, Garcia AE (2004) J Mol Biol 338:419–435
37. Garcia AE, Hummer G (2000) Proteins 38:261–272
38. Rhee YM, Sorin EJ, Jayachandran G, Lindahl E, Pande VS (2004) Proc Natl Acad Sci USA 101:6456–6461
39. Cheung MS, Garcia AE, Onuchic JN (2002) Proc Natl Acad Sci USA 99:685–690
40. Schreiber G (2002) Curr Opin Struct Biol 12:41–47

A Brief Introduction to Krylov Space Methods for Solving Linear Systems

Martin H. Gutknecht

With respect to the "influence on the development and practice of science and engineering in the 20th century," Krylov space methods are considered as one of the ten most-important classes of numerical methods [1]. *Large sparse linear systems of equations* or *large sparse matrix eigenvalue problems* appear in most applications of scientific computing. Sparsity means that most elements of the matrix involved are zero. In particular, discretization of PDEs with the finite element method (FEM) or with the finite difference method (FDM) leads to such problems. In case the original problem is nonlinear, linearization by Newton's method or a Newton-type method leads again to a linear problem. We will treat here systems of equations only, but many of the numerical methods for large eigenvalue problems are based on similar ideas as the related solvers for equations.

Sparse linear systems of equations can be solved by either so-called *sparse direct solvers*, which are clever variations of Gauss elimination, or by iterative methods. In the last thirty years, sparse direct solvers have been tuned to perfection: on the one hand by finding strategies for permuting equations and unknowns to guarantee a stable LU decomposition and small fill-in in the triangular factors, and on the other hand by organizing the computation so that optimal use is made of the hardware, which nowadays often consists of parallel computers whose architecture favors block operations with data that are locally stored or cached.

The *iterative methods* that are today applied for solving large-scale linear systems are mostly *preconditioned Krylov (sub)space solvers*. Classical methods that do not belong to this class, like the successive overrelaxation (SOR) method, are no longer competitive. However, some of the classical matrix splittings, e.g., the one of SSOR (the symmetric version of SOR), are still used for preconditioning. Multigrid is in theory a very effective iterative method, but normally it is now applied as an inner iteration with a Krylov space solver as outer iteration; then, it can also be considered as a preconditioner.

In the past, Krylov space solvers were referred to also by other names such as *semi-iterative methods* and *polynomial acceleration methods*. Some of them can also be used as (fixed) preconditioners, in which case they are known as *polynomial preconditioners*; but in this function they are reportedly not competitive. *Flexible preconditioning* allows us to apply any Krylov space solver as a preconditioner of another Krylov space solver. Such combinations, called *inner–outer iteration* methods, may be very effective; see, e.g., [14].

Krylov space methods for solving $\mathbf{Ax} = \mathbf{b}$ have the special feature that the $N \times N$ matrix $\mathbf{A}$ needs only be given as an operator: for any N-vector $\mathbf{y}$ one must be able to compute $\mathbf{Ay}$; so $\mathbf{A}$ may be given as a function (or procedure or subroutine). We refer to the operation $\mathbf{Ay}$ as *matrix–vector product (MV)*. In practice, it may be much more complicated than the multiplication of a vector by a sparse matrix; e.g., this operation may include the application of a preconditioner, which may also require the solution of a large linear system.

1 From Jacobi Iteration to Krylov Space Methods

The simplest iterative method is *Jacobi iteration*. It is the same as diagonally preconditioned *fixed point iteration*: if the diagonal matrix $\mathbf{D}$ with the diagonal of $\mathbf{A}$ is nonsingular, we can transform $\mathbf{Ax} = \mathbf{b}$ into

$$\mathbf{x} = \widehat{\mathbf{B}}\mathbf{x} + \widehat{\mathbf{b}} \quad \text{with} \quad \widehat{\mathbf{B}} := \mathbf{I} - \mathbf{D}^{-1}\mathbf{A}, \quad \widehat{\mathbf{b}} := \mathbf{D}^{-1}\mathbf{b} \tag{1}$$

and apply the fixed point iteration $\mathbf{x}_{n+1} := \widehat{\mathbf{B}}\mathbf{x}_n + \widehat{\mathbf{b}}$.

It is easy to show by considering the powers of the Jordan canonical form of $\widehat{\mathbf{B}}$ that the following convergence result is valid for Jacobi iteration:

$$\mathbf{x}_n \to \mathbf{x}_\star \text{ for every } \mathbf{x}_0 \iff \rho(\widehat{\mathbf{B}}) < 1, \tag{2}$$

where $\mathbf{x}_\star := \mathbf{A}^{-1}\mathbf{b}$ and $\rho(\widehat{\mathbf{B}}) := \max\{|\lambda| \mid \lambda \text{ eigenvalue of } \widehat{\mathbf{B}}\}$ is the *spectral radius* of $\widehat{\mathbf{B}}$. From the suggested proof of (2) one actually sees that the asymptotic root convergence factor is bounded by $\rho(\widehat{\mathbf{B}})$ (if less than 1), and that this bound is sharp:

$$\limsup_{n \to \infty} \|\mathbf{x}_n - \mathbf{x}_\star\|^{1/n} \leq \rho(\widehat{\mathbf{B}}), \tag{3}$$

and equality holds for some $\mathbf{x}_0$.

Unfortunately, even the simplest examples of boundary value problems, like $u'' = f$ on $(0, 1)$ with $u(0) = u(1) = 0$, discretized by a standard finite difference method, show that Jacobi iteration may converge extremely slowly.

Unless we know the solution $\mathbf{x}_\star$ we cannot compute the nth *error (vector)*

$$\mathbf{d}_n := \mathbf{x}_n - \mathbf{x}_\star. \tag{4}$$

Thus, for checking the convergence we normally use the nth *residual (vector)*

$$\mathbf{r}_n := \mathbf{b} - \mathbf{A}\mathbf{x}_n. \tag{5}$$

Note that $\mathbf{r}_n = -\mathbf{A}(\mathbf{x}_n - \mathbf{x}_\star) = -\mathbf{A}\mathbf{d}_n$.

Assuming $\mathbf{D} = \mathbf{I}$ and letting $\mathbf{B} := \mathbf{I} - \mathbf{A}$ we have

$$\mathbf{r}_n = \mathbf{b} - \mathbf{A}\mathbf{x}_n = \mathbf{B}\mathbf{x}_n + \mathbf{b} - \mathbf{x}_n = \mathbf{x}_{n+1} - \mathbf{x}_n,$$

so we can rewrite the Jacobi iteration as

$$\mathbf{x}_{n+1} := \mathbf{x}_n + \mathbf{r}_n. \tag{6}$$

Multiplying this by $-\mathbf{A}$, we obtain a recursion for the residual

$$\mathbf{r}_{n+1} := \mathbf{r}_n - \mathbf{A}\mathbf{r}_n = \mathbf{B}\mathbf{r}_n. \tag{7}$$

So we can compute $\mathbf{r}_n$ either according to definition (5) or by the recursion (7); either way it requires one MV. From (7) it follows by induction that

$$\mathbf{r}_n = p_n(\mathbf{A})\mathbf{r}_0 \in \mathrm{span}\,\{\mathbf{r}_0, \mathbf{A}\mathbf{r}_0, \ldots, \mathbf{A}^n\mathbf{r}_0\}, \tag{8}$$

where $p_n(\zeta) = (1 - \zeta)^n$ is a polynomial of exact degree n. From (6) we have

$$\mathbf{x}_n = \mathbf{x}_0 + \mathbf{r}_0 + \cdots + \mathbf{r}_{n-1} = \mathbf{x}_0 + q_{n-1}(\mathbf{A})\mathbf{r}_0 \tag{9}$$

with a polynomial q_{n-1} of exact degree $n - 1$. So, $\mathbf{x}_n$ lies in the affine space $\mathbf{x}_0 + \mathrm{span}\,\{\mathbf{r}_0, \ldots, \mathbf{A}^{n-1}\mathbf{r}_0\}$ obtained by shifting the subspace of $\mathbf{r}_{n-1}$.

Computing $q_{n-1}(\mathbf{A})\mathbf{r}_0$ and $p_n(\mathbf{A})\mathbf{r}_0$ requires $n + 1$ MVs, because we need to build up the subspace $\mathrm{span}\,\{\mathbf{r}_0, \mathbf{A}\mathbf{r}_0, \ldots, \mathbf{A}^{n-1}\mathbf{r}_0\}$. This is the main work. We may ask: *Is there a better choice for $\mathbf{x}_n$ in the same affine space?*

The subspace that appears in (8) and (9) is what we call a Krylov space:

Definition 1. *Given a nonsingular $\mathbf{A} \in \mathbb{C}^{N \times N}$ and $\mathbf{y} \neq \mathbf{o} \in \mathbb{C}^N$, the nth Krylov (sub)space $\mathcal{K}_n(\mathbf{A}, \mathbf{y})$ generated by $\mathbf{A}$ from $\mathbf{y}$ is*

$$\mathcal{K}_n := \mathcal{K}_n(\mathbf{A}, \mathbf{y}) := \mathrm{span}\,(\mathbf{y}, \mathbf{A}\mathbf{y}, \ldots, \mathbf{A}^{n-1}\mathbf{y}). \tag{10}$$

Clearly, $\mathcal{K}_1 \subseteq \mathcal{K}_2 \subseteq \mathcal{K}_3 \subseteq \ldots$, and the dimension increases at most by one in each step. *But when does the equal sign hold?* Moreover, it seems clever to choose the nth approximate solution $\mathbf{x}_n$ in $\mathbf{x}_0 + \mathcal{K}_n(\mathbf{A}, \mathbf{r}_0)$. *But can we expect to find the exact solution $\mathbf{x}_\star$ in one of those affine spaces?* These questions are answered in the following lemmas and corollaries that we cite without proof.

Lemma 1. *There is a positive integer $\nu := \nu(\mathbf{y}, \mathbf{A})$, called grade of $\mathbf{y}$ with respect to $\mathbf{A}$, such that*

$$\dim \mathcal{K}_n(\mathbf{A}, \mathbf{y}) = \begin{cases} n & \text{if } n \leq \nu, \\ \nu & \text{if } n \geq \nu. \end{cases}$$

Corollary 1. *$\mathcal{K}_\nu(\mathbf{A}, \mathbf{y})$ is the smallest $\mathbf{A}$-invariant subspace that contains $\mathbf{y}$.*

Lemma 2. *The nonnegative integer ν of Lemma 1 satisfies*

$$\nu = \min\,\{n \mid \mathbf{A}^{-1}\mathbf{y} \in \mathcal{K}_n(\mathbf{A}, \mathbf{y})\}.$$

Corollary 2. *Let $x_\star$ be the solution of $Ax = b$ and let x_0 be any initial approximation of it and $r_0 := b - Ax_0$ the corresponding residual. Moreover, let $\nu := \nu(r_0, A)$. Then*
$$x_\star \in x_0 + \mathcal{K}_\nu(A, r_0).$$

The idea behind Krylov space solvers is to generate a sequence of approximate solutions $x_n \in x_0 + \mathcal{K}_n(A, r_0)$ of $Ax = b$, so that the corresponding residuals $r_n \in \mathcal{K}_{n+1}(A, r_0)$ "converge" to the zero vector o. Here, "converge" may also mean that after a finite number of steps, $r_n = o$, so that $x_n = x_\star$ and the process stops. This is in particular true (in exact arithmetic) if a method ensures that the residuals are linearly independent: then $r_\nu = o$. In this case we say that the method has the *finite termination property*.

It is not so easy to give a formal definition of Krylov space solvers that covers all relevant cases and is not too general in the sense that it no longer grasps some relevant aspects of Krylov space solvers. Here is our proposal:

Definition 2. *A (standard) Krylov space method for solving a linear system $Ax = b$ or, briefly, a Krylov space solver is an iterative method starting from some initial approximation x_0 and the corresponding residual $r_0 := b - Ax_0$ and generating for all, or at least most n, until it possibly finds the exact solution, iterates x_n such that*
$$x_n - x_0 = q_{n-1}(A)r_0 \in \mathcal{K}_n(A, r_0) \tag{11}$$
with a polynomial q_{n-1} of exact degree $n-1$. For some n, x_n may not exist or q_{n-1} may have lower degree.

A similar statement can be made for the residuals, if they exist.

Lemma 3. *The residuals r_n of a Krylov space solver satisfy*
$$r_n = p_n(A)r_0 \in r_0 + A\mathcal{K}_n(A, r_0) \subseteq \mathcal{K}_{n+1}(A, r_0), \tag{12}$$
where p_n is a polynomial of degree n, called nth residual polynomial, which is related to the polynomial q_{n-1} of (11) by
$$p_n(\zeta) = 1 - \zeta q_{n-1}(\zeta). \tag{13}$$
In particular, it satisfies the consistency condition $p_n(0) = 1$.

The vague expression "for all, or at least most n" in Definition 2 is needed because in some widely used Krylov space solvers (e.g., BICG) there may exist exceptional situations, where for some n the iterate x_n and the residual r_n are not defined. In other Krylov space solvers (e.g., CR), there may be indices where x_n exists, but the polynomial q_{n-1} is of lower degree than $n-1$.

There are also *nonstandard Krylov space methods* where the search space for $x_n - x_0$ is still a Krylov space but one that differs from $\mathcal{K}_n(A, r_0)$.

When applied to large real-world problems Krylov space solvers often converge very slowly – if at all. In practice, Krylov space solvers are therefore nearly always applied with *preconditioning*: $\mathbf{Ax} = \mathbf{b}$ is replaced by

$$\underbrace{\mathbf{CA}}_{\widehat{\mathbf{A}}}\mathbf{x} = \underbrace{\mathbf{Cb}}_{\widehat{\mathbf{b}}} \quad \text{or} \quad \underbrace{\mathbf{AC}}_{\widehat{\mathbf{A}}}\underbrace{\mathbf{C}^{-1}\mathbf{x}}_{\widehat{\mathbf{x}}} = \mathbf{b} \quad \text{or} \quad \underbrace{\mathbf{C}_L\mathbf{AC}_R}_{\widehat{\mathbf{A}}}\underbrace{\mathbf{C}_R^{-1}\mathbf{x}}_{\widehat{\mathbf{x}}} = \underbrace{\mathbf{C}_L\mathbf{b}}_{\widehat{\mathbf{b}}}.$$

The first two cases are referred to as *left and right preconditioning*, respectively, while in the last case we apply a *split preconditioner* $\mathbf{C}_L\mathbf{C}_R$. Actually, here $\mathbf{C}$ and $\mathbf{C}_L\mathbf{C}_R$ are *approximate inverses* of $\mathbf{A}$. Often we use instead of $\mathbf{C}$ a preconditioner $\mathbf{M} \approx \mathbf{A}$ with the property that the system $\mathbf{My} = \mathbf{z}$ is easily solved for any $\mathbf{z}$. Then, in the earlier formulas we have to replace $\mathbf{C}$ by $\mathbf{M}^{-1}$ and $\mathbf{C}_L$, $\mathbf{C}_R$ by $\mathbf{M}_L^{-1}$, $\mathbf{M}_R^{-1}$. Applying a preconditioned Krylov space solver just means to apply the method to $\widehat{\mathbf{A}}\widehat{\mathbf{x}} = \widehat{\mathbf{b}}$.

2 The Conjugate Gradient Method

The *conjugate gradient* (CG) *method* is due to Hestenes and Stiefel [7]. It should only be applied to systems that are symmetric positive definite (spd) or Hermitian positive definite (Hpd), and it is still the method of choice for this case. We will assume real data here.

CG is the archetype of a Krylov space solver that is an *orthogonal projection method* and satisfies a minimality condition: the error is minimal in the so-called *energy norm* or $\mathbf{A}$-*norm* of the error vector $\mathbf{d} := \mathbf{x} - \mathbf{x}_\star$,

$$\|\mathbf{d}\|_\mathbf{A} = \sqrt{\mathbf{d}^\mathsf{T}\mathbf{A}\mathbf{d}}.$$

In nature, stable states are characterized by minimum energy. Discretization leads to the minimization of a quadratic function

$$\Psi(\mathbf{x}) := \frac{1}{2}\mathbf{x}^\mathsf{T}\mathbf{A}\mathbf{x} - \mathbf{b}^\mathsf{T}\mathbf{x} + \gamma \qquad (14)$$

with an spd matrix $\mathbf{A}$. Ψ is convex and has a unique minimum. Its gradient is $\nabla\Psi(\mathbf{x}) = \mathbf{A}\mathbf{x} - \mathbf{b} = -\mathbf{r}$, where $\mathbf{r}$ is the residual corresponding to $\mathbf{x}$. Hence,

$$\mathbf{x} \text{ minimizer of } \Psi \iff \nabla\Psi(\mathbf{x}) = \mathbf{o} \iff \mathbf{A}\mathbf{x} = \mathbf{b}. \qquad (15)$$

If $\mathbf{x}_\star$ denotes the minimizer and $\mathbf{d} := \mathbf{x} - \mathbf{x}_\star$ the error vector, and if we choose $\gamma := \frac{1}{2}\mathbf{b}^\mathsf{T}\mathbf{A}^{-1}\mathbf{b}$, it is easily seen that

$$\|\mathbf{d}\|_\mathbf{A}^2 = \|\mathbf{x} - \mathbf{x}_\star\|_\mathbf{A}^2 = \|\mathbf{A}\mathbf{x} - \mathbf{b}\|_{\mathbf{A}^{-1}}^2 = \|\mathbf{r}\|_{\mathbf{A}^{-1}}^2 = 2\Psi(\mathbf{x}). \qquad (16)$$

In summary: *If $\mathbf{A}$ is spd, to minimize the quadratic function Ψ means to minimize the energy norm of the error vector of the linear system $\mathbf{A}\mathbf{x} = \mathbf{b}$. The minimizer $\mathbf{x}_\star$ is the solution of $\mathbf{A}\mathbf{x} = \mathbf{b}$.*

The earlier discussion suggest to find the minimizer of Ψ by descending on the surface representing Ψ by following the direction of steepest descent. If we took infinitely many infinitesimal steps, we would find the minimum by following a curved line. However, each determination of the gradient requires an MV, and therefore we need to take long steps and follow a piecewise straight line. In each step we go to the lowest point in the chosen descent direction.

Since $\mathbf{A}$ is spd, the level curves $\Psi(\mathbf{x}) = \text{const}$ are just concentric ellipses if $N = 2$ and concentric ellipsoids if $N = 3$. As can be seen from a sketch, *even for a 2×2 system many steps may be needed to get close to the solution.*

We can do much better: by choosing the second direction $\mathbf{v}_1$ *conjugate* or $\mathbf{A}$-*orthogonal* to $\mathbf{v}_0$, the first one, i.e., so that $\mathbf{v}_1^T \mathbf{A} \mathbf{v}_0 = 0$, we find the minimum in at most two steps, because for ellipses, any radius vector and the corresponding tangential direction are conjugate to each other.

How does this generalize to N dimensions, and what can be said about the intermediate results? We choose *search directions* or *direction vectors* $\mathbf{v}_n$ that are conjugate ($\mathbf{A}$–orthogonal) to each other:

$$\mathbf{v}_n^T \mathbf{A} \mathbf{v}_k = 0, \qquad k = 0, \ldots, n-1, \tag{17}$$

and define

$$\mathbf{x}_{n+1} := \mathbf{x}_n + \mathbf{v}_n \omega_n, \tag{18}$$

so that $\mathbf{r}_{n+1} = \mathbf{r}_n - \mathbf{A} \mathbf{v}_n \omega_n$. Here, ω_n is again chosen such that the $\mathbf{A}$-norm of the error is minimized on the line $\omega \mapsto \mathbf{x}_n + \mathbf{v}_n \omega$. This means that

$$\omega_n := \frac{\langle \mathbf{r}_n, \mathbf{v}_n \rangle}{\langle \mathbf{v}_n, \mathbf{A} \mathbf{v}_n \rangle}. \tag{19}$$

Definition 3. *Any iterative method satisfying (17)–(19) is called a conjugate direction (CD) method.*

By definition, such a method chooses the step length ω_n so that $\mathbf{x}_{n+1}$ is locally optimal on the search line. But does it also yield the best

$$\mathbf{x}_{n+1} \in \mathbf{x}_0 + \text{span}\{\mathbf{v}_0, \ldots, \mathbf{v}_n\} \tag{20}$$

with respect to the $\mathbf{A}$-norm of the error? By verifying that

$$\Psi(\mathbf{x}_{n+1}) = \Psi(\mathbf{x}_n) - \omega_n \mathbf{v}_n^T \mathbf{r}_0 + \frac{1}{2} \omega_n^2 \mathbf{v}_n^T \mathbf{A} \mathbf{v}_n.$$

one can prove that this is indeed the case [3].

Theorem 1. *For a conjugate direction method the problem of minimizing the energy norm of the error of an approximate solution of the form (20) decouples into $n+1$ one-dimensional minimization problems on the lines $\omega \mapsto \mathbf{x}_k + \mathbf{v}_k \omega$, $k = 0, 1, \ldots, n$. Therefore, a conjugate direction method yields after $n+1$ steps the approximate solution of the form (20) that minimizes the energy norm of the error in this affine space.*

In general, conjugate direction methods are not Krylov space solvers, but with suitably chosen search directions they are. Since

$$\mathbf{x}_{n+1} = \mathbf{x}_0 + \mathbf{v}_0 \omega_0 + \cdots + \mathbf{v}_n \omega_n \in \mathbf{x}_0 + \text{span}\{\mathbf{v}_0, \mathbf{v}_1, \ldots, \mathbf{v}_n\},$$

we need that

$$\text{span}\{\mathbf{v}_0, \ldots, \mathbf{v}_n\} = \mathcal{K}_{n+1}(\mathbf{A}, \mathbf{r}_0), \qquad n = 0, 1, 2, \ldots. \tag{21}$$

Definition 4. *The conjugate gradient (CG) method is the conjugate direction method with the choice (21).*

Theorem 1 yields now the main result on CG.

Theorem 2. *The CG method yields approximate solutions $\mathbf{x}_n \in \mathbf{x}_0 + \mathcal{K}_n(\mathbf{A}, \mathbf{r}_0)$ that are optimal in the sense that they minimize the energy norm ($\mathbf{A}$-norm) of the error (i.e., the $\mathbf{A}^{-1}$-norm of the residual) for $\mathbf{x}_n$ from this affine space.*

Associated with this minimality is the *Galerkin condition*

$$\mathcal{K}_n \perp \mathbf{r}_n \in \mathcal{K}_{n+1}, \tag{22}$$

which implies that the residuals $\{\mathbf{r}_n\}_{n=0}^{\bar{\nu}-1}$ are orthogonal to each other and form an orthogonal basis of $\mathcal{K}_{\bar{\nu}}$. On the other hand, by assumption, the search directions $\{\mathbf{v}_n\}_{n=0}^{\bar{\nu}-1}$ form a conjugate basis of $\mathcal{K}_{\bar{\nu}}$.

Here is a formulation of the standard version of CG due to Hestenes and Stiefel, which is sometimes called ORTHOMIN or OMIN version now.

Algorithm 1 (OMIN form of the CG method) *For solving $\mathbf{A}\mathbf{x} = \mathbf{b}$ let $\mathbf{x}_0$ be an initial approximation, and let $\mathbf{v}_0 := \mathbf{r}_0 := \mathbf{b} - \mathbf{A}\mathbf{x}_0$ and $\delta_0 := \|\mathbf{r}_0\|^2$. Then, for $n = 0, 1, 2, \ldots$, compute*

$$\delta'_n := \|\mathbf{v}_n\|_{\mathbf{A}}^2, \tag{23a}$$
$$\omega_n := \delta_n / \delta'_n, \tag{23b}$$
$$\mathbf{x}_{n+1} := \mathbf{x}_n + \mathbf{v}_n \omega_n, \tag{23c}$$
$$\mathbf{r}_{n+1} := \mathbf{r}_n - \mathbf{A}\mathbf{v}_n \omega_n, \tag{23d}$$
$$\delta_{n+1} := \|\mathbf{r}_{n+1}\|^2, \tag{23e}$$
$$\psi_n := -\delta_{n+1}/\delta_n, \tag{23f}$$
$$\mathbf{v}_{n+1} := \mathbf{r}_{n+1} - \mathbf{v}_n \psi_n. \tag{23g}$$

If $\|\mathbf{r}_{n+1}\| \leq \text{tol}$, the algorithm terminates and $\mathbf{x}_{n+1}$ is a sufficiently accurate approximation of the solution.

The conjugate residual (CR) method is defined in analogy to the CG method, but the 2-norm is replaced by the $\mathbf{A}$-norm. So what is minimized is now the $\mathbf{A}^2$-norm of the error, which is the 2-norm of the residual. Here, the residuals $\{\mathbf{r}_n\}_{n=0}^{\bar{\nu}-1}$ form an $\mathbf{A}$-orthogonal basis of $\mathcal{K}_{\bar{\nu}}$ and the search directions $\{\mathbf{v}_n\}_{n=0}^{\bar{\nu}-1}$ form a $\mathbf{A}^2$-orthogonal basis of $\mathcal{K}_{\bar{\nu}}$. The challenge is to find recurrences so that still only one MV is needed per iteration.

3 Methods for Nonsymmetric Systems

Solving nonsymmetric (or non-Hermitian) linear systems iteratively with Krylov space solvers is considerably more difficult and costly than symmetric (or Hermitian) systems. There are two different ways to generalize CG:

- Maintain the orthogonality of the projection by constructing either orthogonal residuals $\mathbf{r}_n$ ($\leadsto$ *generalized* CG (GCG)) or $\mathbf{A}^\mathsf{T}\mathbf{A}$-orthogonal search directions $\mathbf{v}_n$ ($\leadsto$ generalized CR (GCR)). Then, *the recursions involve all previously constructed residuals or search directions and all previously constructed iterates.*
- Maintain short recurrence formulas for residuals, direction vectors and iterates ($\leadsto$ biconjugate gradient (BICG) method, Lanczos-type product methods (LTPM)). The resulting methods are at best *oblique projection methods. There is no minimality property of error or residual vectors.*

3.1 The Biconjugate Gradient (BICG) Method

While CG (for spd $\mathbf{A}$) has mutually *orthogonal residuals* $\mathbf{r}_n$ with

$$\mathbf{r}_n = p_n(\mathbf{A})\mathbf{r}_0 \in \text{span}\{\mathbf{r}_0, \mathbf{A}\mathbf{r}_0, \ldots, \mathbf{A}^n\mathbf{r}_0\} =: \mathcal{K}_{n+1}(\mathbf{A}, \mathbf{r}_0),$$

BICG constructs in the same spaces residuals that are orthogonal to a dual Krylov space spanned by *"shadow residuals"*

$$\widetilde{\mathbf{r}}_n = \widetilde{p}_n(\mathbf{A}^\mathsf{T})\widetilde{\mathbf{r}}_0 \in \text{span}\{\widetilde{\mathbf{r}}_0, \mathbf{A}^\mathsf{T}\widetilde{\mathbf{r}}_0, \ldots, (\mathbf{A}^\mathsf{T})^n\widetilde{\mathbf{r}}_0\} =: \mathcal{K}_{n+1}(\mathbf{A}^\mathsf{T}, \widetilde{\mathbf{r}}_0) =: \widetilde{\mathcal{K}}_{n+1}.$$

The initial shadow residual $\widetilde{\mathbf{r}}_0$ can be chosen freely. So, BICG requires two MVs to extend $\mathcal{K}_n$ and $\widetilde{\mathcal{K}}_n$: one multiplication by $\mathbf{A}$ and one by $\mathbf{A}^\mathsf{T}$. But there are still short recurrences for $\mathbf{x}_n$, $\mathbf{r}_n$, and $\widetilde{\mathbf{r}}_n$. Now there are two Galerkin conditions:

$$\widetilde{\mathcal{K}}_n \perp \mathbf{r}_n \in \mathcal{K}_{n+1}, \qquad \mathcal{K}_n \perp \widetilde{\mathbf{r}}_n \in \widetilde{\mathcal{K}}_{n+1},$$

but only the first one is relevant for determining $\mathbf{x}_n$.

The residuals $\{\mathbf{r}_n\}_{n=0}^m$ and the shadow residuals $\{\widetilde{\mathbf{r}}_n\}_{n=0}^m$ form *biorthogonal bases* or *dual bases* of $\mathcal{K}_{m+1}$ and $\widetilde{\mathcal{K}}_{m+1}$:

$$\langle \widetilde{\mathbf{r}}_m, \mathbf{r}_n \rangle = \begin{cases} 0 & \text{if } m \neq n, \\ \delta_n \neq 0 & \text{if } m = n. \end{cases}$$

The search directions $\{\mathbf{v}_n\}_{n=0}^m$ and the *"shadow search directions"* $\{\widetilde{\mathbf{v}}_n\}_{n=0}^m$ form *biconjugate bases* of $\mathcal{K}_{m+1}$ and $\widetilde{\mathcal{K}}_{m+1}$:

$$\langle \widetilde{\mathbf{v}}_m, \mathbf{A}\mathbf{v}_n \rangle = \begin{cases} 0 & \text{if } m \neq n, \\ \delta'_n \neq 0 & \text{if } m = n. \end{cases}$$

BICG goes back to Lanczos [8], but was brought to its current, CG-like form later. For a detailed discussion of versions and difficulties such as breakdowns and a possibly somewhat erratic convergence see [6].

3.2 Lanczos-Type Product Methods (LTPMs)

Sonneveld [17] found with the (bi)conjugate gradient squared method (CGS) a way to replace the multiplication with $\mathbf{A}^\mathsf{T}$ by a second one with $\mathbf{A}$. The nth residual polynomial of CGS is p_n^2, where p_n is still the nth BiCG residual polynomial. In each step the dimension of the Krylov space and the search space increases by two. Convergence is nearly twice as fast as for BiCG, but even more erratic.

BiCGStab by van der Vorst [18] includes some local optimization and smoothing, and therefore the residual norm histories tend to be much smoother. The nth residual polynomial is $p_n t_n$, where now t_n satisfies the recursion

$$t_{n+1}(\zeta) = (1 - \chi_{n+1}\zeta)t_n(\zeta)$$

with χ_{n+1} chosen by residual minimization. However, in BiCGStab all zeros of t_n are real (if $\mathbf{A}$, $\mathbf{b}$ are real). If $\mathbf{A}$ has a complex spectrum, it is better to choose two possibly complex new zeros in every other iteration. This is the idea behind BiCGStab2 [5]. Further generalizations of BiCGStab include BiCGStab(ℓ) by Sleijpen and Fokkema [15,16] and GPBI-CG by Zhang [19]; see [6] for references to yet other proposals.

In view of the form of the residual polynomials, this family of methods has been called *Lanczos-type product methods*. These are often the most efficient solvers. They have short recurrences, they are typically about twice as fast as BiCG, and they do not require $\mathbf{A}^\mathsf{T}$. Unlike in GMRes, the memory needed does not increase with the iteration index n.

3.3 Solving the System in Coordinate Space: MinRes, SymmLQ, GMRes, and QMR

There is yet another class of Krylov space solvers, which includes well-known methods like MinRes, SymmLQ, GMRes, and QMR. It was pioneered by Paige and Saunders [10]. Their approach was later adapted to more general cases by Saad and Schultz [12] and Freund and Nachtigal [2].

The basic idea is to successively construct a basis of the Krylov space by combining the extension of the space with Gram-Schmidt orthogonalization or biorthogonalization, and to update at each iteration the approximate solution of $\mathbf{A}\mathbf{x} = \mathbf{b}$ in coordinate space. There are essentially three cases:

- Symmetric Lanczos process $\leadsto$ MinRes, SymmLQ [10]
- Nonsymmetric Lanczos process $\leadsto$ QMR [2]
- Arnoldi process $\leadsto$ GMRes [12]

3.4 Further Reading

For further study we suggest recent books on Krylov space solvers such as those of Greenbaum [4], Meurant [9], and Saad [11], as well as the review article by Simoncini and Szyld [13], which also covers developments of the last ten years. Nonsymmetric Lanczos-type solvers were reviewed in [6].

References

1. J. Dongarra and F. Sullivan. Guest editors' introduction to the top 10 algorithms. *Comput. Sci. Eng.*, 2(1):22–23, 2000
2. R. W. Freund and N. M. Nachtigal. QMR: a quasi-minimal residual method for non-Hermitian linear systems. *Numer. Math.*, 60:315–339, 1991
3. G. H. Golub and C. F. van Loan. *Matrix Computations*. Johns Hopkins University Press, Baltimore, MD, 3nd edition, 1996
4. A. Greenbaum. *Iterative Methods for Solving Linear Systems*. SIAM, Philadelphia, PA, 1997
5. M. H. Gutknecht. Variants of BiCGStab for matrices with complex spectrum. *SIAM J. Sci. Comput.*, 14:1020–1033, 1993
6. M. H. Gutknecht. Lanczos-type solvers for nonsymmetric linear systems of equations. *Acta Numer.*, 6:271–397, 1997
7. M. R. Hestenes and E. Stiefel. Methods of conjugate gradients for solving linear systems. *J. Res. Nat. Bureau Stand.*, 49:409–435, 1952
8. C. Lanczos. Solution of systems of linear equations by minimized iterations. *J. Res. Nat. Bureau Stand.*, 49:33–53, 1952
9. G. Meurant. *Computer solution of large linear systems*, volume 28 of *Studies in Mathematics and its Applications*. North-Holland, Amsterdam, 1999
10. C. C. Paige and M. A. Saunders. Solution of sparse indefinite systems of linear equations. *SIAM J. Numer. Anal.*, 12:617–629, 1975
11. Y. Saad. *Iterative Methods for Sparse Linear Systems*. SIAM, Philadelphia, 2nd. edition, 2003
12. Y. Saad and M. H. Schultz. Conjugate gradient-like algorithms for solving nonsymmetric linear systems. *Math. Comput.*, 44:417–424, 1985
13. V. Simoncini and D. B. Szyld. Recent developments in krylov subspace methods for linear systems. *Numer. Linear Algebra Appl.* See //www3.interscience.wiley.com/cgi-bin/jissue/99019936
14. V. Simoncini and D. B. Szyld. Flexible inner-outer Krylov subspace methods. *SIAM J. Numer. Anal.*, 40(6):2219–2239 (electronic) (2003), 2002
15. G. L. G. Sleijpen and D. R. Fokkema. BiCGstab(l) for linear equations involving unsymmetric matrices with complex spectrum. *Electron. Trans. Numer. Anal.*, 1:11–32, 1993
16. G. L. G. Sleijpen, H. A. van der Vorst, and D. R. Fokkema. BiCGstab(l) and other hybrid Bi-CG methods. *Numer. Algorithms*, 7:75–109, 1994
17. P. Sonneveld. CGS, a fast Lanczos-type solver for nonsymmetric linear systems. *SIAM J. Sci. Statist. Comput.*, 10:36–52, 1989
18. H. A. van der Vorst. Bi-CGSTAB: a fast and smoothly converging variant of Bi-CG for the solution of nonsymmetric linear systems. *SIAM J. Sci. Statist. Comput.*, 13:631–644, 1992
19. S.-L. Zhang. GPBI-CG: generalized product-type methods based on Bi-CG for solving nonsymmetric linear systems. *SIAM J. Sci. Comput.*, 18(2):537–551, 1997

Recent Developments in DNS and Modeling of Rotating Wall-Turbulent Shear Flows with Heat Transfer

Yasutaka Nagano and Hirofumi Hattori

Summary. We have carried out DNSs of various rotation-number channel flows with heat transfer and with arbitrary rotating axes in order to obtain fundamental statistics on these flows. It is found from the DNS results of various rotation-number channel flows with arbitrary rotating axes and with heat transfer that the streamwise rotating channel flow generates counter-gradient turbulent diffusion for spanwise turbulence quantities, so that the linear and the quadratic nonlinear two-equation models can not be applied to calculate streamwise rotating channel flows. As for thermal field, the transport mechanism of thermal turbulence quantities on the suction side, where the laminarization phenomenon occurs remarkably, is elucidated from DNS results. Using the present DNS data, both the cubic and the quadratic two-equation models are then evaluated. It is found that the existing models can not accurately predict turbulent quantities in a streamwise rotating channel flow. Therefore, using the assessment results, we propose a new cubic expression for Reynolds stresses in conjunction with two-equation turbulence modeling. This model satisfactorily predicts rotational channel flows with arbitrary rotating axes. The proposed model can also reproduce anisotropy in turbulent intensities near the wall, and thereby satisfies the wall-limiting behaviour of turbulent quantities in the cases under study. Also, based on the present assessment of turbulent heat-transfer models, we have proposed a new model which can predict complicated heat transfer in rotating channel flows.

1 Introduction

Rotating wall shear flows with heat transfer are encountered in many engineering relevant applications such as a flow in turbomachinery. Recently, a turbulence model has been improved for analysis of a turbulent flow with rotation and heat transfer. In particular, in order to conduct a more precise analysis of a rotating wall-turbulent flow with heat transfer, it is found that nonlinear two-equation heat transfer models are required. On the other hand, direct numerical simulation (DNS) can be effective to explore the complex turbulent transport mechanism of rotating wall shear flows. Although DNS studies on spanwise rotating channel flows have been reported, few of them deal with streamwise and wall-normal rotating channel flows.

In the present study, in order to obtain fundamental statistics on rotating channel flows with heat transfer and arbitrary rotating axes, DNSs have been conducted using the spectral method. By sorting out all the turbulence quantities obtained, we have created a DNS database for the evaluation of turbulence models. DNSs provide detailed turbulence quantities which are extremely difficult to obtain from measurements, and also provide the budget for the turbulent transport equations. In particular, turbulent diffusion and dissipation terms which are essential to construct a turbulence model in the turbulent transport equation can be obtained from DNSs. Based on the present DNSs, existing nonlinear two-equation heat-transfer turbulence models have been evaluated. And using the results of evaluation, we have improved the nonlinear eddy diffusivity model (NLEDM) for the Reynolds stresses and the nonlinear eddy diffusivity for the heat model (NLEDHM) for turbulent heat-fluxes, and constructed two-equation models for rotating wall-turbulent flows with heat-transfer. The proposed model can also reproduce anisotropy in turbulent intensities near the wall, and thereby satisfies the wall-limiting behavior of turbulent quantities in the cases under study. Also, based on the present assessment of turbulent heat-transfer models, we have proposed a model which can predict complicated heat transfer in rotating channel flows.

2 Governing Equations in Rotational Coordinate System

We consider a fully developed turbulent heated channel flow with rotation at a constant angular velocity as shown in Fig. 1. The governing equations for an incompressible rotating channel flow with heat transfer in reference frame rotational coordinates can be described in the following dimensionless forms:

$$u^+_{i,i} = 0, \tag{1}$$

$$Du^+_i/Dt^+ = -p^+_{\text{eff},i} + (1/Re_\tau)u^+_{i,jj} - \epsilon_{ik\ell} Ro_{\tau k} u^+_\ell, \tag{2}$$

$$D\theta^+/Dt^+ = (1/PrRe_\tau)\theta^+_{,jj}, \tag{3}$$

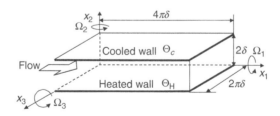

Fig. 1. Rotating channel flow and coordinate system

where D/Dt implies the substantial derivative, the Einstein summation convention is used, and a comma followed by an index indicates differentiation with respect to the indexed spatial coordinate. A temperature does not affect a velocity field, i.e., temperature is treated as a passive scalar, and the rotation number $Ro_{\tau k}$ is defined as $2\Omega_k \delta/u_\tau$. The centrifugal force can be included in the effective pressure p_{eff}^+ ($= p_s^+ - Ro^2 r_c^2/8$), if fluid properties are constant, where p_s^+ is the normalized static pressure, $Ro = (Ro_j Ro_j)^{1/2}$ is the absolute rotation number, and r_c is the dimensionless distance from the rotating axis [1]. To solve the governing equations numerically, the spectral method is used for the DNS [2]. The boundary conditions are nonslip conditions for the velocity field, and different constant temperatures for the thermal field on the walls, $\Delta\Theta = \Theta_H - \Theta_c = $ Const., as well as periodic conditions in the streamwise and spanwise directions.

The Reynolds-averaged equations for the turbulence model can be written as follows:

$$\bar{U}_{i,i} = 0 \tag{4}$$

$$D\bar{U}_i/Dt = -(1/\rho)\bar{P}_{\text{eff},i} + \left(\nu \bar{U}_{i,j} - \overline{u_i u_j}\right)_{,j} - 2\epsilon_{ik\ell}\Omega_k \bar{U}_\ell \tag{5}$$

$$D+\bar{\Theta}/D+t = \left(\alpha \bar{\Theta}_{,j} - \overline{u_j \theta}\right)_{,j} \tag{6}$$

$$\overline{u_i u_j} = (2/3)k\delta_{ij} - 2C_0 \nu_t S_{ij} + \text{High-order terms} \tag{7}$$

$$\overline{u_j \theta} = -\alpha^t_{jk}\bar{\Theta}_{,k} + \text{High-order terms}, \tag{8}$$

where $\nu_t = C_\mu f_\mu (k^2/\varepsilon)$ and $\alpha^t_{jk} = -C_{t0}\overline{u_j u_k}\tau_m$ are eddy diffusivities for momentum and heat, respectively. τ_m is the hybrid/mixed time scale, and $S_{ij} = (\partial \bar{U}_i/\partial x_j + \partial \bar{U}_j/\partial x_i)/2$ is the strain-rate tensor. The transport equations of turbulence quantities, i.e., turbulence energy, k, dissipation-rate of k, ε, temperature variance, $k_\theta (= \overline{\theta^2}/2)$, and dissipation-rate of k_θ, ε_θ which compose the eddy diffusivities, can be given as follows:

$$Dk/Dt = \nu k_{,jj} + T_k + \Pi_k + P_k - \varepsilon, \tag{9}$$

$$D\varepsilon/Dt = \nu\varepsilon_{,jj} + T_\varepsilon + \Pi_\varepsilon + (\varepsilon/k)(C_{\varepsilon 1}P_k - C_{\varepsilon 2}f_\varepsilon \varepsilon) + E + R, \tag{10}$$

$$Dk_\theta/Dt = \alpha k_{\theta,jj} + T_{k_\theta} + P_{k_\theta} - \varepsilon_\theta, \tag{11}$$

$$D\varepsilon_\theta/Dt = \alpha\varepsilon_{\theta,jj} + T_{\varepsilon_\theta} + (\varepsilon_\theta/k_\theta)(C_{P1}f_{P1}P_{k_\theta} - C_{D1}f_{D1}\varepsilon_\theta) \\ + (\varepsilon_\theta/k)(C_{P2}f_{P2}P_k - C_{D2}f_{D2}\varepsilon) + E_\theta + R_\theta, \tag{12}$$

where T_k, T_ε, T_{k_θ}, and T_{ε_θ} are turbulent diffusion terms, Π_k and Π_ε are pressure diffusion terms, and P_k ($= -\overline{u_i u_j}\partial \bar{U}_i/\partial x_j$) and P_{k_θ} ($= -\overline{u_j \theta}\partial \bar{\Theta}/\partial x_j$) are production terms. E and E_θ are extra terms, and R and R_θ are rotation-influenced addition terms [2,3]. Note that, in the rotational coordinate system, the vorticity tensor should be replaced with the absolute vorticity tensor in (7) and (8), i.e., $W_{ij} = \Omega_{ij} + \epsilon_{mji}\Omega_m$ to satisfy the material frame indifference (MFI) [4].

Table 1. Computational conditions

Re_τ		150		Pr		0.71
	case 1: spanwise rotation + heat transfer					
		SPR1	SPR2	SPR3	SPR4	SPR5
$Ro_{\tau 3}$		0.0	0.75	1.5	3.0	5.0
	case 2: streamwise rotation					
		STR1	STR2	STR3	STR4	STR5
$Ro_{\tau 1}$		1.0	2.5	7.5	10.0	15.0
	case 3: wall-normal rotation					
		WNR1	WNR2	WNR3	WNR4	WNR5
$Ro_{\tau 2}$		0.01	0.02	0.05	0.1	0.3

3 Results and Discussion of DNSs

DNSs have been performed under the conditions listed in Table 1. It is well known that a region exists in a spanwise rotating channel (case 1) where the vorticity ratio, $S = -2\Omega/(d\bar{U}/dy)$, becomes $S = -1$, which represents neutral stability [5,6]. This relation yields the following equation:

$$\bar{U}^+ = Ro_\tau (y/\delta) + C. \tag{13}$$

Profiles of mean velocities are shown in Fig. 2, where regions of neutral stability are clearly observed at various rotation numbers. Since the DNS captures exactly the relation $S = -1$ in the region of neutral stability, the numerical accuracy should be sufficient. Distributions of Reynolds shear stress are indicated in Fig. 2. Obviously, the decrease in Reynolds shear stress is observed on the suction side, in which the flow tends to laminarize. Thus, it

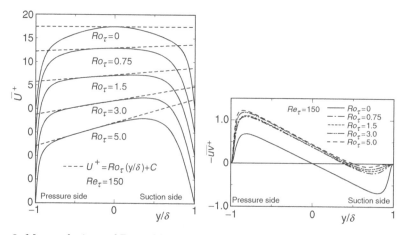

Fig. 2. Mean velocity and Reynolds shear stress profiles in spanwise rotating channel flows at various rotation numbers

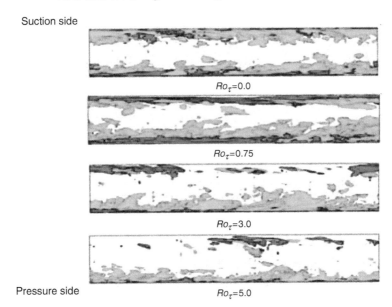

Fig. 3. Streak structures at various rotation numbers (*side view*)

can be seen in Fig. 3, which shows the streaky structures in the side view of a channel, that there are few streaks on the suction side, as the rotation number increases.

In the thermal field of case 1, the mean temperature and turbulent heat-flux at various rotation numbers are shown in Fig. 4. With increasing rotation numbers, a varying mean temperature and turbulent heat-flux are observed. In particular, the region of large gradient for mean temperature expands on the suction side though a laminarization phenomenon that occurs remarkably there. However, the turbulent heat-flux, $\overline{v\theta}^+$, is decreasing here, where the wall-normal velocity fluctuations decrease. Thus, the correlation between the wall-normal velocity fluctuation and temperature fluctuation becomes small. The marked characteristics of the thermal field are an increase in temperature variance on the suction side as shown in Fig. 4c. It can observed that the turbulent diffusion is active at $y^+ > 20$ on the suction side ($Ro_\tau = 5.0$). Since the skewness factor of wall-normal velocity fluctuation becomes positive in the rotating channel flows, passive scalars are transported from the pressure side to the suction side, i.e., the turbulent diffusion for passive scalar becomes active there. From Fig. 5 showing the velocity vector averaged in the streamwise direction and temperature fluctuation in the rotating channel flow, it can be seen that a pair of large-scale vortices occurs across the channel from the heated pressure side to the cooled suction side in the rotating channel flow, and higher temperature fluid is transported by these large-scale motions. Consequently, the region of large gradient for mean temperature expands and the temperature variance increases on the suction side.

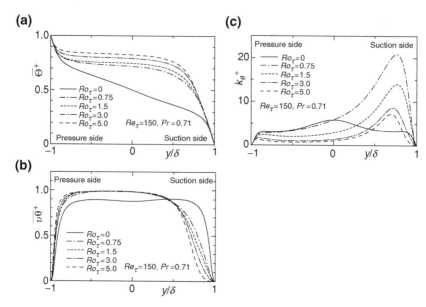

Fig. 4. Turbulent quantities of thermal field in spanwise rotating channel flows at various rotation numbers: (**a**) mean temperature, (**b**) turbulent heat-flux, and (**c**) temperature variance

Next, DNSs of fully developed channel flows with streamwise (case 2) and wall-normal (case 3) rotation are carried out for the 10 cases indicated in Table 1. DNS results of cases for STR1 ∼ STR5 are shown in Fig. 6, respectively, (case 3 is not shown here). In these cases, the spanwise mean velocity appears to be caused by a rotational effect, which increases with the increase in rotation number in both cases. It is found from DNS results that cases of streamwise rotating channel flow (case 2) involve the counter gradient turbulent diffusion in the spanwise direction shown in Fig. 6. In view of turbulence modeling, this fact clearly demonstrates that the linear and the quadratic nonlinear two-equation models cannot be applied to calculate the case of a streamwise rotating channel flow. Thus, the following modeled (14) employed

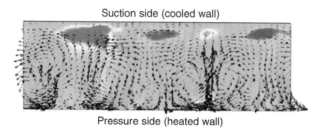

Fig. 5. Streamwise-averaged velocity vectors and temperature fluctuations in non-rotating and rotating channel flows: $Ro_\tau = 5.0$; $\theta/\Delta\Theta = -0.06 \sim +0.06$

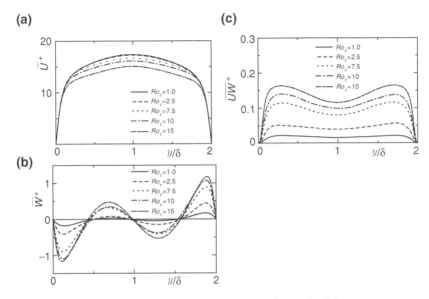

Fig. 6. DNS results of streamwise rotating flows (case 2): (**a**) streamwise mean velocity, (**b**) spanwise mean velocity, and (**c**) Reynolds shear stresses, $\overline{vw}$

in the linear and the quadratic nonlinear two-equation models cannot clearly express a counter gradient turbulent diffusion:

$$-\overline{vw} = \nu_t \bar{W}_{,y}, \qquad (14)$$

where the Reynolds shear stresses of a quadratic nonlinear two-equation model are expressed identical to a linear model. Also, it is noted that the rotational term does not appear in the momentum equation of a fully developed streamwise rotating channel flow, in which the rotational effect is included implicitly in the Reynolds shear stress in the same manner as the spanwise rotational flow [3]. Therefore, a cubic nonlinear two-equation model or Reynolds stress equation model should be used for the calculation of streamwise rotating channel flows.

4 Nonlinear Eddy Diffusivity Models

To reconstruct the proper turbulence models for rotational heated channel flows with arbitrary rotating axes, we evaluated the existing turbulence models of both the velocity and the thermal fields using the present DNSs to know the improvement points.

Using the evaluation results for turbulence model, to predict adequately rotational channel flows with arbitrary rotating axes, we proposed a cubic NLEDM in a two-equation turbulence model, in which a modeling of wall-limiting behavior of Reynolds stress components is also considered. The modeled expression of cubic Reynolds shear stress is given as follows [7]:

$$\begin{aligned}\overline{u_i u_j} &= (2/3)k\delta_{ij} - c_1\nu_t S_{ij} + c_2 k(\tau_{Ro}^2 + \tau_{R_w}^2)(S_{ik}W_{kj} - W_{ik}S_{kj}) \\
&+ c_3 k(\tau_{Ro}^2 + \tau_{R_w}^2)(S_{ik}S_{kj} - S_{mn}S_{mn}\delta_{ij}/3) \\
&+ c_4 k\tau_{Ro}^2 (W_{ik}W_{kj} - W_{mn}W_{mn}\delta_{ij}/3) \\
&+ c_5 k\tau_{Ro}^3 (W_{ik}S_{k\ell}S_{\ell j} - S_{ik}S_{k\ell}W_{\ell j}) \\
&+ c_6 k\tau_{Ro}^3 (W_{ik}W_{k\ell}S_{\ell j} + S_{ik}W_{k\ell}W_{\ell j} - 2S_{\ell m}W_{mn}W_{n\ell}\delta_{ij}/3) \\
&+ c_7 k\tau_{Ro}^3 (S_{ij}S_{k\ell}S_{k\ell} - S_{ij}W_{k\ell}W_{k\ell}).\end{aligned} \qquad (15)$$

The characteristic time-scale τ_{R_w} is introduced in an NLHN model [3] for an anisotropy and wall-limiting behavior of turbulent intensity. For the modeled transport equations of k and ε as indicated in (9) and (10), the turbulent diffusion and the pressure diffusion terms are modeled [3]. The extra term in the ε-equation is adopted as identical to the NLHN model [3]. Finally, a rotation-influenced addition term, R, in the ε-equation (10) is generalized as follows:

$$R = C_\Omega f_\Omega k \epsilon_{ij\ell} W_{ij} d_\ell \Omega_\ell, \qquad (16)$$

where d_ℓ is the unit vector in the spanwise direction.

On the other hand, the transport phenomenon of a spanwise rotating channel flow with heat transfer (case 1) is discussed based on results of DNS. It is obvious that a dissimilarity between the velocity and thermal fields exists in the rotating channel flow under the defined boundary conditions. Therefore, we propose the nonlinear eddy diffusivity for heat model (NLEDHM) as follows [2]:

$$\overline{u_j\theta} = -\alpha_{jk}^t \bar{\Theta}_{,k} + (C_{\theta 1}/f_{RT})\tau_m^2 \overline{u_\ell u_k}\left(C_{\theta 2}S_{j\ell} + C_{\theta 3}W_{j\ell} + 2C_{\theta 1}\epsilon_{jm\ell}\Omega_m\right)\bar{\Theta}_{,k}. \qquad (17)$$

The turbulent diffusion terms in the k_θ- and ε_θ-equations are modeled with a GGDH plus convection by large-scale motions proposed by Hattori and Nagano [8] to reflect the transport of scalars by large-scale vortices in the rotating channel flow. In the modeling of the ε_θ-equation, the following rotation-influenced additional term as introduced in the ε-equation is proposed

$$R_\theta = C_\Omega f_\Omega k_\theta \epsilon_{ij\ell} W_{ij} \Omega_\ell. \qquad (18)$$

5 Performance of NLEDM and NLEDHM

In order to evaluate the present NLEDM and NLEDHM, we have calculated the channel flow with spanwise rotation and heat transfer. Figure 7 show the

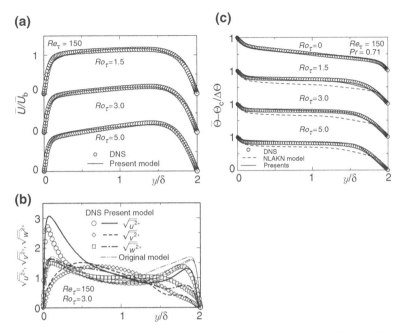

Fig. 7. Predicted turbulent quantities in spanwise rotating channel flows (case 1): (a) mean velocity, (b) Rms velocity ($Ro_\tau = 3.0$), and (c) mean temperature

prediction of turbulence quantities in the rotating thermal field in comparison with the present DNS. It is obvious that the proposed model is improved for predictions of the rotating channel flows with heat transfer at various rotation numbers.

The predictions of fully developed streamwise (case 2) and wall-normal (case 3) rotating channel flows calculated by the present model are shown in Figs. 8 and 9. The results with the NLCLS model [9] are also included in the figure for comparison. It can be seen that the present model gives accurate predictions of all cases. Also, the present model adequately reproduces redistribution of Reynolds normal stresses, and can predict wall-limiting behavior properly as shown in Fig. 9b.

6 Conclusions

We have carried out DNSs of various rotation-number channel flows with heat transfer and arbitrary rotating axes in order to obtain fundamental statistics on these flows. The transport mechanism of thermal turbulence quantities on the suction side, where the laminarization phenomenon occurs remarkably, is elucidated from DNS results. Also, it is found from DNS results that cases of streamwise rotating channel flow involve a counter gradient turbulent diffusion

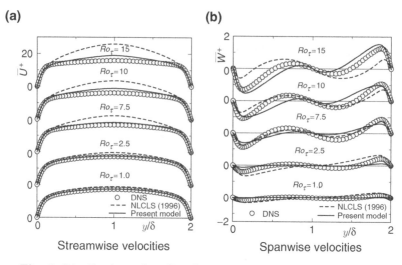

Fig. 8. Distributions of predicted streamwise rotating flows (case 2)

regarding spanwise turbulent quantities, so that the linear and the quadratic nonlinear two-equation models can not be applied to calculate the case of streamwise rotating channel flow. Using the present DNS data, both the cubic and the quadratic two-equation models are then evaluated. It is found that the existing models can not accurately predict turbulent quantities in a rotating channel flow with heat transfer. Therefore, using the assessment results, we proposed a cubic NLEDM and an NLEDHM in a two-equation heat-transfer

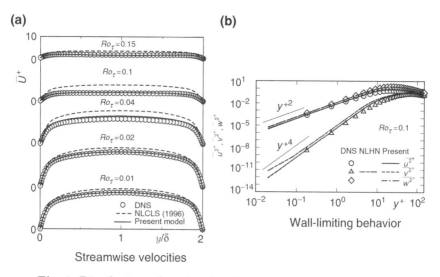

Fig. 9. Distributions of predicted wall-normal rotating flows (case 3)

turbulence model which can satisfactorily predict rotational channel flows with heat transfer and arbitrary rotating axes. The proposed model can also reproduce anisotropy of turbulent intensity near the wall, and thereby satisfies the wall-limiting behavior of turbulent quantities in the cases under study.

Acknowledgment

This research was partially supported by a Grant-in-Aid for Scientific Research (S), 17106003, from the Ministry of Education, Culture, Sports, Science and Technology (MEXT), Japan.

References

1. Wu H, Kasagi N (2004) Phys Fluids, 16:979–990
2. Nagano Y, Hattori H (2003) J Turbulence, 4:1–15
3. Nagano Y, Hattori H (2002) J Turb., 3:1–13
4. Speziale C G, Sarkar S, Gatski T B (1997) J Fluid Mech, 227:245–272
5. Johnston J P, Halleen R M, Lezius D K (1972) J Fluid Mech, 56:533–557
6. Kristoffersen R, Andersson H (1993) J Fluid Mech, 256:163–197
7. Hattori H, Ohiwa N, Nagano Y (2006) Int J Heat Fluid Flow, 27:838–851
8. Hattori H, Nagano Y (1998) Numer Heat Transfer B, 33:153–180
9. Craft T J, Launder B E, Suga K (1996) Int J Heat Fluid Flow, 17:108–115

Contribution of Computational Biology and Structural Genomics to Understand Genome and Transcriptome

Mitiko Go, Kei Yura, and Masafumi Shionyu

Summary. Genome sequencing and structural genomics projects are both proceeded to gain a new perspective of life, that is global views on mechanisms of life with comprehensive and unbiased fashion. We now have genome sequences of human and other species, and are going to have a three-dimensional structure of whole proteins. Those massive pieces of information can only be deciphered with collaboration of computational biology. In this paper, we will discuss the amount of data we have at the moment and one of the new views on mechanisms of cellular function regulation obtained based on the computational analyses of those data.

1 Structural Genomics and Homology Modeling

1.1 Structural Genomics

In 2001, initial sequencing results of the human genome were presented [1]. The human genome was found to consist of around 2,850 million bases and expected to contain around 20,000 protein-coding genes [2]. Those pieces of information are the tip of the global views of genomes and enable us to attack scientific issues on life science in a comprehensive manner.

In a cell, a protein, but not a gene itself, is the macromolecule functioning as an enzyme for a chemical reaction, a scaffold of a cellular structure, a regulator of information flow and so forth. A gene residing in a DNA sequence is transcribed into a messenger RNA (mRNA) and then translated into an amino acid sequence or protein, which assumes a three-dimensional (3D) structure unique to its sequence. Specific locations of atoms enable proteins to facilitate chemical reactions and to interact with other molecules in the cell. The protein 3D structure is determined based on the amino acid sequence, hence in theory, once we obtain the nucleotide sequence of the genome, we could have 3D structures of whole proteins. In reality, however, prediction of protein-coding genes out of a nucleotide sequence has attained to a certain level, but we still do not know the whole genes encoded in the human genome [3], and we still cannot deduce a protein 3D structure from an amino acid sequence with certainty. Therefore, projects to determine 3D structures of proteins in a great scale (structural genomics projects) are going on around the world [4].

1.2 Role of Homology Modeling in Structural Genomics

Most of the structural genomics projects do not aim to determine whole protein 3D structures, but try to determine 3D structures of representative proteins. Determining whole protein 3D structures is impractical in a short period. In the amino acid sequence database, approximately 1.6 million unique amino acid sequences are registered and determination of 3D structures of those proteins will take more than 400 years, even with a speed of 10 structures per day.

Proteins can be classified into, at least, 16,000 groups based on the amino acid sequence similarity [5]. Proteins in each group are likely evolutionary related and they share a similar 3D structure [6]. Therefore, once we have a 3D structure of, at least, one of the proteins in a group, then 3D structures of other proteins in the same group can be computationally deduced. The method to obtain the 3D structures of those proteins in the same group is called homology modeling [7]. Once we have a 3D structure of a single protein for each group, then using the protein 3D structure determined experimentally as a template, we could model the 3D structures of whole proteins. Most of the structural genomics projects are determining the 3D structures of the representative proteins from the groups.

2 Coverage of Whole Amino Acid Sequences with Protein 3D Structures

Since 2002, we have been maintaining FAMSBASE, a database for homology modeled protein 3D structures [8]. A protein 3D structure is modeled with FAMS [9], one of the homology modeling programs, based on amino acid sequence alignments given in GTOP database [10]. The latest update of the database based on protein 3D structures determined by November 2003, resulted in protein 3D structure models for 368,724 amino acid sequences in genomes from 276 species. Genomes of those species encode 734,193 proteins in total, and FAMSBASE contains 3D structures of about 50.2% of those proteins. The fraction is about 52.1% for eubacterial genomes, about 46.6% for archaeal genomes, and about 49.1% for eukaryotic genomes (Fig. 1). At the time we launched FAMSBASE in 2002, the fraction was about 44.3% for eubacterial genomes, 38.7% for archaeal genomes, and 40.7% for eukaryotic genomes. In total it was about 41.8% [8]. In the last three years (between 2001 and 2004), therefore, we observed about 8% increase in coverage of proteins by 3D structures and with this rate, we would soon obtain 3D structures for

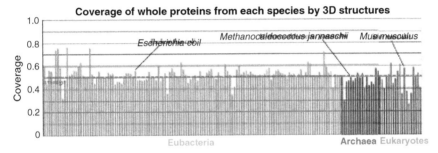

Fig. 1. Coverage of whole amino acid sequences with protein 3D structures on each genome. A vertical bar is the fraction of proteins with 3D structure

the whole proteins and could start to analyze cellular processes with protein 3D structures in a global manner [11].

At the moment, however, we do not have the entire view of protein 3D structures yet. But with a computational technique to build a protein 3D structure and with a reasonable number of template structures, we started to assess the effect of one of the most characteristic phenomena in eukaryotic cell on protein 3D structures.

3 Alternative Splicing: Multiple Proteins from a Single Gene

3.1 What is Alternative Splicing?

A gene structure in a genome of eubacteria and eukaryotes is quite different. A protein-coding region of a gene in eubacteria is encoded as a contiguous nucleotide sequence, whereas a protein-coding region of a gene in eukaryotes is often split by a stretch of a noncoding region named intron (Fig. 2). Because of this difference, maturation process of mRNA in both kingdoms of life is not the same. In the maturation process of mRNA in eukaryotes, introns are cut out and the juxtaposing coding regions, each of which is named exon, are spliced. This splicing process in eukaryotes enables building variations in combinations of exons. Alternative splicing is a process that generates multiple mature mRNAs (splicing isoforms) from a single gene by switching usage of exons during the process of mRNA maturation (Fig. 2). At first, the number of genes undergoing alternative splicing was thought to be small, but recent analyses of genomic data and a large number of mRNA products (transcriptome) reveal that a large fraction of protein-coding genes of multicellular organisms are involved in alternative splicing [12]. In humans, computational and experimental analyses indicate that about 70% of all genes undergo alternative splicing and a single gene generates two to three mature mRNAs on average [13]. Because mature mRNAs generated by alternative splicing can encode

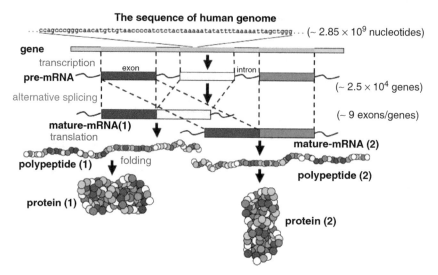

Fig. 2. Schematic description of alternative splicing

different amino acid sequences, alternative splicing is thought to be an important mechanism to expand a repertoire of proteins from limited number of genes.

Biochemical analyses on a limited number of splicing isoforms demonstrate that alternative splicing regulates the features of proteins, such as binding properties, intracellular localization, enzymatic activity, protein stability, and posttranslational modifications [13]. To obtain the comprehensive and unbiased views on the impact of alternative splicing in cellular processes, functional analysis of all splicing isoforms should be performed. However, since the number of splicing isoforms is expected to be so large, experimental analyses of all splicing isoforms are impractical in a short period. We, therefore, have started to assess the possible effects of alternative splicing on the function of splicing isoforms through computational investigation of the 3D structures of the isoforms [14].

3.2 Estimating Functional Differences of Splicing Isoforms by Homology Modeling: Case Study in Acyl-Protein Thioesterase I

Some of Ras superfamily proteins and α subunit of trimeric G protein, which is involved in cell signal transduction, are modified posttranslationally with palmitate to interact with the cell membrane. Acyl-protein thioesterase I (APT-I) is cytoplasmic enzyme that catalyzes depalmitoylation of trimeric G protein α subunit and may promote release of α subunit from the membrane, which impairs interactions of the signal transduction proteins [15]. In humans, at least, two splicing isoforms are produced from APT-I gene and the 3D structure of one of the splicing isoforms is determined by X-ray crystallography [15]. The crystal structure of the isoform shows that APT-I is a homodimer. The active site is buried at the homodimer interface and hence the dimer cannot be an active form. Dissociation of the two subunits is needed to expose the active site to the solvent to perform catalytic function. This isoform evidently regulates its enzymatic activity by association and dissociation of the two subunits [15].

The other splicing isoform of APT-I is a translated product of mature mRNA without the fourth exon that encodes 16 amino acid residues. The functional differences, if any, of the two isoforms remain to be elucidated. The 16 residues were found to be a part of homodimer interface, but they were distant from the catalytic site. Therefore, lack of the 16 residues in the short isoform seemed to result in modulation of homodimer stability, without perturbing the APT-I enzymatic activity. To support this idea, we modeled a 3D structure of the short isoform using the 3D structure of the long isoform as a template. The modeled structure of short isoform apparently showed no disturbance on the overall structure and catalytic residues of a monomer. The homodimer interface area of the long isoforms was about $1,300\,\text{Å}^2$. The 16 residues were a part of the interface with about the size of $400\,\text{Å}^2$, and the interface for the short isoforms was predicted to lose that area. A size of a homodimer interface is empirically correlated with a molecular weight of each subunit [16], and for APT-I (about 23 kDa), $1,300\,\text{Å}^2$ is a typical size for the homodimer interface. The predicted size of the homodimer interface for the short isoforms was less than $1,000\,\text{Å}^2$ and this was far off the empirical rule. The reduction of homodimer interface area may alter the condition of equilibrium and increase monomer population by disfavoring a formation of homodimer. The change in the population, in effect, makes the catalytic site likely have more chance to be exposed to the solvent and more chance to encounter the substrates than the catalytic site of the long isoform. This, in turn, can accelerate depalmitoylation of signal transduction proteins, and may impair interactions of them.

4 Concluding Remarks

3D structures of about half of the proteins predicted from genome sequences are now known by the outcome of structural genomics and by computational method. The fraction is gradually increasing owning to the worldwide effort of structural genomics. Proteins function with their 3D structures, and the knowledge of 3D structures is essential to elucidate protein functions. Here we showed importance of protein 3D structure for regulation of cellular function found in one of the cases of alternative splicing. In this case, a function to modulate signal transduction pathway is likely regulated by modifying 3D structures in two splicing isoforms.

Acknowledgment

This work was supported by Grants-in-Aid for Scientific Research on Priority Areas (C) "Genome Information Science" to KY and MG, and for Scientific Research (B) to MG from the Ministry of Education, Culture, Sports, Science and Technology of Japan (MEXT). The work of MS was supported by JSPS Research Fellowships for Young Scientists.

References

1. International Human Genome Sequencing Consortium (2001) Nature 409:860–921
2. International Human Genome Sequencing Consortium (2004) Nature 431:931–945
3. Brent MR (2005) Genome Res 15:1777–1786
4. Burley SK (2000) Nature Struct Biol 7 Suppl: 932–934
5. Vitkup D, Melamud E, Moult J, Sander C (2001) Nature Struct Biol 8:559–566
6. Chothia C, Lesk AM (1986) EMBO J 5:823–836
7. Baker D, Sali A (2001) Science 294:93–96
8. Yamaguchi A, Iwadate M, Suzuki E, Yura K, Kawakita S, Umeyama H, Go M (2003) Nucleic Acids Res 31:463–468
9. Ogata K, Umeyama H (2000) J Mol Graph Model 18:258–272
10. Kawabata T, Fukuchi S, Homma K, Ota M, Araki J, Ito T, Ichiyoshi N, Nishikawa K (2002) Nucleic Acids Res 30:294–298
11. Yura K, Yamaguchi A, Go M (2006) J Str Func Genomics in press
12. Ast G (2004) Nature Rev Genet 5:773–782
13. Stamm S, Ben-Ari S, Rafalska I, Tang Y, Zhang Z, Toiber D, Thanaraj TA, and Soreq H (2005) Gene 344:1–20
14. Yura K, Shionyu M, Hagino K, Hijikata A, Hirashima Y, Nakahara Eguchi T, Shinoda K, Yamaguchi A, Takahashi K, Itoh T, Imanishi T, Gojobori T, Go M (2006) Gene 380:63–71
15. Devedjiev Y, Dauter Z, Kuznetsov SR, Jones TL, Derewenda ZS (2000) Structure 8:1137–1146
16. Jones S, Thornton JM (1996) Proc Natl Acad Sci USA 93:13–20

Invited Papers

Achromatic and Chromatic Visual Information Processing and Discrete Wavelets

Hitoshi Arai

Summary. In this paper we propose a general scheme of discrete wavelet analysis for studying visual information processing. After surveying our mathematical models of the early vision, we give a computer simulation of the occurrence of an achromatic visual illusion and a chromatic one.

1 Introduction

For the past two decades wavelet analysis has been applied to many research areas. This paper is concerned with applications of discrete wavelets to mathematical study on vision science. In particular, I will describe how discrete wavelets relate to visual information processing, and propose a general scheme of discrete wavelet analysis of vision. So far, several studies on vision have been based on the Gabor or DOG functions. However for the computational study on vision, discrete wavelets have many advantages as is shown later.

The organization of this paper is as follows: In Sect. 2, I will propose a general scheme of wavelet analysis for vision. Mathematical models proposed in [1–4] are regarded as partial realizations of this general scheme. By using these models, I will show in Sect. 3 new computer simulations on the occurrence of visual illusions related to achromatic and chromatic visual information processing.

2 A New Scheme for Discrete Wavelet Analysis of Vision

Let us begin with a brief review of physiological or psychophysical facts on the visual information processing. As is well known, our real visual system encodes incoming images in the retina, and then sends the coded signals through LGN to the striate cortex, V1. In V1, the signals are mapped as the so-called retinotopic maps and processed by cortical cells. There are two different opinions about the role of simple cells in V1. One of them is that they perform as zero-crossing detectors [5]. Another is that they behave like band-pass filters with respect to spatial frequencies [6, 7]. Indeed simple cells have multiple-oscillatory receptive fields and the corresponding spatial frequency functions are tuned to narrow ranges (see [6, Fig. 4.14]). The former opinion is one of bases of Marr's theory of vision [8]. However, also the latter opinion by De Valois is also compelling, because it is compatible with the multiple-channel hypothesis known in psychophysics. It asserts that human's visual system has a number of spatial frequency channels which are tuned to a relatively narrow range, and that the contrast sensitivity function is the envelope of these channels [9, 10]. There is a considerable number of psychophysical, physiological, and anatomical evidence supporting this hypothesis (see [6, Chap.6]). In this paper I intend to study mathematical model based on De Valois' hypothesis.

There is one more important fact from physiology and psychophysics. It is that the visual information is not processed only locally by simple cells, but it is processed also globally by horizontal connection. For example, V1 has a mechanism called "stimulus context-dependent modulation" [11, Chap.10].

After these processing in V1, the encoded biological signals are sent further to higher levels of visual cortex.

Now let us turn to mathematical scheme. A mathematical and computational problem which we have to consider is to construct a mathematical model of the above mentioned visual information processing and to give computer simulations of them. For studying this problem, it should be noted that quantitative receptive field profiles of cells in the retina and in LGN are similar to the so-called Mexican hat function in shape while the receptive fields of simple cells in V1 look like the Gabor function or the DOG (difference of Gaussian) function. From this reason, it has been recognized that these functions are a mathematical realization of the above mentioned cells (see [6, 12]). However in this paper I propose to use discrete, stationary, compactly supported wavelets [13–15] or framelets [4] instead of Mexican hat, Gabor, and DOG functions. There are several reasons why we use such wavelets: One of them is that Gabor and DOG functions have noncompact supports, therefore from viewpoint of computation by a computer we need to make finite impulse response (FIR) filters from these functions by some approximation. However this means that the inversion formula for the approximated FIR filters does not hold rigorously. On the other hand, a number of discrete wavelets or framelets are themselves FIR filters, and moreover stationary filter banks consisting of such wavelets or framelets have the perfect reconstruction property, which is abbreviated to PR. I will mention later why PR is important for our study. Now let me show remaining reasons. The second reason is that stationary wavelet filter banks can be naturally equipped with a multiresolution analysis which is regarded as a good mathematical realization of the multiple-channel hypothesis. This structure is important for studying V1. The third reason is that the multiresolution analysis provides a fast algorithm for computation. The fast computation is practically helpful in order to carry out computer experiments for finding an appropriate model of visual information processing. In addition to these reasons, I would like to note also that by many computer experiments, the stationary filter banks are robust with respect to nonlinear global processing, while the usual decimated filter banks produce unexpected distortion of signals by global processing. The author considers that one cause is that the decimated filter banks do subsampling of data.

From these reasons, wavelets are suitable for encoders of incoming images. However, we need to select carefully wavelets which are fit for our purpose: for instance, since the receptive fields of cells in the retina or LGN are circular, so for modeling of processing in that place we need to employ nonseparable wavelets. On the other hand, simple cells in V1 have orientation selectivity as well as frequency selectivity. Therefore their mathematical realization should be wavelets or framelets having orientation selectivity. After selecting appropriate wavelets, another important problem arises, that is,

(P) How the global processing in V1 and the processing in higher levels of visual cortex are implemented mathematically?

From what we have seen, we can conclude that the mathematical analogue of the global processing of visual information should be implemented as a processing on wavelet coefficients. However it is difficult to find mathematical models of all such processing. The author considers that a way for studying the problem (P) is to find models step by step or one by one.

Let us now return to the question why PR is important for our study. The answer is related deeply to the problem (P). In fact, if the filter bank has not PR, even if one can construct a mathematical model of global visual information processing, one can not judge whether it is appropriate or not, because changes of the outputted image may be due to the failure of the filter bank itself.

Lastly, I would like to note that to simulate the occurrence of visual illusions is necessary in order to know whether a constructed model is appropriate or not, because perceiving of illusions is a feature of human's vision.

3 Mathematical Modeling of a Global Processing

In the previous paper [1], the author gave a mathematical model of one of global processing, which is regarded as a partial answer to problem (P). It is that if an incoming image contains strong contrasts, weak contrast parts of the image are inhibited, but if the image has no strong contrast parts, weak contrast parts are enhanced (see [1] for detail). We called this phenomenon the contrast induction effect. By our computer simulations we found that many so-called lightness illusions are explained by using this effect [1]. By the way, it is known that the classical lateral inhibition theory for the retina can explain the occurrence of many lightness illusions such as Mach band (see [16]).

Fig. 1. Mach ring

However, there are a number of lightness illusions which the lateral inhibition theory cannot explain (see [1, 17]). Hence it has been conjectured that the occurrence of such illusions is a consequence of some postretinal, nonlinear processing. Our model equipped with the contrast induction effect can simulate such lightness illusions as well as illusions to which we can apply the classical theory [1]. These simulations imply that the contrast induction effect is one cause of lightness illusions. In this section, we give new simulations of two illusions which support our above assertion. The two illusions are variations of the classical Mach band which were made by S. Arai for computer experiments. Let me begin with studying one of them (see Fig. 1). We call Fig. 1 Mach ring.

In Fig. 1 there are light discs and dark discs. The graph of the luminance of each disc is approximately a truncated cone (Fig. 2(left)). However, we

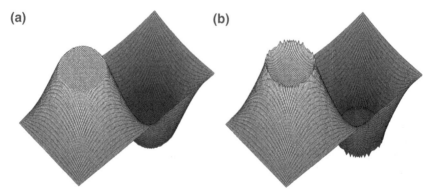

Fig. 2. Luminance of a part of Mach ring (*left*) and of output image (*right*)

 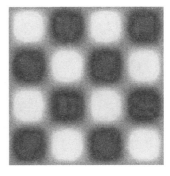

Fig. 3. Colored Mach ring. *Left*: original image, *Right*: output image

perceive a brighter (respectively, darker) ring at the boundary of each highland (respectively, valley). Since the receptive fields of cells in the retina are circular, the lateral inhibition theory can explain this illusion. On the other hand, since simple cells have oval receptive fields, simple cells having one same directional selectivity seem to contribute only partially production of the illusion. However, since simple cells have various directional selectivity, several types of simple cell can produce integrative the illusion. In fact, our model constructed in [1] can simulate the occurrence of Mach ring (see Fig. 2(right)). Therefore our assertion about contribution of the contrast induction effect to the occurrence of lightness illusion is also true for Mach ring.

Now I turn to examine chromatic visual information processing. The model in [1] is for achromatic visual information processing. However, by modifying it, the author and Arai constructed a computational model of color vision, and gave simulations of several illusions related to color [3]. Here we are concerned with a colored Mach ring (Fig. 3 (left)): If we input this image to our computational system for color vision, it outputs Fig. 3(right). Figure 4 depicts a part of the original image in sRGB and Fig. 5 a part of the output image in sRGB. This simulation provides us an evidence that our computational

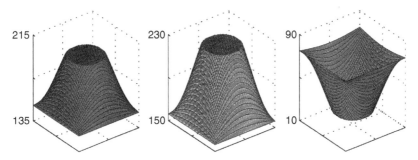

Fig. 4. Colored Mach ring in terms of sRGB

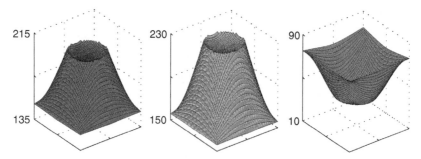

Fig. 5. Output image in terms of sRGB

system proposed in [3] is an appropriate model of chromatic visual information processing.

4 Conclusion

In this paper I proposed a general scheme of discrete wavelet analysis of vision. In this scheme, an important problem is to find mathematical implementation of global nonlinear processing by the brain (see problem (P)). The mathematical models of the contrast induction effect formulated in [1,3] are regarded as a partial answer to problem (P). In fact, by using of these models we were able to give computer simulations on several visual illusions [1,3]. In this paper more new simulations of two illusions related to lightness and color were given.

Acknowledgment

The author is grateful to Shinobu Arai for her considerable assistance. Without her help and discussion with her he could not finish this paper.

References

1. Arai H (2005) Interdisciplinary Inform. Sci. 11: 177–190
2. Arai H, Arai S (2005) VISION, J. Vision Soc. Jpn. 17: 259–265
3. Arai H, Arai S (2006) Wavelets and illusions related to chromatic visual information processing, preprint in preparation
4. Arai H, Arai S (2006) Simple constructions of finite shift-invariant directional framelet filterbanks for visual information processing, preprint
5. Marr D, Hildreth EC (1980) Proc. R. Soc. London, 200: 187–217
6. De Valois RL, De Valois KK (1988) Spatial vision, Oxford University Press, New York

7. De Valois RL, Albrecht DG, and Thorell LG (1982) Vis. Res. 22: 545–559
8. Marr D (1982) Vision: A computational investigation into the human representation and processing of visual information, W.H. Freemann, San Francisco
9. Blakemore C, Campbell FW (1969) J. Physiol. 203: 237–260
10. Campbell FW, Robson JG (1968) J. Physiol. 197: 551–566
11. Fukuda Y, Sato H (2002) Brain and vision (Japanese), Kyoritsu, Tokyo
12. Blakeslee B, McCourt ME (2004) Vision Res. 44: 2483–2503
13. Mallat S (1998) A wavelet tour of signal processing, Academic, San Diego
14. Nason GP, Silverman BW (1995) Lect. Notes Statis. 103: 281–299
15. Coifman RR, Donoho DL (1995) Lect. Notes Statis. 103: 125–150
16. Palmer SE (1999) Vision science, photons and phenomenology, MIT, Cambridge, MA; London
17. Spillmann L (1994) Perception 23: 691–708

On Decaying Two-Dimensional Turbulence in a Circular Container

Kai Schneider and Marie Farge

Summary. Direct numerical simulation of two-dimensional decaying turbulence in a circular container with no-slip boundary conditions are presented. Starting with random initial conditions the flow rapidly exhibits self-organisation into coherent vortices. We study their formation and the role of the viscous boundary layer on the production and decay of integral quantities. The no-slip wall produces vortices which are injected into the bulk flow and tend to compensate the enstrophy dissipation.

1 Introduction

In oceanography two-dimensional turbulence plays an important role, e.g. in the vortex formation in coastal currents. Experiments in rotating tanks, e.g. in [1], leading to quasi two-dimensional geostrophic flows, have shown the formation of long-lived coherent vortices. Only few numerical simulations of two-dimensional turbulence in bounded circular domains have been performed so far. Decaying two-dimensional turbulent flows in circular domains with no-slip boundary conditions have been computed in [7–9], using a spectral method with Bessel functions of the first kind. This pure spectral scheme has a prohibitive numerical cost and therefore these simulations were limited to low Reynolds numbers, $Re < 10^3$, where Re is based on the rms initial velocity and the circle radius. In [3] numerical simulations of forced two-dimensional turbulent flows in circular geometry for Reynolds numbers up to 3,500 using a Tchebycheff–Fourier discretisation have been presented. The aim of the present paper is to present direct numerical simulation (DNS) of two-dimensional decaying turbulence in a circular geometry with higher initial Reynolds-number of 5×10^4 computed at resolution $N = 1,024^2$ and to compare the results with low Reynolds number flows, $Re = 1,000$, computed at resolution $N = 256$.

2 Governing Equations and Numerical Discretisation

The numerical method is based on a Fourier pseudospectral method with semi-implicit time integration and adaptive time-stepping [11]. The circular container Ω of radius $R = 2.8$ is imbedded in a periodic square domain and the no-slip boundary conditions on the wall $\partial\Omega$ are imposed using a volume penalisation method [2]. The Navier–Stokes equations are then solved in a square domain of size $L = 2\pi$ using the vorticity–velocity formulation with periodic boundary conditions. Numerical validations can be found in [5, 11]. The resulting equation in vorticity–velocity formulation reads,

$$\partial_t \omega + \boldsymbol{u} \cdot \nabla \omega - \nu \nabla^2 \omega + \nabla \times \left(\frac{1}{\eta} \chi \boldsymbol{u}\right) = 0,$$

where $\boldsymbol{u}$ is the divergence-free velocity field, i.e. $\nabla \cdot \boldsymbol{u} = 0$, $\omega = \nabla \times \boldsymbol{u}$ the vorticity, ν the kinematic viscosity and $\chi(\boldsymbol{x})$ a mask function which is 0 inside the fluid, i.e. $\boldsymbol{x} \in \Omega$, and 1 inside the solid wall. The penalisation parameter η is chosen to be sufficiently small ($\eta = 10^{-3}$) [11].

Different integral quantities of the flow can be derived [6]. The energy E, enstrophy Z and palinstrophy P are defined as

$$E = \frac{1}{2}\int_\Omega |\boldsymbol{u}|^2 d\boldsymbol{x}, \; Z = \frac{1}{2}\int_\Omega |\omega|^2 d\boldsymbol{x}, \; P = \frac{1}{2}\int_\Omega |\nabla \omega|^2 d\boldsymbol{x}, \tag{1}$$

respectively.

The energy dissipation is given by $d_t E = -2\nu Z$ and the enstrophy dissipation by

$$d_t Z = -2\nu P + \nu \oint_{\partial\Omega} \omega(\boldsymbol{n} \cdot \nabla \omega) ds, \tag{2}$$

where $\boldsymbol{n}$ denotes the outer normal vector with respect to $\partial\Omega$. The surface integral reflects the enstrophy production at the wall involving the vorticity and its gradients.

3 Numerical Results

The numerical simulations are initialised with a correlated Gaussian noise with zero angular momentum and an energy spectrum $E(k) \propto k^{-4}$. In the first simulation presented here, the initial Reynolds number is chosen to be $Re = 2R\sqrt{2E}/\nu = 5 \cdot 10^4$ and the numerical resolution is $N = 1{,}024^2$. We introduce a dimensionless time $\tau = t/t_e$ based on the initial eddy turnover time $t_e = 1/\sqrt{2Z(0)} = 0.061$. The flow has been integrated for $650\,t_e$ corresponding to more than 10^5 time steps. Figure 1 shows a snap shot of the vorticity field (left) at $\tau = 320$ together with the corresponding stream function (right). We observe the formation of vorticity sheets at the wall which roll

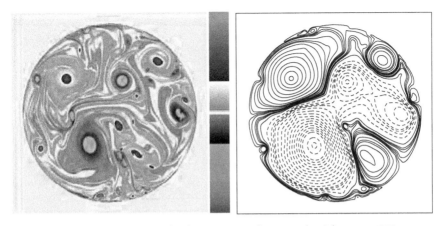

Fig. 1. Vorticity field (*left*) and streamfunction (*right*) at $\tau = 320$

up into coherent vortices. This active unstable strong boundary layer persists throughout the simulation. The resulting continuous injection of vorticity and vorticity gradients into the flow leads to a concomitant increase of the energy dissipation. Where the boundary layer detaches from the wall we observe the formation of dipolar vortices, which then move into the bulk flow and interact with other vortices as observed in rotating tanks [1].

Figure 2 shows vertical cuts of the vorticity, the velocity components at $\tau = 320$, together with the mask function. The cuts illustrate the spiky behaviour of the vorticity and the stiffness of the problem due to the strong gradients to be resolved.

In Fig. 3 we plot the time evolution of energy, enstrophy and palinstrophy. We observe that the kinetic energy slowly decays. At the final instant the

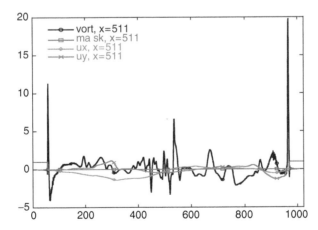

Fig. 2. Vertical cuts of vorticity, velocity components at $\tau = 320$

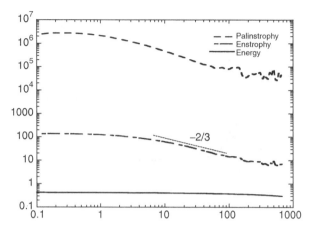

Fig. 3. Time evolution of energy E, enstrophy Z and palinstrophy P in log–log coordinates

energy has lost 71% of its initial value, while the enstrophy has decreased to only 5.1% and the palinstrophy to only 1.5% of their initial values. The enstrophy exhibits a self-similar decay over one decade (from $\tau = 10$ to about 100), proportional to $t^{-2/3}$. Note that this is much slower than in double periodic simulations [10] where typically a slope of -1 is observed for the enstrophy decay. At later times, for $\tau > 150$, we also observe a non monotonous behaviour for Z and P which is due to the generation of vorticity and its gradients at the no-slip wall.

The time evolution of the different terms in (3) for the enstrophy dissipation are shown in Fig. 4. We observe a monotonous decay of all terms up

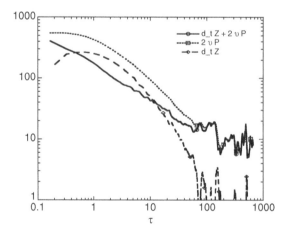

Fig. 4. Time evolution $d_t Z$, $2\nu P$ and $\nu \oint_{\partial \Omega} \omega(\mathbf{n} \cdot \nabla \omega) ds = d_t Z + 2\nu P$

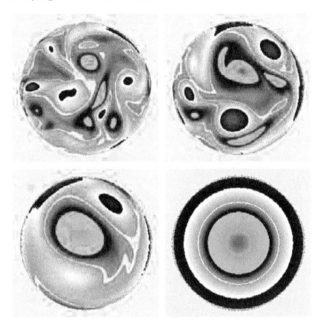

Fig. 5. Vorticity fields at $\tau = 5, 10, 30$ and 200. Resolution $N = 256^2$ and $Re = 800$

to $\tau = 100$. The enstrophy production term at the wall yields a power-law behaviour with slope $-2/3$, and for later times oscillations can be observed. Furthermore, the enstrophy production at the wall ($\nu \oint_{\partial\Omega} \omega(\boldsymbol{n} \cdot \nabla \omega) \mathrm{d}s$) coincides with the term νP for $\tau > 100$. This implies that the enstrophy dissipation $\mathrm{d}_t Z$ becomes negligible and oscillates around zero.

In the second simulation we compute a flow at initial Reynolds number $Re = 1{,}000$ with resolution $N = 256^2$. Four snapshots of the vorticity field at $\tau = 5, 10, 30$ and 200 are shown in Fig. 5. We also observe the formation of coherent vorticites and vortex sheets. However, the boundary layer is less pronounced and the flow rapidly decays. At $\tau = 200$ we observe already a quasisteady state, a negative circular vortex surrounded by a vortex ring of positive vorticity, which corresponds to theoretical predictions.

The evolutions of the integral quantities shown in Fig. 6 confirm the rapid decay of the flow, which in contrast to the high Reynolds number simulation is monotonuous.

4 Conclusion

In conclusion, we have shown, by means of DNS performed in circular geometry, that no-slip boundaries play a crucial role for decaying turbulent flows. At early times we observe a decay of the flow which leads to self-organisation and the emergence of vortices in the bulk flow, similarly to flows in double

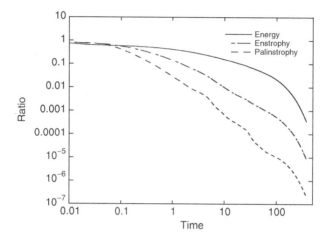

Fig. 6. Time evolution of energy, enstrophy and palinstrophy for $Re = 1,000$. The values have been normalised with the corresponding values at $\tau = 0$

periodic boxes. In the high Reynolds number case, the production of coherent vortices at the boundary compensates the enstrophy dissipation and the flow decay is drastically reduced. This is reflected in the time evolution of enstrophy and palinstrophy which decay in a non monotonous way, while for the low Reynolds number simulation both quantities decay monotonously. More details on the high Reynolds number simulation can be found in [12].

Acknowledgement

We thankfully acknowledge financial support from Nagoya University and from the contract CEA/EURATOM n⁰ V.3258.001.

References

1. J. Aubert, S. Jung and H.L. Swinney. *Geophys. Res. Lett.* **29**, 1876, 2002
2. P. Angot, C.H. Bruneau and P. Fabrie. *Numer. Math.* **81**, 497–520, 1999
3. H.J.H. Clercx, A.H. Nielsen, D.J. Torres and E.A. Coutsias. *Eur. J. Mech. B-Fluids* **20**, 557–576, 2001
4. H.J.H. Clercx, S.R. Maasen and G.J.F. van Heijst. *Phys. Rev. Lett.* **80**, 5129–5132, 1998
5. N. Kevlahan and J.-M. Ghidaglia. *Eur. J. Mech./B*, **20**, 333–350, 2001
6. R.H. Kraichnan and D. Montgomery. *Rep. Progr. Phys.* **43**, 547–619, 1980
7. S. Li and D. Montgomery. *Phys. Lett. A* **218**, 281–291, 1996
8. S. Li, D. Montgomery and B. Jones. *J. Plasma Phys.* **56**, 615–639, 1996

9. S. Li, D. Montgomery and B. Jones. *Theor. Comput. Fluid Dyn.* **9**, 167–181, 1997
10. W.H. Matthaeus, W.T. Stribling, D. Martinez and S. Oughton. *Phys. Rev. Lett.* **66**, 2731–2734, 1991
11. K. Schneider. *Comput. Fluids*, **34**, 1223–1238, 2005
12. K. Schneider and M. Farge. *Phys. Rev. Lett.* **95**, 244–502, 2005

A Study Toward Improving Accuracy of Large Scale Industrial Turbulent Flow Computations

Kazuhiko Suga

1 Introduction

With recent advanced high performance computers and commercial CFD software, it is now possible for industrial engineers to perform very large scale applications. However, near-wall resolution requirements for numerical grids still make sublayer-resolving turbulence models prohibitively expensive in industrial applications. This is particularly true for high Reynolds number and/or high Prandtl number flows which require very high grid resolution near walls. Moreover, for flows over rough surfaces which are common in industrial applications, sublayer-resolving turbulence models are not practical at all since one cannot hope to resolve the details of each small wall-roughness element.

Hence, industrial engineers still rely on so-called "wall-function" (WF) approaches. Since the WF is economical and stable for computation, it has been over used in large scale applications by RANS, TRANS, and LES.[1]

However, the most part of the "standard" log-law WF strategies was proposed in the 1970s with the assumption of semilogarithmic variations of the near-wall velocity and temperature. It is now well known that such assumptions do not apply in flows with strong pressure gradients and separation. Therefore, for improving accuracy of large scale industrial applications, it is particularly important to innovate the WF approaches.

In order to address such industrial requirements, a novel near-wall treatment [1], which integrates simplified momentum and energy equations analytically over the control volumes adjacent to a wall, is generalized for the effects of fine-grain surface roughness [2,3] by the group involving the present author. This paper summarizes and presents the latest development of the analytical WF (AWF) methodology for RANS computations with up-to-date application results by an unstructured grid code.

---

[1] RANS: Reynolds Averaged Navier-Stokes simulation; TRANS: Transient RANS; LES: Large Eddy Simulation

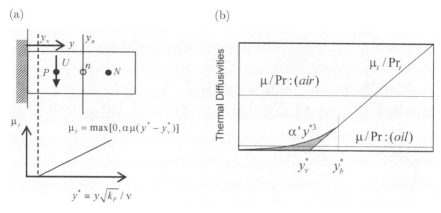

Fig. 1. Near-wall eddy viscosity and thermal diffusivity

2 Analytical Wall-Function

In the AWF, the wall shear stress and heat flux are obtained through the analytical solution of simplified near-wall versions of the transport equation for the wall-parallel momentum and temperature. In case of the forced convection regime, the main assumption required for the analytical integration of the transport equations is a prescribed variation of the turbulent viscosity μ_t. In the context of Fig. 1a, the distribution of μ_t over the wall-adjacent cell P is modeled as in a one-equation turbulence model:

$$\mu_t = \rho c_\mu k^{1/2} \ell = \rho c_\mu k^{1/2} c_\ell y \simeq \alpha \mu y^*, \tag{1}$$

where $y^* \equiv y k_P^{1/2}/\nu$, $\alpha = c_\ell c_\mu$, $c_\ell = 2.55$, and $c_\mu = 0.09$, and ρ, μ, ν, ℓ, y, and k_P are, respectively, the fluid density, the molecular viscosity, the kinematic viscosity, the turbulent length scale, the wall-normal distance and k (turbulence energy) at the node P. In order to consider viscous sublayer effects, instead of introducing a damping function, the profile of μ_t is modeled as:

$$\mu_t = \max\{0, \alpha\mu(y^* - y_v^*)\}, \tag{2}$$

in which μ_t grows from the edge of the viscous sublayer: $y_v^*(\equiv y_v k_P^{1/2}/\nu)$. Then, the simplified forms of the momentum and energy equations become

$$\frac{\partial}{\partial y^*}\left[(\mu + \mu_t)\frac{\partial U}{\partial y^*}\right] = \frac{\nu^2}{k_P}\left[\frac{\partial}{\partial x}(\rho U U) + \frac{\partial P}{\partial x}\right] = C_U, \tag{3}$$

$$\frac{\partial}{\partial y^*}\left[\left(\frac{\mu}{\Pr} + \frac{\mu_t}{\Pr_t}\right)\frac{\partial \Theta}{\partial y^*}\right] = \frac{\nu^2}{k_P}\left[\frac{\partial}{\partial x}(\rho U \Theta) + S_\theta\right] = C_T, \tag{4}$$

where x, U, P, Θ, S_θ, $\Pr$, and $\Pr_t$ are, respectively, the wall parallel direction, the x-component of the mean velocity, the mean pressure, the mean temperature, the local heat source, Prandtl number and a prescribed turbulent Prandtl number, taken as 0.9. The further assumption made is that convective transport and the wall-parallel pressure gradient $\partial P/\partial x$ do not change across the wall-adjacent cell which is a standard treatment in the finite volume method. Thus, the right hand side (rhs) terms C_U and C_T of (3) and (4) can be treated as constant. Then, the equations can be integrated analytically over the wall-adjacent cell giving:

if $y^* < y_v^*$,

$$dU/dy^* = (C_U y^* + A_U)/\mu, \qquad (5)$$
$$d\Theta/dy^* = \Pr(C_T y^* + A_T)/\mu, \qquad (6)$$

if $y^* \geq y_v^*$,

$$dU/dy^* = (C_U y^* + A_U')/[\mu\{1 + \alpha(y^* - y_v^*)\}], \qquad (7)$$
$$d\Theta/dy^* = \Pr(C_T y^* + A_T')/[\mu\{1 + \alpha_\theta(y^* - y_v^*)\}], \qquad (8)$$

where $\alpha_\theta = \alpha \Pr/\Pr_t$. Further integration to obtain U and Θ is also straightforward. The integration constants A_U, A_T etc. are determined by applying boundary conditions at the wall, y_v and the point n. The values at n are determined by interpolation between the calculated node values at P and N, while at y_v a monotonic distribution condition is imposed by ensuring that U, Θ and their gradients should be continuous at $y = y_v$. Notice that the obtainable logarithmic velocity equation by integrating (7) includes C_U. This implies that the velocity profile has sensitivity to the pressure gradient since C_U includes $\partial P/\partial x$.

Consequently, the wall shear stress and wall heat flux can be expressed as:

$$\tau_w = \mu \left.\frac{dU}{dy}\right|_{wall} = \mu \frac{k_P^{1/2}}{\nu}\left.\frac{dU}{dy^*}\right|_{wall} = \frac{k_P^{1/2} A_U}{\nu}, \qquad (9)$$

$$q_w = -\frac{\rho c_p \nu}{\Pr}\left.\frac{d\Theta}{dy}\right|_{wall} = -\frac{\rho c_p \nu}{\Pr}\frac{k_P^{1/2}}{\nu}\left.\frac{d\Theta}{dy^*}\right|_{wall} = -\frac{\rho c_p k_P^{1/2} A_T}{\mu}, \qquad (10)$$

where c_p is the specific heat capacity at constant pressure. For the $k - \varepsilon$ models, the local generation rate of k, $P_k(= \nu_t(dU/dy)^2)$, is written as:

$$P_k = \begin{cases} 0 & , \text{ if } y^* < y_v^* \\ \dfrac{\alpha k_P}{\nu}(y^* - y_v^*)\left(\dfrac{C_U y^* + A_U'}{\mu\{1 + \alpha(y^* - y_v^*)\}}\right)^2 & , \text{ if } y^* \geq y_v^* \end{cases} \qquad (11)$$

which can then be integrated over the wall-adjacent cell to produce an average value $\overline{P_k}$ for use in solving the k equation. For the dissipation rate ε, the following model is employed:

$$\varepsilon = \begin{cases} 2\nu k_P/y_\varepsilon^2, & \text{if } y < y_\varepsilon \\ k_P^{1.5}/(c_\ell y), & \text{if } y \geq y_\varepsilon. \end{cases} \qquad (12)$$

The characteristic dissipation scale y_ε can be defined as $y_\varepsilon^* = 2c_\ell$ to ensure a continuous variation of ε at y_ε. Thus, the cell averaged value is obtained as:

$$\bar{\varepsilon} = \begin{cases} 2k_P^2/(\nu y_\varepsilon^{*2}) & , \text{ if } y_\varepsilon^* > y_n^* \\ \dfrac{1}{y_n}\left(y_\varepsilon \dfrac{2\nu k_P}{y_\varepsilon^2} + \displaystyle\int_{y_\varepsilon}^{y_n} \dfrac{k_P^{1.5}}{c_\ell y}dy\right) = \dfrac{k_P^2}{\nu y_n^*}\left[\dfrac{2}{y_\varepsilon^*} + \dfrac{1}{c_\ell}\ln\left(\dfrac{y_n^*}{y_\varepsilon^*}\right)\right], & \text{ if } y_\varepsilon^* \leq y_n^*. \end{cases} \quad (13)$$

The optimized value of the constant y_v^* for smooth walls is 10.7.

2.1 For Rough-Wall Flows

For rough-wall turbulence, the AWF strategy involves the use of the dimensionless roughness height $h^* = hk_P^{1/2}/\nu$. In a rough-wall turbulence, y_v^* is no longer fixed and is modeled to become smaller. This provides that the modeled distribution of μ_t shown in Fig. 1a shifts toward the wall depending on h^*. At a certain value of the dimensionless roughness height, it is expediently allowed to have a negative value of y_v^* to give a positive value of μ_t at the wall. The optimized form is:

$$y_v^* = y_{vs}^*\{1 - (h^*/70)^m\}, \quad (14)$$
$$m = \max[\{0.5 - 0.4(h^*/70)^{0.7}\}, \{1 - 0.79(h^*/70)^{-0.28}\}], \quad (15)$$

where y_{vs}^* is the viscous sublayer thickness in the smooth wall case ($y_{vs}^* = 10.7$). Unlike in a sublayer over a smooth wall, the total shear stress includes the drag force from the roughness elements in the inner layer. This implies that the convective and pressure gradient contributions should be represented somewhat differently across the inner layer, below the roughness element height. Hence, the practice of simply evaluating the rhs of (3) in terms of nodal values needs modifying. It is thus assumed that the total shear stress remains constant across the roughness element height. Then,

$$C_U = \begin{cases} 0 & , \text{ if } y^* \leq h^* \\ \dfrac{\nu^2}{k_P}\left[\dfrac{\partial}{\partial x}(\rho UU) + \dfrac{\partial P}{\partial x}\right], & \text{ if } y^* > h^*. \end{cases} \quad (16)$$

In the energy equation, $\Pr_t$ is also no longer constant over the wall-adjacent cell. The reason for this is that, since the fluid trapped around the roots of the roughness elements forms a thermal barrier, the turbulent transport of the thermal energy is effectively reduced relative to the momentum transport. Thus, a functional profile is assumed in the roughness region of $y \leq h$ as:

$$\Pr_t = \Pr_t^\infty + C_0 \max(0, 1 - y^*/h^*), \quad C_0 = \dfrac{5.5}{1 + (h^*/70)^{6.5}} + 0.6. \quad (17)$$

Over the rest of the field ($y > h$), $\Pr_t = \Pr_t^\infty = 0.9$ is applied.

2.2 For High Pr Flows

Since the theoretical wall-limiting variation of μ_t is proportional to y^3, in the viscous sublayer, the AWF methodology does not count a certain amount of turbulent viscosity. Despite that, its effect is not serious for flow field prediction since the contribution from the molecular viscosity is more significant in the sublayer. This is also true for the thermal field prediction of fluids whose Prandtl number is less than 1.0. However, in high Prandtl number fluid flows such as oil flows, since the effect of the molecular thermal diffusivity becomes very small as in Fig. 1b, it is then necessary to consider the contribution from the turbulent thermal diffusivity inside the sublayer.

Thus, to correct the profile of μ_t it is assumed that its linear profile is connected to a function: $\alpha' y^{*3}$ at the point y_b^*, as illustrated in Fig. 1b. By referring to DNS data, the value of y_b^* is optimized as $y_b^* = 11.7$ and thus $\alpha' = \alpha(y_b^* - y_v^*)/y_b^{*3}$ is obtainable. Then, the simplified energy equation over the wall adjacent cell can be rewritten as:

$$\frac{\partial}{\partial y^*}\left[\left(\frac{\mu}{\mathrm{Pr}} + \frac{\alpha' \mu}{\mathrm{Pr}_t}y^{*3}\right)\frac{\partial \Theta}{\partial y^*}\right] = C_T, \quad \text{for } y^* \leq y_b^*, \tag{18}$$

$$\frac{\partial}{\partial y^*}\left[\left(\frac{\mu}{\mathrm{Pr}} + \frac{\alpha \mu}{\mathrm{Pr}_t}(y^* - y_v^*)\right)\frac{\partial \Theta}{\partial y^*}\right] = C_T, \quad \text{for } y^* > y_b^*. \tag{19}$$

By analytically integrating (18) and (19), one can easily obtain the integration constant appearing in (10).

3 Application Results

Figure 2a shows the superior performance of the AWF to that of the log-law WF (LWF) for the friction velocity in a recirculating flow over a sand dune

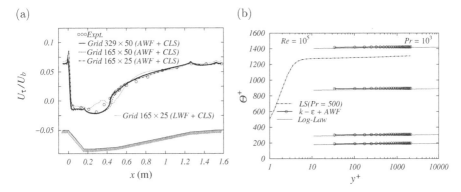

Fig. 2. Examples of the evaluation tests of the AWF; (a) friction velocity over a sand dune; (b) temperature distribution in high Pr turbulent channel flows

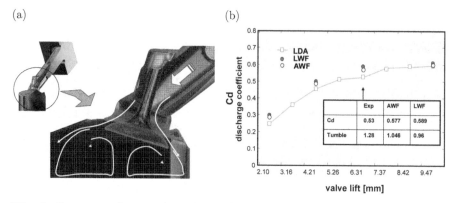

Fig. 3. Geometry of port-valve-cylinder flow and gas discharge coefficients by the standard $k - \varepsilon$ model (by courtesy of Dr. Basara of AVL LIST GMBH, Graz, Austria)

which is a rough-wall turbulent flow at Re = 175,000 with $h^+ \simeq 75$ at the inlet. In fact, in combination with the cubic nonlinear $k - \varepsilon$ model (CLS) [4], the AWF performs well with any grid resolution tested.

For thermal fields of high Prandtl number flows, Fig. 2b shows normalized temperature distribution in turbulent (smooth) channel flows at high Prandtl numbers. It is obvious that although the low Reynolds number $k - \varepsilon$ (LS) model [5] predicts too high a level of log-law temperature distribution, the AWF produces reliable distribution with the same grid resolution.

The final more complicated case is an IC Engine port-valve-cylinder flow. Figure 3a shows the geometry and the mesh used for this case. The flow comes through the intake-port and the valve section into the cylinder at Re $\simeq 10^5$. As shown in Fig. 3b, although it is not very significant, the distribution of the gas discharge coefficient Cd is in fact improved by the AWF. The tumble ratio is also better predicted by the AWF.

4 Concluding Remarks

(1) The AWF applies well to smooth and rough-wall turbulent heat transfer as well as high Pr fluid flows. (2) The AWF is fairly insensitive to mesh resolutions and works well with a mesh for the standard WF with almost the same computation cost as that for the standard WF. (3) Thus, the AWF improves large scale industrial applications without increasing the computation cost.

References

1. Craft TJ, Gerasimov AV, Iacovides H, Launder BE (2002) Progress in the generalization of wall-function treatments. Int J Heat Fluid Flow 23:148–160

2. Suga K, Craft TJ, Iacovides H (2005) Extending an analytical wall-function for turbulent flows over rough walls. In: Rodi W, Muras M (eds) Engineering Turbulence Modelling and Experiments 6. Elsevier, New York, pp. 157–166
3. Suga K, Craft TJ, Iacovides H (2005) An analytical wall-function for turbulent rough wall heat transfer. In: Proceedings of the 4th International Symposium on Turbulence and Shear Flow Phenomena, Williamsburg, USA, pp. 1083–1088
4. Craft TJ, Launder BE, Suga K (1996) Development and application of a cubic eddy-viscosity model of turbulence. Int J Heat Fluid Flow 17:108–115
5. Launder BE, Sharma BI (1974) Application of the energy-dissipation model of turbulence to the calculation of flow near a spinning disc. Lett Heat Mass Transfer 1:131–138

Direct Numerical Simulations of Turbulent Premixed Flames with Realistic Kinetic Mechanisms

Mamoru Tanahashi, Yuzuru Nada, Nobuhiro Shiwaku, and Toshio Miyauchi

Summary. Direct numerical simulation (DNS) of turbulent combustion with realistic kinetic mechanisms is one of the most difficult problems in computational science. In this paper, three-dimensional DNS of hydrogen-air and methane-air turbulent premixed flames propagating in homogeneous isotropic turbulence are conducted to investigate local flame structures in turbulent premixed flames. Detailed analyses of DNS results show that proper decompositions of strain rates and curvature at the flame front are important to understand the mechanism of local extinction of turbulent premixed flames.

1 Introduction

Turbulent combustion is one of the most difficult problems in computational science, because this phenomenon includes very complicated interaction of turbulence and chemical reactions. In combustion chemistry, very fast elementary reactions cause quite small time and spatial scales, which is well known as stiffness problem. Direct numerical simulations (DNS) of turbulent flows have been used for turbulence research in last few decades, whereas the first three-dimensional DNS of turbulent combustion with a detailed kinetic mechanism has reported in 2000 [1]. The combustion can be classified into premixed flames and nonpremixed (diffusion) flames. The characteristics of the turbulent premixed flames have been classified by the ratio of the laminar burning velocity (S_L) to the turbulence intensity (u'_{rms}) and the ratio of the laminar flame thickness (δ_F) to the turbulence length scale (l), which is called as turbulent combustion diagram [2]. In the turbulent combustion diagram, the flame structure split into four regimes: wrinkled flamelets, corrugated flamelets, thin reaction zones, and broken reaction zones. In the wrinkled flamelets and the corrugated flamelets regimes, local flame structure is considered to be laminar flame with small curvature under strain field, whereas characteristics of the flame elements in the thin reaction zones and the broken reaction zones are supposed to be quite different from that of laminar flame. However, actual

Table 1. Numerical parameters for DNS of turbulent premixed flames

	mixture	Re_λ	Re_l	u'/S_L	l/δ_F	D/δ_F
HR37LL	hydrogen–air	37.4	144	0.85	169	44.1
HR37MM	hydrogen–air	37.4	144	1.70	84.3	30.3
HR37HS	hydrogen–air	37.4	144	3.41	42.2	17.9
HR60HM	hydrogen–air	60.8	203	3.39	90.1	18.7
HR97HM	hydrogen–air	97.1	515	5.78	122	10.6
MR37HS	methane–air	37.4	144	5.80	24.7	5.69

flame structure in turbulence has not been clarified yet [3]. In this study, DNS of hydrogen–air and methane–air turbulent premixed flames are conducted to investigate local flame structures in turbulent premixed flames.

2 DNS of Turbulent Premixed Flames

Details of DNS can be found in Tanahashi et al. [1, 4, 5]. Numerical parameters of DNS are listed in Table 1. In Table 1, the ratios of the most expected diameter of the fine scale eddy (D) of turbulence to δ_F are presented. D is eight times Kolmogorov length (η) in turbulence [6]. As for hydrogen–air turbulent premixed flames, a detailed kinetic mechanism which includes 12 species and 27 elementary reactions is used and all of the DNS were conducted for $\phi = 1.0$, 700 K and 0.1 MPa. The maximum Reynolds number based on Taylor micro scale (Re_λ) is 97.1. For methane–air turbulent premixed flames, a reduced kinetic mechanism which consists of 23 chemical species and 19 step reactions is used [5]. A methane–air mixture in unburned side is set to be $\phi = 1.0$ at 0.1 MPa and 950 K. In the turbulent combustion diagram proposed by Peters [2], HR37HS, HR97HM, and MR37HS are classified into the thin reaction zones. HR37LL is classified into the wrinkled flamelets and other cases are into the corrugated flamelets.

3 Local Flame Structure of Turbulent Premixed Flames

Figure 1 shows contour surfaces of the heat release rate with the axes of the fine scale eddies for HR97HM and MR37HS. The contour level of the heat release rate is $\Delta H/\Delta H_L \geq 1.0$, where ΔH_L denotes the maximum heat release rate of the laminar flame. The thickness of the axis is drawn to be proportional to the square root of the second invariant of the velocity gradient tensor on the axis. The second invariant (Q) is normalized by u'_{rms} and η in the unburned side. A thicker eddy possesses stronger swirling motion around the eddy. Note that the most expected diameter of these fine scale eddies is 8η and the maximum azimuthal velocity reaches to 3–$4u'_{rms}$ [7]. The number density of the fine

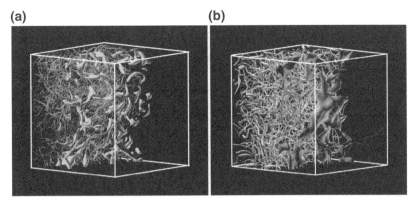

Fig. 1. Contour surfaces of heat release rate with the axes of the coherent fine scale eddies for HR97HM (**a**) and MR37HS (**b**)

scale eddies in the unit volume of the integral length scale (l^3) increases with the increase of Re_λ [8]. For high Reynolds number hydrogen–air turbulent premixed flames, the wrinkling of the flame surfaces also increases and appearance of three-dimensional structure increases because of high probability of the fine scale eddies which possess azimuthal velocity faster than S_L [9].

Differences between hydrogen–air and methane–air turbulent premixed flames can be observed in distribution of the heat release rate. In Fig. 1b, there is a large region of low heat release rate (see circle in Fig. 1b). This has never been observed in hydrogen–air turbulent premixed flames. The low heat release rate area shows about $0.3\Delta H_L$ although temperature is relatively uniform and still high. There is a possibility that this region represents the local extinction in turbulent premixed flames.

4 Statistical Characteristics of Local Flame Structure

To clarify effects of turbulence characteristics on local flame structure quantitatively, the statistical characteristics of flame elements are investigated. The flame fronts are defined as points with the local maximum temperature gradient. Probability density functions of local heat release rate in hydrogen–air and methane–air turbulent premixed flames are shown in Fig. 2. The local heat release rate is the maximum heat release rate in a flame element and normalized by ΔH_L. In the case of hydrogen–air turbulent premixed flames, the maximum heat release rate increases with the increase of turbulence intensity or the decrease of turbulence length scale, and reaches to $1.3\Delta H_L$ for the same Reynolds number. For the high Reynolds number case, probability of high local heat release rate increases compared with low Reynolds number cases. The variance of local heat release rate in methane–air turbulent premixed flame is much larger than those in hydrogen–air turbulent premixed

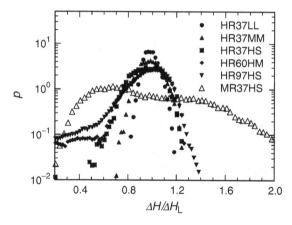

Fig. 2. Probability density functions of local heat release rate

flames, and there are lots of flame elements that show high heat release rate of $2.0\Delta H_\mathrm{L}$ and low heat release rate of $0.5\Delta H_\mathrm{L}$ along the flame front. For hydrogen–air turbulent premixed flames, probability density functions show their peaks at about $1.0\Delta H_\mathrm{L}$. For methane–air turbulent premixed flame, however, probability density function shows a weak peak at $0.7\Delta H_\mathrm{L}$. This value is lower than the maximum heat release rate of the laminar case.

In this study, the correlation between local heat release rates and geometrical shapes of flame front is examined by using principal curvatures (k_1 and k_2) [5,9]. Principal curvatures are defined as $k = k_1 + k_2 (k_1 < k_2)$, where k denotes mean curvature. Figure 3 shows the joint pdfs of principal curvatures of flame front with conditional mean local heat release rate for HR60HM and MR37HS. The flame elements convex toward the burned side are defined to have positive value and the principal curvatures are normalized by η. From

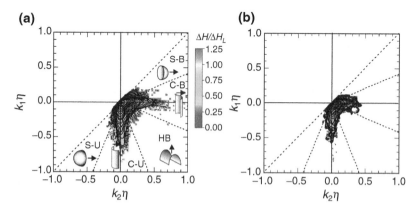

Fig. 3. Joint probability density functions of principal curvatures of flame front with conditional mean local heat release rate for HR60HM (**a**) and MR37HS (**b**)

two principal curvatures, flame shape can be classified into spherical surface convex toward the burned side (S-B), cylindrical surface convex toward the burned side (C-B), hyperboloidal surface (HB), cylindrical surface convex toward the unburned side (C-U) and spherical surface convex toward the unburned side (S-U). The conditional mean local heat release rate is shown in Fig. 3 by colors. For hydrogen–air turbulent premixed flames, number of flame elements in C-B and S-B regimes increases with the increase of Re_λ. Flame elements in S-B attribute to the spire and handgrip structures [10], and those in C-B regime are mainly created by fine scale eddies parallel to the flame front. In the case of HR60HM, the contributions of C-B and S-B regimes to the total heat release rate are clearly high and make up about 50% of the total heat release rate. These results suggest that C-B and S-B flame elements would dominate the total heat release rate and the turbulent burning velocity in high Reynolds number hydrogen–air turbulent premixed flames. Heat release rates tend to increase when the regime changes from S-U to S-B as shown in Fig. 3a, while the correlation between heat release rate and curvatures for methane–air turbulent premixed flame is not clear compared with hydrogen–air cases.

The strain rate tangential to the flame front has been discussed by $a_t = t_1 t_1 : \nabla u + t_2 t_2 : \nabla u$, where t_1 and t_2 represent unit vectors tangential to the flame front and are satisfying a relation of $t_1 \cdot t_2 = 0$ [11]. In this study, we introduce the minimum and the maximum strain rate on the flame surface to investigate strain rate effect correctly [5, 9]. Figure 4 shows joint pdfs of the minimum and maximum strain rate with conditional mean local heat release rate for HR60HM and MR37HS. a_{t1} and a_{t2} denote the minimum and maximum strain rates and are normalized by u'_{rms}/λ. From a_{t1} and a_{t2}, the tangential strain rate on the flame surface can be classified into three types: stretching in the two directions (S–S), stretching and compression in each

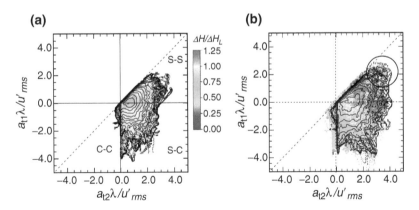

Fig. 4. Joint probability density functions of the minimum and maximum strain rate with conditional mean local heat release rate for HR60HM (**a**) and MR37HS (**b**)

direction (S–C), and compression in two direction (C–C). The most expected strain field is simple two-dimensional strain rate of the order of u'_{rms}/λ ($a_{t1} \approx 0$ and $a_{t2} \approx u'_{rms}/\lambda$). For both turbulent premixed flames, the flame elements in S–S regime are more than 50% and almost all of flame elements are under the stretching in one direction at least. Flame elements in C–C regime are scarcely observed and number of flame element in S–C regime increases with the increase of Re_λ. It should be noted that flame elements in S–S regime could be approximated by laminar flames observed in counterflow flame, whereas it is not the case for those in S–C regime.

In the cases of hydrogen–air turbulent premixed flames, correlations between heat release rates and tangential strain rates are hardly observed. However, in the case of methane–air turbulent premixed flame, they show relatively strong correlation. The heat release rates tend to increase from C–C regime to S–S regime. In S–S regime, the heat release rate also increases with the increase of the tangential strain rate. However, in S–S regime, the flame elements that are stretching strongly in two directions (see circle in Fig. 4b) tend to show low heat release rate. Most of these flame elements exist in the vicinity of low heat release rate region in Fig. 1b and they are considered to be closely related to local extinction of turbulent premixed flames.

5 Conclusions

In this study, DNS of three-dimensional hydrogen–air and methane–air turbulent premixed flames propagating in homogeneous isotropic turbulence are conducted to investigate the effects of turbulence characteristics on the local flame structure. The three-dimensional flame structures such as the handgrip and spire structure are important to determine the total heat release rate and turbulent burning velocity in high Reynolds number hydrogen–air turbulent premixed flames. In methane–air turbulent premixed flames, local heat release rate is well correlated with tangential strain rate at the flame front. The most expected strain rate is simple two-dimensional stretching of the order of u'_{rms}/λ. The flame elements stretched into two tangential directions shows high heat release rate. However, excessive stretching leads to the decrease of the heat release rate. This result indicates that proper decomposition of tangential strain rates is important to understand the mechanism of local extinction of turbulent premixed flames.

References

1. Tanahashi M, Fujimura M, Miyauchi T (2000) Proc. Combust. Inst. 28:529–535
2. Peters N (1999) J. Fluid Mech. 384:107–132
3. Tanahashi M, Nada Y, Ito Y, Miyauchi T (2002) Proc. Combust. Inst. 29: 2041–2049

4. Miyauchi T, Tanahashi M, Sasaki K, Ozeki T (1996) Transport Phenomena in Combustion 1095–1105
5. Tanahashi M, Kikuta S, Nada Y, Shiwaku N, Miyauchi T (2005) Proc. Fourth Int. Symp. Turbul. Shear Flow Phenom. 2:775–780
6. Tanahashi M, Miyauchi T, Ikeda J (1997) Proc. 11th Symp. Turbul. Shear Flow 1:4-17–4-22
7. Tanahashi M, Iwase S, Miyauchi T (2001) J. Turbul. 2:6
8. Tanahashi M, Iwase S, Uddin A, Takada N, Miyauchi T (2000) Therm. Sci. Eng. 8:29–38
9. Tanahashi M, Nada Y, Miyauchi T (2004) Proc. 6th World Cong. Comput. Mech. CD-ROM 181
10. Nada Y, Tanahashi M, Miyauchi T (2004) J. Turbul. 5:16
11. Candel S, Poinsot T. (1990) Combust. Sci. Technol. 70:1–15

Variational Monte Carlo Studies of Superconductivity in κ-BEDT-TTF Salts

Tsutomu Watanabe, Hisatoshi Yokoyama, Yukio Tanaka, and Jun-ichiro Inoue

Summary. In order to consider superconductivity (SC) arising in organic superconductors, κ-(BEDT-TTF)$_2$X, we study a Hubbard model on an anisotropic triangular lattice, using a variational Monte Carlo method. We reveal that a first-order superconductor-to-insulator transition takes place irrespective of strength of frustration, and robust d-wave SC appears for moderate strength of frustration and close to the transition. The origin of SC is an antiferromagnetic spin correlation.

1 Introduction

Organic compounds, κ-BEDT-TTF salts (κ-(ET)$_2$X) [1], are very intriguing in some points: (i) They have quasi-two-dimensional conducting planes, whose structure is an anisotropic triangular lattice, which has strong magnetic frustration. (ii) First-order superconductor (SC)-to-insulator (so-called Mott) transitions take place by applying (chemical) pressure (the critical temperature of SC (T_c) is around 10 K). The connectivity of the ET dimers can be properly modeled by an anisotropic triangular lattice with a hopping integral t in two directions and t' in the third direction ($t'/t = 0.5$–1.1) [2], where a hole exists on each unit dimer (half-filling). The system is quite similar to that of high-T_c cuprate SC, which has a two-dimensional square lattice ($t' = 0$) and gives rise to a SC-to-insulator transition by carrier doping. Therefore, it is important to clarify whether, in κ-(ET)$_2$X, a new superconducting (SC) mechanism is realized or an antiferromagnetic (AF) spin correlation causes superconductivity (SC) as in the cuprates [3].

So far many theoretical studies have considered this issue: mean-field [4] and fluctuation exchange (FLEX) [5] approximations. However, these methods could not correctly treat the SC-to-insulator transition, because they are approximations for weak Coulomb interaction of electrons. Therefore, our purpose is to reveal the mechanism of SC arising close to the transition in κ-(ET)$_2$X, using a variational Monte Carlo (VMC) method, which can study the physics irrelevantly to the strength of Coulomb interaction.

2 Method

As a model of the conducting plane in κ-(ET)$_2$X, we consider a Hubbard model on an anisotropic triangular lattice:

$$H = -\sum_{\langle i,j\rangle\sigma} t_{ij}\left(c^\dagger_{i\sigma}c_{j\sigma} + \text{H.c.}\right) + U\sum_i c^\dagger_{i\uparrow}c_{i\uparrow}c^\dagger_{i\downarrow}c_{i\downarrow}, \quad (1)$$

where $t_{ij} = t$ for nearest-neighbor-site pairs $\langle i,j\rangle$ in two of lattice directions, $t_{ij} = t'$ in the remaining direction, and $t_{ij} = 0$ otherwise. $c_{i\sigma}(c^\dagger_{i\sigma})$ exhibits annihilation (creation) operator of an electron with spin σ on i site, and $U(>0)$ is strength of onsite Coulomb repulsion. In this model, decrease of U/t corresponds to increase of pressure applied, because pressure tends to increase the band width (t). Throughout this paper, we fix the electron density at half-filling. We apply a VMC method to this model [6], which can treat the whole parameter space of U/t. As a variational trial wave function, we use a two-body Jastrow type: $\Psi = P\Phi$, where Φ is a one-body (Hartree-Fock) part expressed as a Slater determinant, and P is a correlation factor. As for P, we adopt a product of a onsite (Gutzwiller) factor $P_G(g)$ and an intersite factor $P_Q(\mu,\mu')$. Here, $P_G(g)$ controls the onsite Coulomb repulsion by parameter g, and $P_Q(\mu,\mu')$ binds a doublon (doubly occupied site) to an adjacent holon (empty site) in the direction of $t(t')$ by parameter $\mu(\mu')$. P_Q is a crucial factor to describe a Mott transition [7].

Regarding Φ, we study two states: (i) a BCS function with a $d_{x^2-y^2}$-wave singlet (or SC) gap $\Delta_\mathbf{k} = \Delta_d(\cos k_x - \cos k_y)$, as a SC state: $\Phi_d(\Delta_d, \zeta, \alpha)$ and (ii) a Hartree-Fock-type AF state: $\Phi_{AF}(\Delta_{AF})$. Here, Δ_d, ζ, and Δ_{AF} are variational parameters to be optimized, which correspond to the d-wave gap, chemical potential, and AF gap, respectively, in the Hartree-Fock theory. Furthermore, α is also a parameter to be optimized, which deforms the quasi Fermi surface in the d-wave state due to the effect of electron correlation.

The variational wave functions we treat here have up to six parameters. In this study, the optimization is carried out with millions of samples. We use the systems of $N_s = L \times L$ ($L = 10, 12,$ and 14) sites with periodic–antiperiodic boundary conditions.

3 Result

As an indicator of SC, we estimate a long-range average of a pair correlation function,

$$P_d(\mathbf{r}) = \frac{1}{4N_s}\sum_{\mathbf{j}}\sum_{\tau,\tau'=x,y}(-1)^{1-\delta(\tau,\tau')}\left\langle \Delta^\dagger_\tau(\mathbf{j})\Delta_{\tau'}(\mathbf{j}+\mathbf{r})\right\rangle, \quad (2)$$

where $\Delta^\dagger_\tau(\mathbf{j}) = (c^\dagger_{\mathbf{j}\uparrow}c^\dagger_{\mathbf{j}+\tau\downarrow} + c^\dagger_{\mathbf{j}+\tau\uparrow}c^\dagger_{\mathbf{j}\downarrow})/\sqrt{2}$. The long-range ($|\mathbf{r}| > 3$) average of the function (P^{ave}_d) corresponds to a possibility that a d-wave singlet Cooper

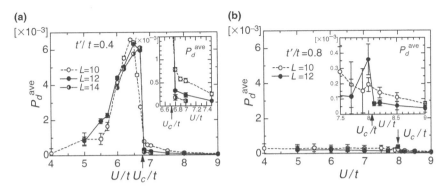

Fig. 1. The long-range average of a pair correlation function as a function of U/t for (a) $t'/t = 0.4$ and (b) 0.8. The insets are the close-ups near U_c/t. Data of different system sizes ($L = 10$–14) are compared

pair of SC within nearest-neighbor sites can move the long distance. In Fig. 1, we depict P_d^{ave} as a function of U/t. First, we can find that for $t'/t = 0.4$ (0.8), P_d^{ave} abruptly decreases at $U = U_c \sim 6.75t$ (8.05t). U_c is a critical value of a Mott transition of first order, because behaviors of charge structure factor and momentum distribution function indicate that a gap opens in the charge degree of freedom for $U > U_c$ (not shown). Therefore, P_d^{ave} abruptly drops to almost zero in the insulating regime ($U > U_c$), as naturally expected. P_d^{ave} will completely vanish in the thermodynamic limit ($L \to \infty$), because the system size dependence is appreciable.

For a moderately frustrated case ($t'/t = 0.4$), P_d^{ave} is very small for small U/t, whereas it is remarkably enhanced as U/t approaches the Mott critical point U_c/t (~ 6.75) with negligible system size dependence. In contrast with this case, for a strongly frustrated system ($t'/t = 0.8$), we cannot find an enhancement of P_d^{ave} below U_c/t (~ 8.05). Thus, robust SC does not appear for a strongly frustrated (or nearly isotropic) case.

To trace the above SC to its origin, we study a spin correlation function,

$$S(\mathbf{q}) = \frac{1}{N_s} \sum_{ij} e^{i\mathbf{q}\cdot(\mathbf{i}-\mathbf{j})} \langle S_\mathbf{i}^z S_\mathbf{j}^z \rangle, \qquad (3)$$

where $S_i^z = 1/2(c_{i\uparrow}^\dagger c_{i\uparrow} - c_{i\downarrow}^\dagger c_{i\downarrow})$ is a spin operator. $S(\mathbf{q})$ has a sharp peak at the AF wave number $\mathbf{q} = (\pi, \pi)$, if robust AF spin correlation appears in the system. In Fig. 2, $S(\mathbf{q})$ is plotted for two values of t'/t. In the moderately frustrated case ($t'/t = 0.4$), in which robust SC appears, $S(\pi, \pi)$ steadily grows as U/t increases in the conductive regime ($U < U_c$). In contrast, in the strongly frustrated case ($t'/t = 0.8$), in which SC does not develop, $S(\pi, \pi)$ does not exhibit a special enhancement. These results indicate that whenever P_d^{ave} is appreciably enhanced, $S(\mathbf{q})$ has an evident peak at $\mathbf{q} = (\pi, \pi)$. Therefore, we can conclude that the SC in this model is induced by the AF spin correlation,

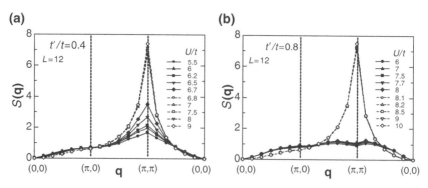

Fig. 2. Spin correlation function $S(\mathbf{q})$ for (a) $t'/t = 0.4$ and (b) 0.8. The *solid* (*open*) symbols denote the points of $U < U_c$ ($U > U_c$). The systems are of $L = 12$

as is the high-T_c cuprates. On the other hand, in the insulating states ($U > U_c$), the conspicuous peaks of $S(\pi,\pi)$ are restored for both $t'/t = 0.4$ and 0.8, as seen in Fig. 2. However, the AF long-range order is always absent, which is confirmed by a staggered magnetization: $\sum_i \langle (-1)^i S_i^z \rangle = 0$. Thus, the insulating states is spin-liquid (nonmagnetic) states, which have basically the same character as Anderson's RVB state [3].

Our results agree with main aspects of experiments in κ-(ET)$_2$X. (1) A compound with a smaller value of t'/t tends to have higher T_c: for X=CuN(CN)$_2$Br ($t'/t = 0.68$), Cu(NCS)$_2$ (0.84), and Cu$_2$(CN)$_3$ (1.06), $T_c = 11.6$, 10.4, and 3.9 K (under pressure of 0.4 GPa), respectively [2]. However, according to our result, robust SC does not arise for $t'/t \gtrsim 0.8$. This fact seemingly contradicts particularly for the compound X=Cu$_2$(CN)$_3$. (2) SC-insulator transitions (Mott transition) of first order are also caused by pressure in our model, which has not been treated within mean-field [4] and FLEX [5] approximations.

References

1. R. H. McKenzie, Science **278** (1997) 820
2. T. Komatsu et al., J. Phys. Soc. Jpn **65** (1996) 1340
3. P. W. Anderson, Science **235** (1987) 1196
4. H. Kino, H. Fukuyama, J. Phys. Soc. Jpn **65** (1996) 2158
5. H. Kino, H. Kontani, J. Phys. Soc. Jpn **67** (1998) 3691; H. Kondo, T. Moriya, J. Phys. Soc. Jpn **67** (1998) 3695
6. C. J. Umrigar et al., Phys. Rev. Lett. **60** (1988) 1719
7. H. Yokoyama, H. Shiba, J. Phys. Soc. Jpn **56** (1987) 1490

Simulation for Measurements of Electric Properties of Surface Nanostructures

Satoshi Watanabe, Ryoji Suzuki, Tomofumi Tada, and Hideomi Totsuka

Summary. We have been simulating quantities obtained from several measurements of electric properties of surface nanostructures to understand the relation between the measured quantities and properties of the nanostructures. In this article, we show two of our recent results on this topic: electron conduction between two probe tips through surface states and local barrier height measurement using scanning tunneling microscope.

1 Introduction

Recent advances in techniques of fabrication and measurements on the nanometer scale have enabled us to measure electric properties of surface nanostructures, i.e., atoms and molecules on surface and nanoscale regions of solid surfaces. Experimentally, many attempts of such measurements have been performed. Interesting results have already been reported and further more are expected to be achieved in the near future. However, interpretation of the results of such experiments is not necessarily straightforward, because of several reasons such as the strong influence of applied electric field on the electronic states of nanostructures.

On the other hand, analyses of such measurements give a challenge to theoretical calculations, because conventional methods, which are powerful to examine properties of bulk crystals and isolated molecules, are often insufficient to treat such measurements involving nonequilibrium steady states under an applied bias voltage. One may think that this issue can be treated within the conventional approach using finite cluster model or thin slab model, but these models do not have enough reservoir to supply charge induced by applied strong electric fields, and thus may involve artificial charge redistribution inside them.

Bearing this in minds, we have been simulating quantities obtained from several measurements of such kind to understand the relation between the measured quantities and properties of the objects, and have also been developing computational methods and programs for the simulations. In this article, we describe our recent results on the analysis of electron conduction between two probe tips through surface states, which is the first step toward the understanding of electric property measurements of surface nanostructures using, for example, micro-four-point probe methods (Sect. 2) and the analysis of local barrier height measured using scanning tunneling microscope (STM) (Sect. 3).

2 Electron Conduction Between Two Probe Tips Through Surface States

Direct measurement of electron conduction through surface states using microscopic methods such as the micro-four-point scanning tunneling microscopy (STM) probe method [1] is expected to be a powerful tool to examine electric properties of the surface nanostructures. However, there are only few theoretical works on electron transport through surface states [2,3]. Therefore, it would be useful to simulate measurements with the micro-multi-point probes using tight-binding calculations to clarify, for example, the effects of probes on surface states and interaction among probes on the measured data.

As the first step toward this, we have developed a new tight-binding program within Hückel approximation [4]. The program can treat models consisting of several STM tips and a flat surface either with or without surface nanostructures. In the model, the surface is represented by a box and five semi-infinite regions connected to the box. The five regions are included to take account of electron propagation into bulk and along the surface. The STM tips are also extended semi-infinitely. For this model, first the wave functions in the semi-infinite surface and tip regions can be obtained by solving the Schrödinger equation with giving an electron energy. Next, we derive simultaneous equations for the wave functions in the whole region, and determine wave functions in the surface box together with the unknown reflection and transmission coefficients. The bulk and scattering region are both simple cubic lattice. These model and method are essentially the same as those adopted by Kobayashi [3]. Only the size of the box is different, that is, we adopted a box of $31 \times 31 \times 8$, which is smaller than Kobayashi's one. We, however, confirmed that there is no significant change in the calculated results. The tight-binding parameters used in the present calculation are also the same as those adopted by Kobayashi [3], which involve Tamm surface states.

The calculated spectrum of the transmission rate from the first to second tips has two resonant peaks at the incident electron energies of -1.10 and $2.91\,\mathrm{eV}$, which is consistent with Kobayashi's result [3]. These peaks appear regardless of the box size. Figure 1 shows the electron density distribution in the topmost layer at the respective resonant energies. These distributions exhibit a striking contrast [4,5]. The electric current in the surface also shows similar contrast [4].

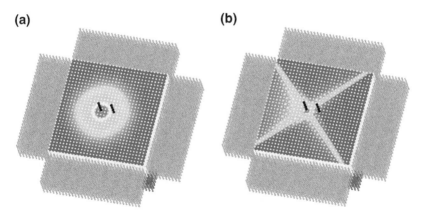

Fig. 1. Electron density distribution in the topmost layer at the incident energies of (a) −1.10 eV and (b) 2.91 eV

Figure 2 shows cross-section of current distribution. We can easily see that the way current flows at one energy is quite different from the one at the other. At the incident energy of −1.10 eV, injected electrons first penetrate into bulk, then some go up to surface and others flow into bulk states. On the other hand, at 2.91 eV, almost all of the injected electrons flow through surface states. These results suggest that measured resistance may not be purely that of surface states.

From the analysis using Green's function method [4], we found that the incident energy of −1.10 eV corresponds to the lower band edge of the Tamm surface states where the wave vector (k_x, k_y) is $(0,0)$, while that of 2.91 eV to the center of the surface state band where $\cos k_x a + \cos k_y a = 0$. Here a is the lattice constant.

It is noted that all the above results were calculated non-self-consistently because we simulated at zero bias limit. For the calculations of the situations with bias voltages applied, self-consistent calculation method such as

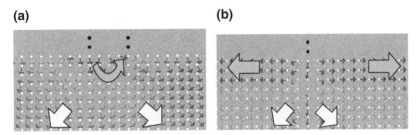

Fig. 2. Cross-section of current distribution at the incident energies of (a) −1.10 eV and (b) 2.91 eV. *Dark arrow* corresponds to large current

density functional based tight-binding method is desirable. Development of an improved program to do this is in progress.

3 Local Barrier Height Measurements

Local barrier height (LBH) is a barrier height for tunneling electrons. Experimentally, the LBH, ϕ_{LBH}, can be estimated from the change of tunneling current with the variation of tip-sample separation in STM using the following formula:

$$\phi_{\text{LBH}} = 0.952 \left(\frac{\mathrm{d}\ln I_t}{\mathrm{d}s}\right)^2, \tag{1}$$

where I_t and s are the tunneling current and the tip-sample separation, respectively. Since the LBH provides useful information on the surface nanostructures, it has attracted the attention of experimentalists recently. However, because the above formula is based on simple one-dimensional square potential and the Wentzel–Kramers–Brillouin method, the physical meaning of the LBH, in particular the site dependence of the LBH, which is usually the most important information obtained from measurements, is not necessarily clear. Therefore, we have simulated the LBH, focusing on its site dependence [6,7].

Firstly, we examined the difference between normal and defect sites [6]. For the normal site, an STM tip and sample surface were represented by an Al atom attached to a semi-infinite jellium electrode and two layers of Al(100) attached to another semi-infinite jellium electrode, respectively. In the model for the defect one, four Al atoms were missing in the layer next to the Al(100) surface. Although this model structure is unstable and thus unrealistic, we expect that the physical meaning of the LBH can be clearly seen by using such an artificial model.

In the analysis, the LBH evaluated using (1) with approximating the derivative by a difference was compared with the barrier height defined as a maximum value of the calculated potential on a straight line perpendicular to the surfaces, which penetrates the tip atom. Hereafter, we call the latter barrier height the maximum barrier height (MBH). In the evaluation of the LBH and MBH, a sample bias voltage of -2 V is applied.

The calculated values of MBH and LBH are shown in Fig. 3. We can see that the value of the LBH is much larger than that of the MBH in both $d = 11$ a.u. and 16 a.u. As Lang [8] has already pointed out, this feature can be understood from the energy increase due to the lateral confinement of tunneling electrons. The most interesting feature seen in Fig. 3 is that the LBH at the normal site is larger than that of the defect one, while the values of the MBH are almost the same. Therefore, we can say that the LBH evaluated using (1) does not directly correspond to the barrier height for tunneling electrons. The above results can be understood from the fact that the decay of the wave functions at the defect site is slower than that at the normal one,

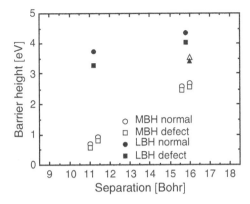

Fig. 3. Calculated maximum barrier height (MBH) and local barrier height (LBH) for the normal and defect sites

because the breaking of the lateral periodicity in the former makes amplitudes of the components having small lateral momenta large.

Secondly, we examined the site dependence within the perfect lattice model of the Al(100) surface [7]. In this calculation, a sample bias voltage of $-50\,\mathrm{mV}$ was applied. The notable feature seen in the calculated result is that the contrast of the LBH is opposite that of the MBH: In the case of the tip-sample distance of 5.9 Å, the calculated LBH's are 3.82 and 3.72 eV for the on-top and hollow sites, respectively, while the MBH's are 1.40 and 1.63 eV. In fact, the site dependence of the LBH is similar to that of the tunneling current in the sense that the on-top sites are imaged brighter than the hollow ones. This similarity between the LBH image and STM image is often observed experimentally.

The similarity between the LBH and STM images appears strange, because this means that the tunneling probability increases with the increase of the local work function, if the LBH corresponds to the local work function. On the other hand, the opposite contrast between the MBH and STM images can be easily understood from the same argument. Obviously, these results suggest that it is difficult to interpret LBH as an atomic-scale work function.

From the detailed analysis of the calculated results, we found that the site dependence of the tunneling current distribution is clearly seen for the tip-sample separation of 5.9 Å: The tunneling current distribution in the on-top case is much narrower than that in the hollow case.

When the local lowering of the barrier occurs, the electrons flowing through the hole need additional energy corresponding to the confinement of the transverse motion. This makes the ABH higher than the corresponding MBH as pointed out by Lang [8]. From the lateral current distribution, we can see that this contribution to the LBH is largest in the on-top case and smallest in the hollow case. Therefore, the site dependence of the LBH can be understood in terms of the lateral confinement of tunneling electrons at least partly.

4 Concluding Remarks

As examples of simulation for measurements of electric properties of surface nanostructures to understand the relation between the measured quantities and properties of the objects, in this article we describe our recent results on electron conduction between two probe tips through surface states and local barrier height measurement using STM. In the case of electron conduction, we demonstrate that novel phenomena can be observed in such nanoscale measurements. On the other hand, we can give a plausible explanation for the apparent inconsistency often seen between the local barrier height and STM images.

Our simulations are still insufficient in many aspects. For example, in the case of electron conduction, we have to consider the effects of bias voltages. Definitely, further development of simulations for nanoscale electrical property measurements and methodology for this is strongly desirable.

References

1. Hasegawa S, Shiraki I, Tanabe F, Hobara R, Kanagawa T, Tanikawa T, Matsuda I (2003) Surf. Rev. Lett. 10: 963
2. Chan YS, Heller EJ (1997) Phys. Rev. Lett. 78: 2570
3. Kobayashi K (2002) Phys. Rev. B 65: 035419
4. Suzuki R, Noda M, Tada T, Watanabe S (2006) Jpn. J. Appl. Phys. 45: 2136
5. Niu Q, Chang MC, Shih CK (1995) Phys. Rev. B 51: 5502
6. Totsuka H, Gohda Y, Furuya S, Watanabe S (2002) Jpn. J. Appl. Phys. 41: L1172–L1174
7. Totsuka H, Furuya S, Watanabe S (2006) Phys. Rev. B (submitted)
8. Lang ND (1988) Phys. Rev. B 37: 10395

Ultra-Fast Dynamics in Nanocarbon Explored by TDDFT-MD Simulations

Yoshiyuki Miyamoto

Summary. Methods of identifying defects and removing impurities in carbon nanotubes using electronic excitation are proposed. The present proposal is based on the first-principles calculations for the molecular dynamics (MD) simulation under electronic excitation. We first show the way of identification of a geometrical defect in a graphitic bond network of carbon nanotubes by using light-induced local vibration. Next, we propose an efficient method to remove oxygen impurities from carbon nanotubes by creating two holes in the C–O bonding state using the Auger process. We expect that all these ideas will be applied in future experiments. We note here that all phenomena in the present simulations are ultra-fast events which take less than 200 fs (1 fs = 10^{-15} s).

1 Introduction

The discovery of a variety of chirality in carbon nanotubes [1] attracted scientific and industrial attentions. Since carbon nanotubes are tough despite their small diameters, they are a promising alternative to semiconductor and metal nanowires. Unfortunately, the actual performance of nanotube-based devices is far behind theoretical predictions because of imperfections remaining in carbon nanotubes.

To date, defects and impurities have been ubiquitous in carbon nanotubes grown by chemical vapor deposition. The use of alcohol [2] or water [3], combined with a low growth temperature, causes a risk of potential geometrical defects or oxygen impurities. To improve nanotube quality, these defects need to be clearly identified and the oxygen impurities need to be completely removed.

Some geometrical defects (five- and seven-member rings) can be identified with use of a transmission electron microscope (TEM) [4], due to the changing shape of a nanotube. On the other hand, the presence of the Stone–Wales defect [5], which consists of one pair each of five- and seven-member rings, is widely believed to exist. However, this has not been revealed with a TEM because a Stone–Wales defect does not change the shape of carbon nanotubes. Scanning tunneling microscopy can be used to identify defects, but this method demands a great deal of patience and time, because finding a Stone–Wales defect requires many scans.

In contrast to the Stone–Wales defects, the presence of oxygen impurities can be identified [6] using X-ray absorption fine structure spectroscopy. The most common structure of an impurity is a C–O–C complex. Removing oxygen efficiently is difficult, because carbon atoms accompany oxygen atoms. This is demonstrated by CO or CO_2 emissions during heat treatments [7] and is due to the fact that C–O chemical bonds are stronger than C–C bonds. Therefore, we need an efficient method for removing oxygen atoms from carbon nanotubes without damaging the C–C bond network.

Methods for identifying Stone–Wales defects efficiently and for removing oxygen impurities using electronic excitation are presented. These methods can be alternatives to conventional thermal processes. First-principles calculations for molecular dynamics (MD) simulations based on the time-dependent density functional theory (TDDFT) have been performed using a computer code called FPSEID (éf-psái-dí:), which stands for first-principles simulation tool for electron ion dynamics [8]. Section 2 describes the computational method and Sect. 3 presents the detailed results for identifying Stone–Wales defects and for removing oxygen impurities in nanotubes.

2 Computational Methods

Within the DFT, we can approximately obtain an excited state with a certain atomic configuration by manually assign occupations of electronic state. In case of mimicking single excitation from state n to state m, which are respectively fully occupied and empty, we reassign these states as half-occupied and compute the charge density according to this electronic occupation. This method is called constraint DFT. We analyze characteristics of the wave functions and determine a proper pair of states n and m according to the optical-selection rule. However, when atomic positions are allowed to move, levels frequently alternate among differently occupied states. This makes constraint DFT difficult to perform during MD simulations and may run the risk of misassignment of the occupied level.

Instead of using constraint DFT, real-time propagation of a Kohn–Sham wave function is used throughout the MD simulation. (The constraint DFT is used only at the beginning of the simulation.) The merit of using real-time propagation is avoiding the misassignment of the occupation levels, since we can trace change of many states from t to $t + \delta t$ regardless to the sequential order in their energy levels. Thanks to the TDDFT [9], we are able to handle many-body systems with time-evolving single particle wave functions. The real-time propagation within TDDFT was used for computing optical spectrum [10], but had not been coupled with MD simulation up to few hundreds femtoseconds (1 fs = 10^{-15} s) prior to a literature [8].

The time-dependent Kohn–Sham equation [9] is

$$i\hbar \frac{d}{dt}\psi_n(\boldsymbol{r},t) = H_{KS}(\boldsymbol{r},t)\psi_n(\boldsymbol{r},t), \qquad (1)$$

where $\hbar$ is the Plank constant divided by 2π. $H_{KS}(\boldsymbol{r},t)$ is the Kohn–Sham Hamiltonian and $\psi_n(\boldsymbol{r},t)$ is the time-dependent Kohn–Sham wave function with subscript n indicating the single-electron level. The real-time propagation of (1) is performed using the "Suzuki-Trotter" split operator method [11]. More details of the computational method can be found in the former publication [8].

In these calculations, the plane wave basis sets with kinetic cutoff energy of 40 and 60 Ry are used for cases without and with oxygen, respectively, in the system. Norm-conserving nonlocal pseudopotentials [12] are used to express electron–ion interaction, and a functional form [13] fitted to the numerical data [14] for electron gas is used for exchange-correlation potential within the local density approximation of DFT.

3 Applying of TDDFT-MD Simulation to Carbon Nanotubes

3.1 Identifying the Stone–Wales Defects

An efficient method of identifying a Stone–Wales defect is described below. The left panel of Fig. 1a shows the structure of a Stone–Wales defect on a (3,3) carbon nanotube. Possible excitation was found among the states localized at a Stone–Wales defect with corresponding excitation energies of 0.6 (infrared light) and 6.0 eV (ultraviolet light).

The right panel of Fig. 1a shows the time-evolution of the C–C bond length centered at a Stone–Wales defect upon the excitations as mentioned above. The corresponding frequency of the C–C stretching motion is 1,962 cm^{-1} in both excitations, which is higher than the highest frequency in graphite and nanotube 1,580 cm^{-1}. This frequency (1,962 cm^{-1}) may be distinctive of a Stone–Wales defect. Better sensitivity for this frequency can be achieved by measuring the difference in vibration spectrums with and without photoexcitation.

The existence of a Stone–Wales defect in a boron nitride (BN) tube has been suggested using MD simulation under mechanical strain [15]. We have explored possibility of identifying a Stone–Wales defect in a BN tube by testing the optical excitation of N–N $\sigma \rightarrow \sigma^*$ with 12 eV. Figure 1b shows an N–N bond variation as a function of time. In contrast to a Stone–Wales defect in a carbon nanotube, the N–N bond stretching shows rapid decay. Therefore, identifying a Stone–Wales defect in a BN nanotube is more difficult than in a carbon nanotube. The rapid decay in a BN nanotube is probably due to its

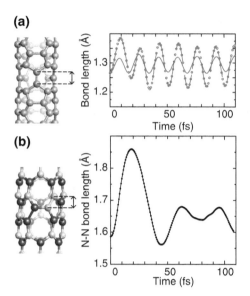

Fig. 1. (**a**) (*Left*) Atomic configuration of Stone–Wales defect in (3,3) nanotube, and (*Right*) time evolution of C–C bond length centered at Stone–Wales defect, denoted by *arrows* in the left panel, induced by infrared (smaller amplitude) and by ultraviolet (larger amplitude) light. (**b**) (*Left*) Atomic configuration of Stone–Wales defect in boron nitride nanotube, and (*Right*) time-evolution of N–N length, denoted by *arrows* in the left panel, by optical excitation with 12 eV. Figures first appeared in [16]

polar nature and a stronger electron–phonon coupling than that of a carbon nanotube.

More details of this calculation can be found in a previous publication [16].

3.2 Eliminating Oxygen Impurities

An efficient method of extracting an oxygen atom is described below. Figure 2a shows the electronic structure of an oxygen impurity, which makes a C–O–C complex in a carbon nanotube. The states localized at the C–O–C complex are labeled "O 2s" and "O 2p," and their energy levels are located below and above the Fermi level, respectively .

The "O 2s" state is hybridized with 2p orbital of the neighboring carbon atom in the bonding phase, while the "O 2p" state is hybridized with the 2p orbital of the neighboring carbon atom in the antibonding phase. A single excitation from the "O 2s" level to the "O 2p" level might be expected to weaken the local C–O–C chemical bond. However, a TDDFT-MD simulation revealed that a single excitation can not completely remove an oxygen atom from carbon nanotube, instead, induces a lattice vibration (see Fig. 2b).

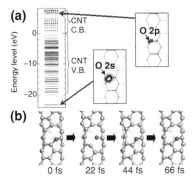

Fig. 2. (a) Electronic structure of C–O–C complex in carbon nanotubes. States labeled as "O 2s" and "O 2p" are located below and above Fermi level, respectively. (b) Snapshots of O-vibration induced by a single excitation from "O 2s" level to "O 2p" level. Figures first appeared in from [17]

Then we change our idea to excite O 1s core level into the "O 2p" level, which causes spontaneous Auger process leaving "two" holes in the "O 2s" level (see Fig. 3a). When we set this Auger final state as the initial condition for the TDDFT-MD simulation, spontaneous oxygen emissions from the carbon nanotubes can be seen. Figure 3b shows snapshots of the TDDFT-MD simulation. Immediately after the emission of an oxygen atom, two neighboring carbon atoms are pushed away to make a large vacancy. However, the this is soon closed by these two carbon atoms which return and form a new C–C bond.

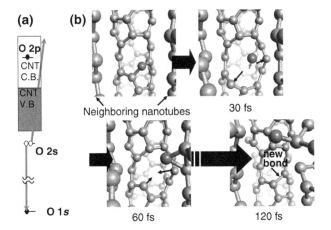

Fig. 3. (a) Schematic of Auger process initiated by "O 1s" core excitation to "O 2p" level leaving "two" holes in the "O 2s" level. (b) Snapshots of spontaneous emission of oxygen atom from carbon nanotube. Figures first appeared in [17]

The emitted oxygen atom attacks the other side of the carbon nanotube. This reoxidation can be prevented by introducing an H_2 molecule to capture the emitted oxygen atom. More details can be found in a previous publication [17]. The combination of core excitation and H_2 introduction may be an efficient post-fabrication process for purifying nanotube-based devices.

Acknowledgments

The author is indebted to Professor Osamu Sugino for developing the computer code, used in this research, FPSEID. The MPI version of FPSEID has been made with a help of Mr. Takeshi Kurimoto and CCRL, MPI-team, at NEC Europe (Bonn). The present research were performed in a collaboration with Professor Angel Rubio and Professor David Tománek. Calculations were performed on the SX5 Supercomputer system in the NEC Fuchu plant, and the Earth Simulator.

References

1. S. Iijima (1991) Nature(London) 354: 56–58; and S. Iijima, T. Ichihashi (1993) Nature(London) 363: 603–605
2. S. Maruyama et al. (2002) Chem. Phys. Lett. 360: 229–234
3. K. Hata et al.(2004) Science 306: 1362–1364
4. S. Iijima, T. Ichihashi, and Y. Ando (1992) Nature (London) 356: 776–778
5. A. J. Stone and D. J. Wales (1986) Chem. Phys. Lett. 128: 501–503
6. A. Kuznetsova et al. (2001) J. Am. Chem. Soc. 123: 10699–10704
7. E. Bekyarova et al. (2002) Chem. Phys. Lett. 366: 463–468
8. O. Sugino and Y. Miyamoto (1999) Phys. Rev. B 59: 2579–2586; O. Sugino and Y. Miyamoto (2002) Phys. Rev. B 66: 89901–89901
9. E. Runge and E. K. U. Gross (1984) Phys. Rev. Lett. 52: 997–1000
10. K. Yabana and J. Bertsch (1996) Phys. Rev. B 54: 4484–4487
11. M. Suzuki (1993) J. Phys. Soc. Jpn 61: 3015–3019
12. N. Troullier and J. L. Martins (1991) Phys. Rev. B 43: 1993–2006
13. J. P. Perdew and A. Zunger (1981) Phys. Rev. B 23: 5048–5079
14. D. M. Ceperley and B. J. Alder (1980) Phys. Rev. Lett. 45: 566–569
15. H. F. Bettinger et al. (2002) Phys. Rev. B 65: 041406(-1)–041406(-4)
16. Y. Miyamoto et al. (2004) Phys. Rev. B 69: 121413(R)–121416(R)
17. Y. Miyamoto, A. Rubio, and D. Tománek (2004) Phys. Rev. 70: 233408(-1)–233408(-4)

Heat Transport in Nanoscale Objects: Classical to Quantum

Takahiro Yamamoto, Naoaki Kondo, Kazuyuki Watanabe, Yingwen Song, and Satoshi Watanabe

Summary. We give a short review of our recent works on the heat transport through carbon nanotubes (CNTs) with vacancy defects. While the thermal conductivity of CNTs at room temperatures is strongly affected by the vacancy defects, the low-temperature heat transport in CNTs is undisturbed by the vacancy because long wavelength phonons are not scattered by the vacancy. Consequently, at cryostatic temperatures, the CNTs are predicted to exhibit the quantization of thermal conductance even in the presence of vacancy defects.

1 Introduction

Miniaturization of electronic devices will arrive at nanoscale in the near future. As the system size decreases down to nanoscale, the heat dissipation and heat transport become extremely important more and more. Carbon nanotubes (CNTs) are potential candidates for heat sink that dissipates the heat from the devices, because thermal conductivity of CNTs at room temperatures is as high as that of good thermal conductors such as diamond and graphite [1]. There is, however, concern that high thermal conductivity diminishes owing to defects generated during the synthesis processes [2]. The first aim of this paper is to clarify the effect of defects on heat transport in CNTs at room temperatures. To achieve the aim, we perform the nonequilibrium molecular dynamics (NEMD) simulations of heat transport in defective CNTs with the vacancy (Sect. 2).

On the other hand, at low temperatures, heat transport in CNTs is predicted to exhibit quantum features. Indeed, recent sophisticated experiments have succeeded in measuring the quantized thermal conductance in the CNTs [3]. Despite the experimental activity, little theoretical attention has been paid to heat transport at low temperatures [4]. This is probably because it is impossible to treat heat transport phenomena including

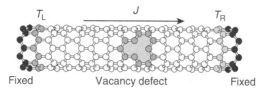

Fig. 1. Schematic diagram of (8,0) CNT with a vacancy defect. The carbon atoms indicated by *black spheres* in each end layer are fixed. The heat thermal current J flows through the CNTs from left to right layers with the temperature T_L and $T_\mathrm{R}(<T_\mathrm{L})$, respectively

quantum effects using conventional approaches such as classical MD methods. Therefore, of course, the influence of defects on the low-temperature heat transport in CNTs has never been studied thus far. In this paper, we investigate the low-temperature heat transport in CNTs with the vacancy using a quantum mechanical approach so-called Landauer formalism (Sect. 3).

2 Room-Temperature Heat Transport in CNTs

2.1 Nonequilibrium Molecular-Dynamics Simulations

To understand the effect of vacancy defect on the thermal conductivity of CNTs at room temperature, we perform the NEMD simulations with the Tersoff–Brenner bond-order potential [5] that accurately reproduces several physical properties of various carbon materials. In our NEMD simulations, we assign different temperatures, T_L and $T_\mathrm{R}(<T_\mathrm{L})$, to the left- and right-end layers of CNTs to generate a heat current through the CNTs (see Fig. 1). A Gaussian thermostat is used to maintain the end-layer temperatures constant [6], and the fixed boundary condition is imposed, i.e., the edge atoms of CNTs are fixed rigidly. In our simulations, we solve Hamilton's classical equations of motion using second-order operator splitting integrators [7] with the MD time step of 0.05 fs. All physical quantities calculated in this work are averaged over ~1.0 ns. The thermal conductivity λ is calculated from Fourier's law,

$$J = -\lambda \frac{\mathrm{d}T}{\mathrm{d}z}, \qquad (1)$$

where the heat current J is defined as the amount of heat across a surface of unit area in a unit time. The temperature gradient $\mathrm{d}T/\mathrm{d}z$ along the tube axis (z-axis) is determined by the slope of temperature profile in the central region of CNTs. To estimate the thermal conductivity λ for the CNTs using Fourier's law (1), we calculate the heat current J and the temperature gradient

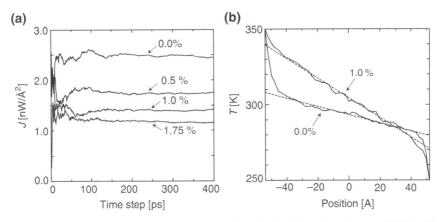

Fig. 2. (a) Time evolution of heat current for the (8,0)-semiconductor CNT with vacancy concentrations of 0, 0.5, 1.0, and 1.75%. The tube length is assumed to be 108 Å including 25 unit cells. (b) Position-dependent temperature profiles of CNTs including 0% and 1.0% vacancy concentrations. *Dashed lines* represent the linear fitting of the temperature profiles

dT/dz for the CNTs by performing our NEMD program improved for long-time simulations.

2.2 Numerical Results

Figure 2a shows the time evolution of heat current for the (8,0)-CNT with the vacancy concentrations of 0, 0.5, 1.0, and 1.75%. Here, the tube length is chosen to be 108 Å including 25 unit cells (800 carbon atoms), and the temperatures at the left- and right-end layers of the CNT are assigned as $T_L = 350$ K and $T_R = 250$ K, respectively. The heat currents for all vacancy concentrations attain steady states in approximately 200 ps and consequently become constant. Steady-state heat current J decreases monotonically with increasing vacancy concentration ρ. For $\rho = 0$, 0.5, 1.0, and 1.75%, $J = 2.43$, 1.74, 1.40, and 1.17 nW Å^{-2}, respectively.

Figure 2b shows the position-dependent temperature profiles of the CNTs with $\rho = 0$ and 1.0% as typical examples. The temperature profiles for $\rho = 0$ and 1.0% exhibit linear behavior with the gradients of $dT/dz = -0.276$ and -0.662 K Å^{-1} in the central region of the CNTs, respectively.

Substituting the obtained J and dT/dz into (1), we obtain the thermal conductivity λ. As shown in Fig. 3, λ decreases dramatically as vacancy concentration ρ increases. Surprisingly, it decreases by approximately 75% with only 1.0% vacancy concentrations. This result is in qualitatively good agreement with previous results from the equilibrium MD simulations by Che et al. [8]. Therefore, high-quality CNTs are needed to maintain the intrinsic high thermal conductivity at room temperatures.

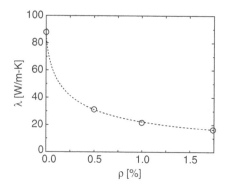

Fig. 3. Thermal conductivity as function of vacancy concentration

3 Low-Temperature Heat Transport in CNTs

3.1 Landauer Formula for Heat Transport

As a classical MD method is not applicable to low-temperature heat transport exhibiting quantum features, our discussion begins here by introducing the Landauer formula of heat transport [9]. Within the framework of Landauer formalism, the thermal conductance is described as

$$\kappa(T) = \frac{k_B}{h} \int_0^\infty d\epsilon \left(\frac{\epsilon}{k_B T}\right)^2 \frac{e^{\epsilon/k_B T}}{\left(e^{\epsilon/k_B T} - 1\right)^2} \zeta(\epsilon), \qquad (2)$$

where k_B and h are the Boltzmann constant and Planck constant, respectively. $\zeta(\epsilon)$ is the phonon transmission function for incident phonon with the energy ϵ, which can be expressed by the Green's functions [10]. If $\zeta(\epsilon)$ is unity for all acoustic modes in the low-temperature limit $T \to 0$, (2) is given as a form of an elementary integration that can be analytically performed and the thermal conductance is quantized as $M\kappa_0 = M(\pi^2 k_B^2 T/3h)$, where M is the number of acoustic modes.

3.2 Numerical Results

The calculated $\zeta(\epsilon)$ for (8,8)-CNT with/without the vacancy is shown in Fig. 4. The dashed curve is $\zeta_p(\epsilon)$ for the perfect (8,8)-CNT without any defects and displays a clear stepwise structure that gives the number of phonon channels. At low-energy region below 2.4 meV being the energy gap of the lowest optical modes, the dashed curve shows $\zeta_p(\epsilon) = 4$ indicating the number of acoustic branches corresponding to a longitudinal, a twisting, and doubly degenerated transverse acoustic modes. On the other hand, the solid curve describes $\zeta_{vac}(\epsilon)$ for the (8,8)-CNT with the vacancy. Although $\zeta_{vac}(\epsilon)$ is dramatically

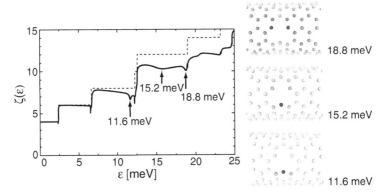

Fig. 4. Phonon transmission function $\zeta(\epsilon)$ for the (8,8)-CNTs. *Solid (dashed)* curve represents $\zeta_{\rm vac(p)}(\epsilon)$ for the (8,8)-CNT with (without) the vacancy. The *right panels* describe the phonon density around the vacancy at 11.6 meV, 15.2 meV, and 18.8 meV indicated by the *arrows*, respectively. The *shading* on the atom spheres indicates the phonon density

deformed from $\zeta_{\rm p}(\epsilon)$ owing to the defect scattering particularly at high energies, it remains unchanged in the low-energy region. This is because the long wavelength acoustic phonons in CNTs are not scattered by the vacancy. The $\zeta_{\rm vac}(\epsilon)$ shows some dips at particular energies indicated by the arrows in Fig. 4, which are clearly distinguished from the dips originating from the van Hove singularity of the optical phonon branches. The right panels in Fig. 4 show the phonon density around the vacancy at the dip positions of 11.6, 15.2, and 18.8 meV. We can see that these phonon densities are highly localized around the vacancy. The incident phonons from the heat bath are critically scattered by the localized phonons.

Substituting the obtained $\zeta(\epsilon)$ into (2), we can obtain the thermal conductance $\kappa(T)$ as a function of temperature T. The low-temperature behaviors of $\kappa(T)$ normalized to the universal quantum κ_0 for (6,6)-, (8,8)- and (10,10)-CNTs with/without defects are shown in Fig. 5a. The dashed and solid curves represent $\kappa_{\rm p}/\kappa_0$ for the perfect CNT and $\kappa_{\rm vac}/\kappa_0$ for defective CNT, respectively. Reflecting the perfect transmission for all acoustic modes at low-energy region, $\kappa(T)$ keeps four universal quanta, $4\kappa_0 = 4(\pi^2 k_{\rm B}^2 T/3h)$, in the limit of $T \to 0$ even in the presence of vacancy.

Finally, we describe the CNT-diameter dependence of vacancy scattering on thermal conductance for moderate temperature up to 300 K. Figure 5b shows the ratios $\kappa_{\rm vac}/\kappa_{\rm p}$ for (6,6)-, (8,8)- and (10,10)-CNTs as a function of T. All curves of $\kappa_{\rm vac}/\kappa_{\rm p}$ decrease rapidly with increasing temperature and become nearly independent of the temperature at $\sim$300 K. The terminal value of $\kappa_{\rm vac}/\kappa_{\rm p}$ also decreases as the CNT gets thin. In other words, the influence of defect scattering in *thin* CNTs on the thermal conductance is more significant than that in *thick* CNTs.

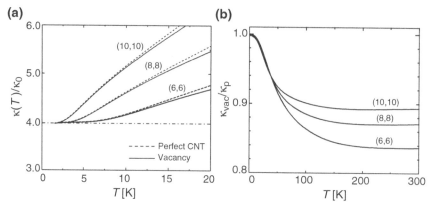

Fig. 5. (a) Low-temperature thermal conductances in (6,6)-, (8,8)- and (10,10)-CNTs. The *solid* (*dashed*) curves represent $\kappa_{\text{vac(p)}}(T)/\kappa_0$ for the CNTs with (without) vacancy. (b) The temperature dependence of the ratio $\kappa_{\text{vac}}/\kappa_{\text{p}}$ for (6,6)-, (8,8)- and (10,10)-CNTs, respectively

Acknowledgments

This work was supported in part by the "Academic Frontier" Project of MEXT (2005-2010). Part of the numerical calculations was performed on the Hitachi SR11000s at ISSP, The University of Tokyo.

References

1. J. Hone, M. Whitney, C. Piskoti, A. Zettl: Phys. Rev. B **59**, R2514 (1999)
2. N. Kondo, T. Yamamoto, K. Watanabe: e-J. Surf. Sci. Nanotech. **4**, 239 (2006)
3. H.-Y. Chiu, et al., Phys. Rev. Lett. **95**, 226101 (2006)
4. T. Yamamoto, S. Watanabe, K. Watanabe: Phys. Rev. Lett. **92**, 075502 (2004)
5. D.W. Brenner: Phys. Rev. B **42**, 9458 (1990)
6. F. Zhang: J. Chem. Phys. **106**, 6102 (1997)
7. M. Tuckerman, B.J. Berne, G.J. Martyna: J. Chem. Phys. **97**, 1990 (1992)
8. J. Che, T. Çağin, W.A. Goddard III: Nanotechnology **11**, 65 (2000)
9. L.G.C. Rego and G. Kirczenow: Phys. Rev. Lett. **81**, 232 (1998)
10. T. Yamamoto and K. Watanabe: Phys. Rev. Lett. **96**, 255503 (2006)

Aromaticity Driven Rupture of CN and CC Multiple Bonds

Cherumuttathu H. Suresh and Nobuaki Koga

Over the last few years, advances in computational hardware and quantum chemical simulation methods have made it possible to construct remarkably accurate computer models of complicated chemical processes providing valuable scientific insights at an unprecedented level of detail. Among the simulation methods, the density functional theory (DFT) based methods have been accepted as efficient and reasonably accurate methods and they have become a powerful tool in molecular modeling, particularly for the ground state properties of organotransition metal complexes [1]. However, many reactions in organotransition metal chemistry are quite complicated and intriguing to explain in simple terms, and understanding the mechanism of such reactions still represents a major challenge for experimental as well as theoretical chemists. In fact, our knowledge on the reactive possibility of organotransition complexes are far from complete and many people believe that these complexes are the most important source of new reactions in organic synthesis. For instance, the transition metal-mediated cleavage of CX type bonds (X= H, first and second row main elements), particularly the CH and CC bonds has been of great interest because of its applicability to organic synthesis [2–4]. A number of noteworthy achievements, especially for the CC single bond cleavage are made in the last two decades [5–9]. However, only a few successful examples are known for the cleavage of CC double, CC triple, and CN triple bonds [10–14]. In this scenario, the reaction reported by Takahashi et al. [15] given in Scheme 1 attains great importance because it describes an unprecedented double CC bond cleavage of Cp ligand in a titanacyclopentadienyl complex.

This is a classic example of a reaction in which cleaved fragments change into useful organic molecules. The pyridine formation in Scheme 1 is possible only if the CN triple bond of at least one Ph-CN molecule is completely cleaved. Such CN bond activation is most likely due to the interaction of an intermediate titanium complex with Ph-CN. However, proposing a mechanism is not very easy from the intuitive chemical knowledge alone. In such cases, the information which can be obtained from good quantum chemical modeling is very valuable and, as a matter of fact, we have recently explored a plausible

Scheme 1

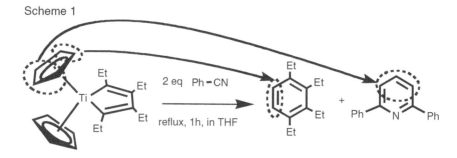

and complete mechanism of this unique reaction using DFT method [16] by using model molecules $Cp_2Ti(C_4H_4)$ and CH_3CN [17]. Here, we will briefly review the results of the calculations [17], which are summarized in Figs. 1–3.

The carbon atoms 1 and 4 in the 16–electron titanocene metallacycle $Cp_2Ti(C_4H_4)$ (**1**) are at the same distance of 2.807 Å from 5 and 6, respectively. This distance is well within the van der Waals' radii of two carbons (3.40 Å) [18] and therefore a small twist or turn in Cp ligand and butadiene moiety can lead to a migration or an insertion type reaction between them. The DFT calculation indeed gives a transition state (TS) for such a process (**TS1**) and at this TS, the Ti–C_1 bond is nearly broken and strong bonding interaction takes place between the Cp carbon C_5 and C_1 at a distance of 1.584 Å. The transformation of **1** to **2** via **TS1** requires activation energy (E_{act}) of 24.66 kcal mol^{-1} and this process is regarded as the migratory insertion of C_5 into the Ti–C_1 bond. The bond length values of **2** are in favor of the schematic structure **2a** and its 14 electron configuration accounts for the high endothermicity of 19.23 kcal mol^{-1} for **1** to **2** conversion.

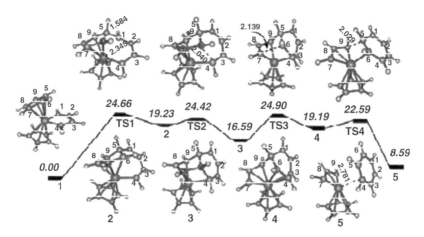

Fig. 1. Migratory insertion of C_5 into the Ti–C_1 bond of **1** and subsequent cleavage of the Cp ring CC bonds. Aromaticity-driven step is **4** → **TS4** → **5**

In complex **2**, the α-carbon C_4 is 2.678 Å close to the nearest Cp carbon and as a result, **2** quickly passes through **TS2** in which bonding interaction between carbon 4 and 6 develops at a distance of 2.040 Å. **TS2** is only 5.19 kcal mol^{-1} higher in energy than **2** and the corresponding product **3** has a relative energy of 16.59 kcal mol^{-1}. In **3**, Ti is bound to an allylic three-carbon unit ($C_7C_8C_9$) as well as a C_4–C_3 double bond (Fig. 1) meaning that Ti possesses a 14 electron configuration. Comparing the schematic structures **2a** and **3a**, it can be easily seen that the conversion of **2** to **3** is the insertion of C_6–C_7 bond into the Ti–C_4 bond and TiC_8C_9 cyclopropane is converted into Ti-allyl with C_7. In **3**, the two-carbon unit composed of C_5 and C_6 of Cp ring is connected to the butadiene-1,4-diyl moiety via σ bonds and therefore the six-membered cyclic region ($C_1C_2C_3C_4C_5C_6$) can be visualized as a precursor structure for benzene formation.

In the next step, C_6–C_7 bond of **3** is cleaved with E_{act} of 8.31 kcal mol^{-1} (Fig. 1) and the corresponding **TS3** shows a C_6–C_7 distance of 2.139 Å. The product complex **4** is 2.60 kcal mol^{-1} less stable than **3**. The structure of **4** depicted in Fig. 1 consists of an allylic $C_3C_4C_6$ unit and a Ti–C_7 bond with significant double bond character. In this structure, the C_8–C_9 bond donates its electrons to Ti atom and therefore, complex **4** is considered as a d^0 system possessing a 14 electron configuration. The next step is the C_5–C_9 bond cleavage (**4** → **TS4** → **5**). Its E_{act} is only 3.40 kcal mol^{-1} and it is exothermic by 10.60 kcal mol^{-1}. The exothermic character is due to the formation of the loosely Ti-bound benzene complex **5** that brings an aromatic stabilization energy of around 30 kcal mol^{-1} [19]. Further, the low E_{act} is due to the partial aromatization of the six-membered ring in **TS4**. In other words, **4** to **5** conversion is an aromaticity driven process. Moreover, the lengths of 1.890, 1.890, 1.441, and 1.437 Å observed for the Ti–C_7, Ti–C_9, C_7–C_8, and C_8–C_9 bonds, respectively, in the Ti$C_7C_8C_9$ metallacycle

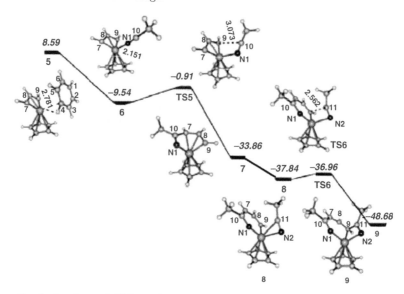

Fig. 2. Activation of CN triple bonds in MeCN molecules. **6** → **TS5** → **7** is aromaticity-driven step

[20a] of **5** suggest delocalized π-electrons in it via the resonance structures **5a** and **5b**.

In the next step, the loosely bound benzene in **5** is displaced by an incoming acetonitrile, which coordinates to the Ti atom in an end-on fashion via its nitrogen lone pair to produce **6** (Fig. 2). This exothermic process releases the energy of 18.13 kcal mol^{-1}. A slight amount of slippage of CH$_3$CN from its end-on coordination leads to the early transition state **TS5** (E_{act} = 8.43 kcal mol^{-1}) that shows weak bonding interaction between C$_9$ and acetonitrile carbon C$_{10}$ at a distance of 3.073 Å. The product that connects **TS5** and **6** is a quite stable metallaheterocycle **7**, which is placed at a relative energy of −33.86 kcal mol^{-1} as compared to complex **1**. The bond lengths of Ti–C$_9$, Ti–N$_1$, N$_1$–C$_{10}$, C$_{10}$–C$_7$, C$_7$–C$_8$, and C$_8$–C$_9$ are 1.958, 1.817, 1.325, 1.441, 1.443, and 1.383 Å, respectively, suggesting significant double bond character to all of them. Therefore, a resonance combination of two structures **7a** and **7b** is assigned for **7**. The higher stability of complex **7** is thus accounted by the delocalized nature of the six π-electrons in the metallaheterocycle, which means significant aromatic character to **7** [20b]. Therefore, **6** to **7** conversion is also considered as aromaticity driven process. The highlight of this part of the reaction is the substantial activation of the CN triple bond by consuming only 8.43 kcal mol^{-1} energy.

At this stage, a second acetonitrile molecule coordinates to complex **7** in a side-on fashion utilizing its CN π-electrons (Fig. 2). The complex **8** thus formed is further stabilized by 3.98 kcal mol^{-1} as compared to complex **7**. Interestingly, unlike complex **7**, the metallaheterocycle unit TiN$_1$C$_{10}$C$_7$C$_8$C$_9$ is planar and it shows though a lesser degree the delocalized nature of the six

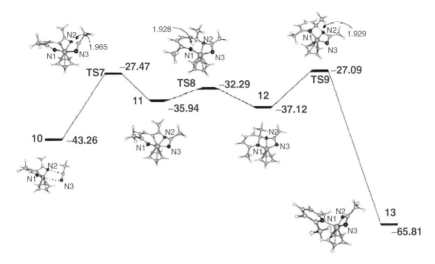

Fig. 3. Rupture of CN triple bond via the reaction of three MeCN molecules to titanium. Aromaticity driven step is **12 → TS9 → 13**

π-electrons [20b]. A slight tilt in the CH$_3$CN moiety in **8** toward the Ti–C$_9$ bond gives transition state **TS6**, which has almost the same relative energy as that of **8**. This TS corresponds to a barrier-less process for the insertion of the Ti–C$_9$ bond to Ti–C$_{11}$ bond. The resulting product **9** is quite stable at the relative energy of -48.68 kcal mol^{-1}. The C$_{11}$–N$_2$ triple bond in complex **8** is clearly activated to a C$_{11}$–N$_2$ double bond in complex **9** with almost no cost of energy. The reaction of a third CH$_3$CN to the most stable intermediate **9** gives **11** through the metastable intermediate **10** and **TS7** (Fig. 3). The E_{act} relative to **9** is found to be 21.41 kcal mol^{-1}. From **11** the formation of the six-membered heterocycle system **12** is very easy (see **TS8**. $E_{act} = 3.65$ kcal mol^{-1}). The next step (**12 → TS9 → 13**) gives pyridine, which is an aromaticity driven process as it shows low activation energy ($E_{act} = 10.0$ kcal mol^{-1}) and high exothermicity of 28.69 kcal mol^{-1}. In this step, the original C$_{11}$–N$_2$ triple bond of second MeCN is completely cleaved.

The mechanism as illustrated in Figs. 1–3 suggests that aromaticity is a major driving force for the unique organometallic reaction discovered by Takahashi et al. [15]. It is perhaps the first example of an aromaticity driven cleavage of strong CC double and CN triple bonds. The bottom line is that the phenomenon of aromaticity commonly observed in organic chemistry can be cleverly utilized as an important catalytic element for designing organometallic systems that are capable of breaking strong CX type bonds.

Acknowledgment

We acknowledge financial support from Ministry of Education, Science, Sports, and Culture, Japan and Council of Scientific and Industrial Research (CSIR), India.

References

1. Seminario JM, Politzer P (eds), (1995). Modern Density Functional Theory. A Tool for Chemistry. Elsevier, Amsterdam
2. (a) Murakami M, Ito Y (1999). In: Activation of Unreactive Bonds and Organic Synthesis. Murai S (ed) Springer, Berlin Heidelberg New York, pp. 97–129; (b) Rybtchinski B, Milstein D (1985). Angew. Chem. Int. Ed. 38:870; (c) Jenning PW, Johnson LL (1994). Chem. Rev. 94:2241; (d) Crabtree RH (1985). Chem. Rev. 85:245
3. Deckers PJW, Hessen B (2002). Organometallics 21:5564–5575
4. Jun CH (2004). Chem Soc Rev 33:610–618
5. Crabtree RH, Dion RP (1984). J Chem Soc, Chem Commun 1260
6. Suggs JW, Jun C-H (1986). J Am Chem Soc 108:4679
7. Hartwig JF, Andersen RA, Bergman RG (1989). J. Am. Chem. Soc. 111:2717
8. Gozin M, Weisman A, Ben-David Y, Milstein, D (1993). Nature 364:699
9. Nishimura T, Araki H, Maeda Y, Uemura S (2003). Org. Lett. 5:2997
10. Moriarty RM, Penmasta R, Awasthi AK, Prakash I (1988). J. Org. Chem. 53:6124
11. Sawaki Y, Inoue H, Ogata Y (1983). Bull. Chem. Soc. Jpn 56:1133
12. Früstner A, Mathes C, Lehmann CW (2001). Chem. Eur. J. 7:5299
13. Ananikov VP, Musaev DG, Morokuma K (2001). Organometallics 20:1652–1667
14. Lim SG, Jun CH (2004). Bull. Korea Chem. Soc. 25:1623–1624
15. Xi Z, Sato K Gao Y, Lu J, Takahashi T (2003). J. Am. Chem. Soc. 125:9568
16. All calculations were carried out using the Gaussian 03 program (Frisch MJ et al. Gaussian 03, revision C.02, Gaussian, Inc.: Wallingford, CT, 2004) using B3LYP/[6-31G* for H, C, and N and LanL2DZ for Ti]
17. Suresh CH, Koga N (2006). Organometallics 25:1924
18. Bondi A (1964). J. Phys. Chem. 68:441
19. (a) Schleyer PvR, Puhlhofer, F (2002). Org. Lett. 4:2873; (b) Suresh CH, Koga N (2002). J. Org. Chem. 67:1965
20. (a) For calculations on titanocyclobutadiene, see Lugo A, Fischer J, Lawson DB (2004). J. Mol. Struct. (THEOCHEM) 674:139; (b) For calculations on metallabenzenes, see Iron MA, Martin, JML, van der Boom ME (2003). J. Am. Chem. Soc. 125:13020

Protein Function Prediction in Proteomics Era

Daisuke Kihara, Troy Hawkins, Stan Luban, Bin Li,
Karthik Ramani, and Manish Agrawal

Summary. The increasing number of genome sequences has become an essential data for biology of this century and function annotation to genes in those genomes is basis of the all biological research. To overcome the limitation of conventional homology-based function annotation methods, here we introduce two types of approaches: A sequence-based approach and a protein surface tertiary shape based approach. The structure-based approach is aimed to predict function of proteins whose tertiary structure was solved. The need of predicting function of proteins from their tertiary structure has emerged by the structural genomics projects, which solve an increasing number of protein structures of unknown function.

1 PFP: Extended Sequence-Based Approach

When a new genome is sequenced, one of the first important bioinformatics tasks is function assignment (prediction) to genes in the genome. Usually function of known homologous genes to a new gene found by a database search method, such as BLAST [1], is transferred to the new gene. Different types of sequence-based methods could be also used, which include a Hidden Markov Model database search [2] or a local motif search [3]. One of the shortcomings of these conventional methods is their limited coverage. A BLAST-based annotation typically covers only up to a half of genes in a genome, hence the rest are remained unknown. The small coverage of function in a genome would prevent us from taking full advantage of omics type experiments, e.g., microarray gene expression analysis, protein identification by mass spectrometry analysis, protein–protein interaction analysis. These omics data deduce relationships between genes, but function annotation is crucial to draw biological conclusion from them.

1.1 Algorithm of PFP

A typical way to use BLAST is to set up a predefined threshold for the statistical significance score (i.e., E-value, a commonly used threshold value is

0.01) and only use highly similar sequences which obtained a score above that threshold value [4]. A more careful way is to identify orthologous genes to a new gene in multiple genomes [5], but again a certain threshold is used for identifying orthologous genes.

Unlike these conventional ways, our novel function prediction methods named Protein Function Prediction (PFP) rather uses even low-scoring sequence hits by a PSI-BLAST search [6] more extensively [7]. PFP uses Gene Ontology (GO) [8] as the vocabulary for gene functions. For an input sequences, PFP outputs ten most probable functions in three GO categories, namely, molecular function (MF), which is essentially biochemical function of proteins, biological process (BP), which indicates pathways where the gene belongs, and cellular component (CC), which specifies localization of the genes in a cell. PFP has the following two key features in the algorithm (1) GO terms associated to all the retrieved sequences by a PSI-BLAST search up to an E-value of 100 are ranked according to their frequency of occurrence in those sequences and the degree of similarity of the originating sequence to the query and (2) PFP also incorporates function associations, i.e., associations between pairs of GO terms found in UniProt database compiled in the form of the function association matrix (FAM). The FAM describes the frequency at which two GO terms occur together in the same context by quantifying the co-occurrence of each pair of annotations within UniProt sequences. Figure 1 gives a graphical representation of FAM. GO terms in the three functional categories are aligned on the two axis of the matrix, and the association between GO pairs is shown in a gray scale.

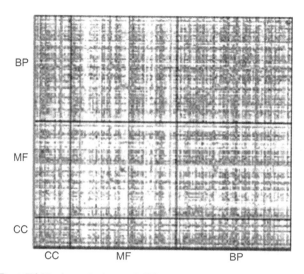

Fig. 1. UniProt FAM. Associations of GO terms within and across the three categories. Darker spots indicate a higher association score.

The PFP score given to a function f_a is computed as follows:

$$s(f_a) = \sum_{i=1}^{N}\sum_{j=1}^{M}((-\log(E\_value(i))+b)P(f_a|f_j)), \quad (1)$$

where N is the number of sequences retrieved by PSI-BLAST, M is the number of GO terms associated to the sequence i, $E\_value(i)$ is the E-value given to the sequence i, f_j is a GO term assigned to the sequence i, $P(f_a|f_j)$ is the conditional probability that f_a is associated with f_j computed by the FAM matrix, and b is a parameter. Following (1), GO terms which repeatedly occur in a PSI-BLAST search will stand out in the PFP score.

1.2 Results

We benchmarked PFP on a 2,000 nonredundant sequences randomly selected from UniProt database. In order to examine PFP's ability to mine correct function from low-scoring sequences in a PSI-BLAST search, the most significant hits using several E-value cutoffs are ignored. We use the default E-value cutoff ($-e$ 10) and E-value threshold for inclusion in multiple iterations ($-h$ 0.005), and set the maximum number of iterations to three ($-j$ 3). Figure 2 shows the fraction of test sequences whose biological process (BP) is correctly predicted by PFP in the top five scoring prediction. A prediction is counted as correct if the common parental node of a predicted GO term and the correct GO term is equal to or deeper than the depth of two in the GO hierarchical tree. The E-value cutoff value (x-axis) represents the minimum similarity for sequences used in the benchmark analysis: Retrieved sequences with the E-value cutoff or smaller (more significant) are not used for function prediction by PFP and PSI-BLAST. PSI-BLAST annotations are transferred from

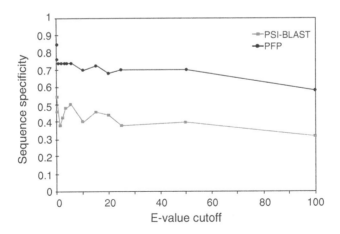

Fig. 2. Sequence-level specificity of PFP and conventional PSI-BLAST

the most similar sequence scoring above each E-value cutoff. It is evident that PFP outperforms PSI-BLAST in the entire range of the E-value cutoff with an accuracy which is almost double of that of PSI-BLAST. It is remarkable that the accuracy does not drop much even when only sequences of an E-value of 10 or larger are used.

PFP was ranked the best in a protein function prediction competition held at the Automated Function Prediction Special Interest Group (AFP-SIG) meeting in the 13th Annual International Conference on Intelligent Systems for Molecular Biology (ISMB) in June, 2005. See the recent paper for our performance at the AFP-SIG meeting and further benchmark results [7]. PFP is publicly available as a Web server at http://dragon.bio.purdue.edu/pfp/.

2 Toward Tertiary Structure-Based Function Prediction

In this section our protein surface shape-based function prediction method is outlined. Generally, a structure-based function prediction works in two steps (1) identifying potentially functionally important sites on a protein surface; and (2) comparing the identified site to known sites stored in a database. And if a similar site is found in this search, function associated to the known site (e.g., ATP binding) is transferred to the query protein. Each of them is described later.

2.1 Identification of Active Sites of Proteins

This step scans the surface of a protein to identify several candidates of active sites characterized by their local landscape, distribution of electrostatic potential, and/or residue conservation among the family. The key feature of a functionally important site is its local landscape, i.e., cavities and protrusions. For enzyme proteins, identifying cavities in a surface is of a special importance because their active sites bind a ligand molecule to catalyze chemical reaction. We have developed a novel algorithm for fast protein pocket region detection by checking *visibility* on the protein surface.

The algorithm called VisGrid first projects a query protein structure on to a three-dimensional grid (voxelization). A cavity is identified as a group of voxels occupied by the protein volume which have less visibility, that is, those which are surrounded by a certain number of empty voxels. The visibility of a voxel on the protein surface is defined as the percentage of viewable directions from the voxel. Voxels at the bottom of a pocket have a low visibility and those at the top of a protrusion have a high visibility. After identifying voxels with a low visibility on the surface, they are clustered into several local sites. Then the identified sites are ranked by their size (i.e., the number of voxels). VisGrid can identify not only pockets on the surface of a protein, but also hollows which are almost buried inside of a protein. This is important because there are often cases where a ligand molecule is held internal core of a protein.

We tested VisGrid on a benchmark dataset used by An et al. [9]. This dataset (LP_SET) consists of 5,561 protein structures taken from PDB, which include naturally occurring hetero molecules of a reasonable size. VisGrid identifies 79.1% of the active site residues (defined as the residues which locate closer than 4.5 Å to the ligand molecule) as the first rank; 87.6% of the binding site residues within the top three rank; and 91.4% among all the sites predicted (sensitivity). The detail of VisGrid will be published elsewhere.

2.2 Comparison of Local Sites of Proteins

Once characteristic local sites are identified in a query protein structure by VisGrid, the next step is to compare the sites with known functionally important sites in proteins. Here representation of local sites and comparison are the crucial steps which are intertwined with each other. We developed a representation of a local site which has following components (1) identifying *critical points* (2) representing the surrounding region of each critical point as *histograms* and (3) representing the entire site surface as a *graph*.

Initially a surface of the local site is represented by triangle tessellae computed by MSMS algorithm [10]. For each vertex of the tessellae, the local curvature is computed using a neighboring area within a certain radius R. Then vertices which has either the local maximum (i.e., a protrusion) or the local minimum (i.e., a pocket) curvature within the radius R is identified as *critical points* of the site. The entire local site can be represented as a complete graph (a graph with fully connected nodes) of the critical points. Now the local landscape of a critical point is described by a histogram of distances between all pairs of vertices in the area of the radius R (Fig. 3).

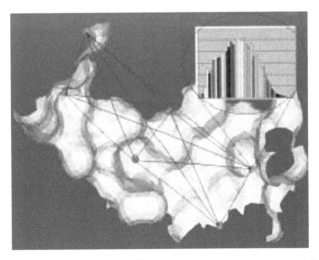

Fig. 3. Local site representation. The entire site is represented as the complete graph of the critical points shown in pink

Hence comparison of two local sites becomes a graph comparison problem. Since we want to identify global as well as local similarity between two sites represented as graphs, finding maximal common regions in the two graphs will be an appropriate measure of similarity. The similarity of two graphs reflects matching of the length of corresponding edges and of corresponding nodes. For matching of nodes, currently only histograms which capture their local geometry information is compared, but physicochemical properties or residue conservation information of critical points will be implemented. Comparison of two graphs comes down to the clique detection problem when an association graph is constructed from the two graphs.

3 Concluding Remarks

We have overviewed sequence- and structure-based function prediction methods currently being developed in our group. Rapid developments of new omics experimental techniques and the progress of structural genomics projects have seriously demanded renovation of bioinformatics tools – and here are a part of our responses to meet their needs.

Acknowledgment

This work was supported in part by the National Institute of General Medical Sciences of the National Institutes of Health (GM-075004), Purdue Research Foundation, and Purdue Alumni Association.

References

1. Altschul SF et al. (1990) J Mol Biol 215:403–410
2. Bateman A et al. (2002) Nucleic Acids Res 30:276–280
3. Hulo N et al. (2004) Nucleic Acids Res 32:D134–D137
4. Pearson WR (1996) Methods Enzymol. 266:227–258
5. Tatusov RL et al. (2003) BMC Bioinformatics 2003, 4:41
6. Altschul SF et al. (1997) Nucleic Acids Res 25:3389–3402
7. Hawkins T, Luban S, Kihara D (2006) Protein Sci In press 15:1550–1556
8. Harris MA et al. (2004) Nucleic Acids Res 32:D258–D261
9. An J, Totrov M, Abagyan R (2005) Mol Cell Proteomics. 4:752–761
10. Sanner M, Olson AJ, Spehner JC (1995) Proceedings of 11th ACM Symposium on Computational Geometry C6–C7

What We can Learn about Protein Folding from Recent Progress in Structure Prediction

George Chikenji, Yoshimi Fujitsuka, and Shoji Takada

Protein folding and protein structure prediction are long-standing fundamental problems in biophysics. Inherently they are very close and so their studies should progress vis-à-vis, but in reality, today's structure prediction technique is highly pragmatic so that its relevance to folding problem has been obscured. This paper contributes to filling this gap by focusing on "what we can learn about protein folding from the recent progress in structure prediction." In particular, we analyzed why the fragment assembly method, currently the most successful method in *de novo* structure prediction, is so powerful by designing a "chimera" protein experiments. In the chimera proteins, local structural preference is specific to the target sequence, while nonlocal interaction is only sequence-independent compaction force. We found that these chimera proteins can find the native fold of the intact sequences with high probability suggesting dominant roles of the local interactions. From these results, we suggest some "principles of protein folding." (1) For small proteins, compact structures that are fully compatible with local structural preference are surprisingly few, one of which is the native fold. (2) These local biases shape up the funnel-like energy landscape of proteins making Gō-like model used for folding study a good approximation of real proteins.

1 Introduction

Natural proteins fold into its native state within biologically relevant timescale [1]. One of the most stringent tests on how much we understood the mechanism may be to predict protein tertiary structures by simulating processes that are analogous to folding, which is often termed *de novo* structure prediction. Recently, significant progress has been made in *de novo* structure prediction, in which the most successful method is the fragment assembly (FA) method developed by David Baker's group and others [2–4]. FA methods showed considerable promise for new fold targets of recent CASPs, the community-wide blind tests of structure prediction [5]. In FA method, simulation is separated into two stages: First, we prepare structural candidates for every short segments of the target sequence retrieving them in structural database. The second stage is to generate tertiary structure that has the lowest approximate energy by assembling these fragments.

Simple questions arose are why the FA method is so successful and what we can learn about protein folding from the success of the FA method. The latter is particularly important because, we believe, some truth must be hidden in a practically successful method. According to Baker et al. [6], FA method is based on the experimental observation of protein folding that local sequence biases but does not uniquely decide the local structure of a protein. To what extent does the modest local bias influences the tertiary structures generated? How is the FA method related to recently developed protein folding theory [7]? In this report, we address these questions.

For the purpose, we need structure prediction software that uses the FA method. Here, we use the in-house developed software, SimFold. The SimFold has its own energy function based on a coarse-grained protein representation [8, 9] and conformational sampling is performed by the FA method [10, 11]. In the recent blind test on structure prediction, CASP6, our group Rokko (as well as the server Rokky) used the SimFold software as a primary prediction tool, showing one of the top level prediction performances in new fold category (elite-club-membership by the assessor, B. K. Lee [12]).

2 Lesson from Structure Prediction in the Past

Let us start with observations on comparisons of structures predicted in the past, which provide us important insight. The observation is from the recent CASPs. In the nonhomology modeling category of CASP6 [12], most prediction groups of high score used FA-based methods. Although each group used its own scoring (i.e., energy) function, predicted structures were sometimes similar to each other. Obviously, well predicted structures are similar. Interestingly, sometimes predicted structures with wrong fold are of some similarity. A nice example is the target T0201 from the new fold category of CASP6. Shown in Fig. 1 are prediction models of three independent groups, Rokko, Baker_ROBETTA, and Jones_UCL using FA methods. Apparently, they have wrong fold, but are very similar to each other. This suggests that the prediction performance of FA method is robust against difference in scoring functions.

3 The Chimera Experiment

Motivated by these observations, we design a chimera experiment for addressing robustness of FA methods and impact of local interactions for protein folding. Here, we note that local and nonlocal interactions are treated explicitly

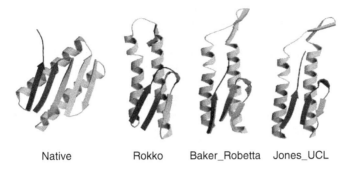

Fig. 1. Examples of structure predictions from the new fold category in CASP6. The native structure of target T0201 (PDB ID code 1s12), the model 1 from group Rokko, the model 6 from group Baker_Robetta, and the moodel 1 from group Jones_UCL are shown. The graphic of protein structures is generated by Molscript [23] and Raster3D [24]

in separate stages in the FA method; local structural propensity is taken into account when fragment candidates are listed up in the first stage, whereas nonlocal interactions play central roles in assembling/folding these fragments in the second stage.

Taking advantage of this separation, we design the following chimera experiment. Let us use immunoglobulin-binding domain of protein G (56-residue $\alpha + \beta$ protein according to the SCOP definition) as an example. First, we prepare fragment candidates of every overlapping eight residues, as usual, for protein G. Second, in the folding stage, we use the energy function for 56-residued polyvaline (poly-Val) sequence instead of the sequence of protein G and perform multicanonical MC simulations of assembling. Thus, the simulation is conducted for the chimera interactions; local structural propensity is for protein G and nonlocal interaction is for poly-Val. For the latter, we choose valine nearly by chance as a representative of hydrophobic residue. Hydrophobic residues tend to attract each other, thus introducing nonspecific force of compaction (collapse). We note that we are not interested in poly-Val folding itself. We tested the same experiment using polyisoleucin and polyleucine finding essentially the same results (data not shown) and we confirmed that the choice of valine is not essential.

Now we discuss the results of experiments. Figure 2 plot energies and RMSD of the conformations generated by the simulations for chimera protein of protein G. We clearly see strong correlation between RMSD and lower envelope of energies. Next, as usual in structure prediction, we perform cluster analysis [13] of low energy structures. Figure 2 show the structures of the largest , the second largest, and the third largest cluster centers for the chimera protein. (The native structure of protein G is depicted in the left hand side of Fig. 2 as a reference.) Surprisingly, the structure of the largest cluster center of the chimera protein has the correct fold (RMSD is

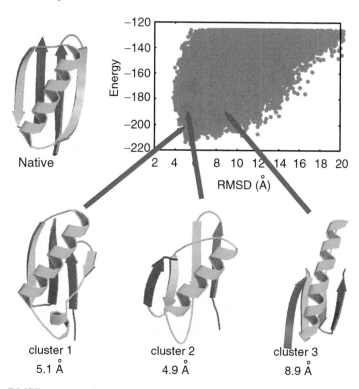

Fig. 2. RMSD-energy plot of "chimera" protein G. The structures of the largest (cluster 1), the second largest (cluster 2), and the third largest cluster (cluster 3) center are shown in bottom of the figure. The native structure of protein G is also shown as a reference

5.1 Å), although the architecture of secondary structure elements is slightly perturbed; the second β strand is too short. The sizes of the largest cluster are larger than 20% of generated low energy structures. The largest cluster size is reasonably large in the standard of *de novo* structure prediction. This indicates that the structure prediction performances using FA is unexpectedly robust with respect to the energy function used for assembling. Essentially the same results can be obtained for other proteins of similar length, as well: We also performed the same experiment for an all α-protein, 434 C1 repressor, and an all β-protein Hex1.These results can be found in [14].

The above chimera experiment showed that local structural preferences specific to the wild-type sequence and nonspecific compaction force together drive the protein into its native fold with reasonably high probability for small proteins. In other words, given the fragment structure ensemble that reflects local structure preferences of the target sequence, compact structures that are compatible with these fragments are few, one of which is the native fold.

4 Why do Gō-like Models Work so well for Describing Folding Reactions?

What do these results tell us about protein folding? Recently it was established that ensemble of folding pathways can often be well-described by the funnel theory [15, 16]. Common to these models is an assumption that only native contacts are favorable and nonnative contacts are neutral or repulsive, which leads to the so-called Gō potential [17]. It is unclear, however, how the smooth and funnel-shaped energy landscape is organized from numerous atomic interactions. Important and interesting questions related to this are why Gō model is a good approximation for describing folding reactions and how we can justify ignorance of nonnative interactions.

By now, it is obvious that the perspective that there are only limited numbers of compact structures fully compatible with local structure preference clears up the mystery: If there are only limited numbers of possible compact structures in the conformational space of real proteins and a native structure is one of them, energetically stable misfolded structures does not exist (because stable misfolded structures must be compact in general), resulting in funnel-like energy landscape.

5 Discussion on Denatured States of Proteins

The finding that local structural preference and compaction force together lead the protein into the native fold may have some links to characteristics of denatured states of proteins, which recently attract considerable attention [18–20]. In simulations of a β-hairpin confined in a pore, Thirumalai and his coworkers found that, under highly denatured condition, the confined β-hairpin possesses partially native-like order [21]. This is in contrast to the denatured ensemble of a bulk β-hairpin, which is extended and has little native-like order. Thus, compaction force alone induces native-like order [22]. The current work suggests that this is quite probable by the help of local structural preferences.

Possibly related to this topic is that Shortle and his collaborators reported a mutant of the staphylococcal nuclease oriented in strained polyacrylamide gels has native-like global structure even in highly denatured state by measuring residual dipolar couplings [19]. The interpretation is now under intensive debates. Since the gels introduce compaction force, one conceivable explanation would be that native-like order may arise by this compaction force accompanied by local structural bias.

6 Conclusions

Although it has been well studied how proteins fold on their funnel-shaped energy landscape, it has not been elucidated what microscopic interactions make protein energy landscape funnel-like. This study addresses this point motivated by recent success of the FA method in structure prediction. We reached a principle of protein folding: (1) There are only limited numbers of "compact" conformations that are fully compatible with local structural preferences, and the native fold is one of them, and thus (2) local structure preferences strongly shape up the protein folding funnel.

References

1. Fersht, A. R. (1999) *Structure and Mechanism in Protein Science: A Guide to Enzyme Catalysis and Protein Folding* (W H Freeman, New York), ISBN: 0716732688
2. Simons, K. T., Kooperberg, C., Huang, E., & Baker, D. (1997) *J. Mol. Biol.* **268**, 209–225
3. Bowie, J. U. & Eisenberg, D. (1994) *Proc. Natl. Acad. Sci. USA* **91**(10), 4436–4440
4. Jones, D. T. (1997) *Proteins* **29**(Suppl 1), 185–191
5. Vincent, J. J., Tai, C., Sathyanarayana, B. K., & Lee, B. K. (2005) *Proteins* **61**(Suppl 7), 67–83
6. Rohl, C., Strauss, C. E. M., Misura, K. M. S., & Baker, D. (2004) *Methods. Enzymol.* **383**, 66–93
7. Onuchic, J. N. & Wolynes, P. G. (2004) *Curr. Opin. Struct. Biol.* **14**(1), 70–75
8. Takada, S., Luthey-Schulten, Z. A., & Wolynes, P. G. (1999) *J. Chem. Phys* **110**, 11616–11629
9. Fujitsuka, Y., Takada, S., Luthey-Schulten, Z. A., & Wolynes, P. G. (2004) *Proteins* **54**, 88–103
10. Chikenji, G., Fujitsuka, Y, & Takada, S. (2003) *J. Chem. Phys.* **119**, 6895–6903
11. Chikenji, G., Fujitsuka, Y, & Takada, S. (2004) *Chem. Phys.* **307**, 157–162
12. http://predictioncenter.org/casp6/meeting/presentations/NF_assessment.pdf.
13. Shortle, D., Simons, K. T., & Baker, D. (1998) *Proc. Natl. Acad. Sci. USA* **95**, 11158–11162
14. Chikenji, G., Fujitsuka, Y. & Takada, S. (2006) *Proc. Natl. Acad. Sci. USA*, in press
15. Clementi, C., Nymeyer, H., & Onuchic, J. N. (2000) *J. Mol. Biol.* **298**, 937–953
16. Koga, N. & Takada, S. (2001) *J. Mol. Biol.* **313**, 171–180
17. Gō, N. (1983) *Ann. Rev. Biophys. Bioeng.* **12**, 183–210
18. Hammarström, P. & Carlsson, U. (2000) *Biochem. Biophys. Res. Comm.* **276**, 393–398
19. Shortle, D. & Ackerman, M. S. (2001) *Science* **293**, 487–489

20. Zagrovic, B. & Pande, V. S. (2003) *Nature Struct. Biol.* **10**, 955–961
21. Klimov, D. K., Newfield, D., & Thirumalai, D. (2002) *Proc. Natl. Acad. Sci. USA* **99**(12), 8019–8024
22. Chan, H. S. & Dill, K. A. (1990) *Proc. Natl. Acad. Sci. USA* **87**(16), 6388–6392
23. Kraulis, P (1991) *J. Appl. Crystallogr.* **24**, 946–950
24. Merrit,E. A., & Bacon,D. J. (1997) *Methods. Enzymol.* **277**, 505–524

Brownian Dynamics Approach to Protein Folding

Tadashi Ando and Ichiro Yamato

Summary. Understanding of protein folding mechanism and prediction of a protein structure are crucial importance not only for biology but also for computational science. In order to address this protein folding problem, we have developed an atomistic Brownian dynamics (BD) simulation that enables us to simulate a protein for long time compared with a conventional molecular dynamics simulation. Using our BD method, folding events of a 13-mer α-helical peptide and a 12-mer β-hairpin peptide were observed many times in canonical simulations. In this model, the driving energy contribution toward folding came from both electrostatic and van der Waals interactions for the α-helical peptide and from van der Waals interactions for the β-hairpin peptide. Although many non-native structures having the same or lower energy than that of native structure were observed, the folded states formed the most populated cluster when the structures obtained by the BD simulations were subjected to the cluster analysis based on distance-based root mean square deviation of side-chains between different structures. This result indicates that we can predict the native structures from conformations sampled by BD simulation. In these respects, the BD method would be a possible approach to protein folding.

1 Introduction

One of the outstanding characteristics of a living system is the ability to self-assemble with remarkable precision and fidelity. The folding of proteins into their compact three-dimensional structures is the most fundamental example of biological self-organization. This folding process is the final and crucial step in the transformation of genetic information into specific biological function, which is called as "the second genetic code" [1]. In addition to this biological importance, protein folding is intrinsically interesting system from the viewpoint of physical chemistry. Proteins are unbranched long-chain polymers consisting of just 20 kinds of amino acids, but unlike most polymers each chain has a well-defined three-dimensional structure and a function. Therefore, understanding the mechanism of the self-organization of a protein and predicting the folded structure from its amino acid sequence will dramatically affect not only biology but also physics and chemistry.

Molecular simulation is a powerful tool that can analyze dynamics of proteins quantitatively. However, when the protein folding problem is approached using this method, we face two difficult problems. The first is the quantitative uncertainty of the free energy function describing both protein's intramolecular interactions and intermolecular interactions with solvent for arbitrary conformations. The second is the insufficiency of simulation time that should be necessary for following the whole process of folding: one nanosecond simulation of a protein-solvent system necessitates about a week of computation time even with up-to-date computers, though protein folding takes place from microseconds to seconds.

In order to overcome the difficulty of these two obstacles, we have developed an atomistic Brownian dynamics (BD) simulation with multiple time step algorithm and a new implicit solvent model to describe the protein folding process at atomic resolution [2–5].

2 Methods

2.1 Brownian Dynamics Simulation Algorithm

By treating the effects of solvent as a dissipative random force, the Langevin equation can be expressed as

$$m_i \frac{d^2 \mathbf{r}_i}{dt^2} = -\zeta_i \frac{d\mathbf{r}_i}{dt} + \mathbf{F}_i + \mathbf{R}_i. \tag{1}$$

Here, m_i and $\mathbf{r}_i$ represent the mass and position of atom i, respectively. ζ_i is a frictional coefficient and is determined by the Stokes' law, that is, $\zeta_i = 6\pi a_i^{\text{Stokes}} \eta$ in which a_i^{Stokes} is a Stokes radius of atom i and η is the viscosity of water. $\mathbf{F}_i$ is the systematic force on atom i. $\mathbf{R}_i$ is a random force on atom i having a zero mean $<\mathbf{R}_i(t)> = 0$ and a variance $<\mathbf{R}_i(t)\mathbf{R}_j(0)> = 2\zeta_i k_\text{B} T \delta_{ij} \delta(t) \mathbf{I}$ where $\mathbf{I}$ is 3×3 unit tensor; this derives from the effects of solvent.

For the overdamped limit (the solvent damping is large and the inertial memory is lost in a very short time), we set the left side of (1) to zero,

$$\zeta_i \frac{d\mathbf{r}_i}{dt} = \mathbf{F}_i + \mathbf{R}_i. \tag{2}$$

Integrated equation of (2) is called Brownian dynamics [2];

$$\mathbf{r}_i(t+h) = \mathbf{r}_i(t) + \frac{\mathbf{F}_i(t)}{\zeta_i} h + \sqrt{\frac{2k_\text{B} T}{\zeta_i}} h \boldsymbol{\omega}_i, \tag{3}$$

where h is a time step and ω_i is a random noise vector obtained from Gaussian distribution.

Time step of 10 fs was used for single time step BD simulation. For multiple time step (MTS) algorithm, short time step of 5 fs and long time step of 40 fs were used [3]. Cut-off method was not used. All bond lengths were constrained with LINCS algorithm [6]. Coordinates and energies were recorded every 100 ps during the simulation. For analysis, the structures collected for first 10 ns were removed.

2.2 Force Field

We used the AMBER91 united-atom force field with an angle-dependent, 12–10 hydrogen-bond potential [4]. To reproduce the solvation effects, three implicit solvent models were used: distance-dependent dielectric model (DD), solvent-accessible surface area model (SA), and effective charge model (EC) [4]. In the DD model, $\varepsilon = 2r_{ij}$ was used. The atomic solvation parameters used in the SA model were $\sigma(C) = 12\,\text{cal mol}^{-1}\,\text{Å}^{-2}$, $\sigma(O, N) = -116\,\text{cal mol}^{-1}\,\text{Å}^{-2}$, $\sigma(S) = -18\,\text{cal mol}^{-1}\,\text{Å}^{-2}$, and $\sigma(O^-/N^+) = -280\,\text{cal mol}^{-1}\,\text{Å}^{-2}$ [5]. The EC model was introduced by us to represent the shielding effect of oriented water molecules around a point charge [4], in which atomic charge of atom i, q_i, is neutralized as a function of solvent-accessible surface area of the atom, $\text{SA}_i(\mathbf{r}^N)$:

$$q'_i = q_i \left[\frac{1 - \text{SA}_i(\mathbf{r}^N)/S_i}{\alpha_{\text{int}}} + \frac{\text{SA}_i(\mathbf{r}^N)/S_i}{\alpha_{\text{ext}}} \right]. \tag{4}$$

Here q'_i is the effective charge of atom i, S_i is the total solvent-accessible surface area of isolated atom i, α_{int} is a shielding parameter against interior of the solute (wherein α_{int} is set at unity), and α_{ext} is a shielding parameter for exterior water. In this study, $\alpha_{\text{ext}} = 5$ was used.

3 Results and Discussion

3.1 Folding Simulations of α-Helical and β-Hairpin Peptides

We performed 400 ns BD simulations of an α-helical, peptide III (13 residues), and a β-hairpin, BH8 (12 residues), peptides five times using different random seeds for each peptide at 298 K from the fully extended conformations; each simulation requires 15 hours on a Pentium4 processor.

Folding Trajectories

Figure 1 shows the fraction of native contacts (Q) of the two peptides as a function of simulation time. We defined nine backbone hydrogen bonds (the $O\cdots H$ distance is smaller than 3 Å) between residue i and $i + 4$ as native

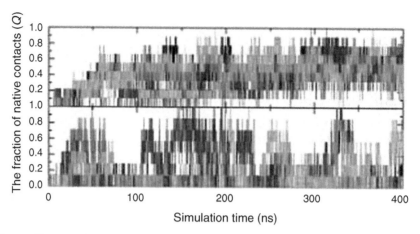

Fig. 1. Time evolutions of the fraction of native contacts during BD simulations of peptide III (*upper*) and BH8 (*lower*). Five trajectories (black, red, green, blue and sky blue) obtained by the BD simulations using different random number seeds are shown

contacts of peptide III. Four interstrand backbone hydrogen bonds (Ile3 NH-Tyr10 CO, Ile3 CO-Tyr10 NH, Val5 NH-Lys8 CO, and Val5 CO-Lys8 NH) and three interstrand side-chain interactions (Ile3-Tyr10, Thr4-Thr9, and Val5-Lys8; distances between geometrical centers of side-chains are smaller than 7 Å) were used for native contacts of BH8. For peptide III, although there were few states having $Q > 0.8$ due to lack of hydrogen bonds at C terminus, the peptide reached the folded states from the extended states within 400 ns in all simulations. Because the formation of perfect helix ($Q = 1.0$) accompanies with large entropic cost of conformation, this state is not expected to exist in a significant amount. In the simulations of BH8, the peptide also folded from the extended structure in all trajectories.

Energy Components

The average effective energy (effective energy is the intraprotein energy plus solvation free energy) and its components (van der Waals term, E_{vdW}, electrostatic term including the effects of DD/SA/EC implicit solvent models, E'_{elec}) of the two peptides as a function of Q are shown in Fig. 2. The total effective energy showed downhill profile for both peptides. The negative gradient of the total effective energy of peptide III was much larger than that of BH8. However, since variances of the total energies were too large, there were many non-native structures having lower effective energy than the energy of the native structure in both systems. This result indicates that it is impossible to predict the native states of the peptides based on the energy alone. For peptide III, the average values of E_{vdW} and E'_{elec} decreased with Q. For

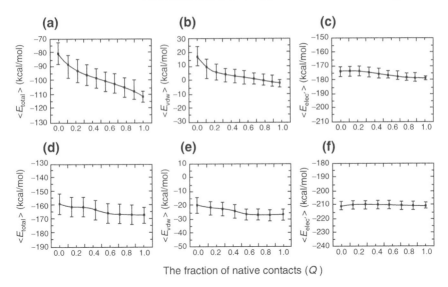

Fig. 2. Energy plots of the 2.0×10^4 conformations sampled during five simulations of (**a–c**) peptide III and (**d–f**) BH8 at 298 K. (**a, d**) Average of total effective energy, (**b, e**) average energy of van der Waals term and (**c, f**) average energy of effective electrostatic term as a function of Q are shown

BH8, although the average value of E_{vdW} decreased with Q, the slope of E'_{elec} is quite flat. These results indicate that the effective driving energy contributions to the folding of the peptides are concluded to be derived from both van der Waals and electrostatic terms for the α-helical peptide, peptide III, and from van der Waals term for β-hairpin peptide, BH8.

Cluster Analysis

Next we performed a cluster analysis based on structural similarity [5] using about 20,000 structures obtained by the simulations of each peptide. The structures of the centers of the three most populated clusters for both peptides are shown in Fig. 3. Interestingly, the most populated cluster had higher average value of Q than that of other clusters and the folded structures belonged to these most populated clusters for both peptides. The central structure of cluster 1 of peptide III had a helical conformation throughout the peptide. For BH8, the central structure of the most populated cluster was a β-hairpin conformation that had side-chains of Ile3, Val5, Lys8, and Tyr10 protruding on the same side of plane of the strands, which is consistent with the NMR data. An important point is that the cluster analysis makes it possible to predict the native folded states from the structures obtained by the BD simulations.

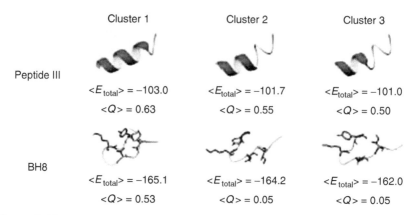

Fig. 3. Ribbon representations of the central structures of the three most populated clusters for peptide III (*upper*) and for BH8 (*lower*). The values of total effective energy (E_{total} in kcal mol^{-1}) and the fraction of native contacts, Q, averaged over the cluster are listed under each central structure. For BH8, residues of Ile3, Val5, Lys8, and Tyr10 are shown in sticks. The figures are generated with MOLMOL [7]

4 Conclusion

Now the BD method made it possible to simulate the folding of the key secondary structure elements in proteins as described above. Adding to the conclusion of this report, we would like to state a future direction of our study. For understanding folding mechanism and predicting the folded structures of proteins from their amino acid sequences, the folding simulations of proteins having more than 50 amino acid residues should be essential. To tackle this problem, we have launched the protein structure prediction server named "TANPAKU" (http://issofty17.is.noda.tus.ac.jp/), in which many BD simulations are performed on distributed computing platform. This method might have a potential to enhance the efficiency of conformation sampling greatly. We believe that these studies presented in this report will become a key step not only to deciphering of "the second genetic code" but also to starting in silico biology.

Acknowledgments

T. Ando acknowledges support from the "Academic Frontier" Project (2005) from MEXT (Ministry of Education, Culture, Sports, Science and Technology of Japan). This work was supported by the "Open Research Center" Project (2002) from MEXT (I. Yamato).

References

1. Duan Y, Kollman PA (2001) IBM Systems J 40: 297–309
2. Ando T, Meguro T, Yamato I (2002) J. Comput. Chem. Jpn 1: 103–110
3. Ando T, Meguro T, Yamato I (2003) Mol. Simul. 29: 471–478
4. Ando T, Meguro T, Yamato I (2004) J. Comput. Chem. Jpn 3: 129–136
5. Ando T, Yamato I (2005) Mol. Simul. 31: 683–693
6. Hess B, Bekker H, Berendsen HJC, Fraaije JGEM (1997) J. Comput. Chem. 18: 1463–1472
7. Koradi R, Billeter M, Wuthrich K (1996) J. Mol. Graphics 14: 51–55

A Fourth Type of Secondary Structure Breaker

Kenichiro Imai, Masashi Sonoyama, and Shigeki Mitaku

Summary. Secondary structure breakers impose strong limitations on the tertiary structures of proteins. Previously, we examined the distribution of various physicochemical properties in a set of 1,031 proteins and found that there are at least three types of secondary structure breakers: clusters of proline, glycine, and amphiphilic residues, respectively (Imai and Mitaku, BIOPHYSICS, 2005). In this new study, we found that clusters of small polar residues function as a fourth type of secondary structure breaker. We used this discovery to improve our prediction system for secondary structure breakers, SOSUIbreaker. Based on the same dataset of 1,031 proteins, predicted breakers now cover 80% of loop regions with accuracies as high as 94%.

1 Introduction

The amino acid sequences of proteins contain various signals for structural stability and function. However, the physical mechanisms underlying secondary structure formation remain somewhat unclear. Consequently, the accuracy of current secondary structure prediction methods remains below 80% in spite of a long history of investigation and advancement in information technology [1]. In a recent paper we focused on secondary structure breakers, which drive secondary structure to loop structural changes. We found that clusters of proline, glycine, and amphiphilic residues can all be secondary structure breakers depending on certain environmental factors, such as hydrophobicity and helical periodicity. On the basis of those factors, we developed a novel secondary structure breaker prediction system (SOSUIbreaker). Using this system on a set of 1,031 proteins, predicted breakers were observed in almost two-thirds of the total loops with accuracies as high as 90% [2]. However, the mechanism for breaking the secondary structures in the remaining loop regions remained unclear.

In this work, we studied clusters of small polar residues (serine, threonine, asparagine, and aspartic acid) in order to ascertain their role as a possible fourth type of secondary structure breaker. Clusters of small polar residues are potential factors in breaking secondary structures because these residues tend to be located at the termini of secondary structures [3, 4]. Similar to the previous study, the dependence of this fourth type of secondary structure breaker on specific environmental conditions was investigated. Such conditions included the average hydrophobicity, the helical periodicity, the presence of tryptophan and tyrosine clusters, and the density of glycine residues. The same dataset of 1,031 proteins was used for analysis. Modification of SOSUIbreaker to include clusters of small polar residues resulted in predicted breakers being observed in 80% of all loop regions. Predicted total secondary structure breakers were located in breaking regions with an accuracy of 70%. When both the breaking region and the loop core together were taken into consideration, 94% of the predicted breakers were located in the loop regions.

2 Method

We improved the SOSUIbreaker prediction system by modifying the previous method, which was based only on clusters of proline, glycine and amphiphilic residues [2], to include clusters of small polar residues as secondary structure breakers as well. In this study, the same previously described dataset of 1,031 proteins with known tertiary structure was used for analysis [2]. The definitions of three regions around the termini of secondary structures, namely the breaking region, the secondary structure core, and the loop core, were also as previously described [2].

Predicting which clusters of small polar residues were secondary structure breakers was based on two steps. First, candidates were enumerated, and then the breakers were predicted, discriminating the candidates by the environmental factors.

Small polar residues were defined as breaker candidates by the function $\langle\langle SP(k)\rangle\rangle$, the double average of $SP(i)$, where $SP(i)$ indicates the presence or absence of a small polar residue at the ith residue. A value of 1 for $SP(i)$ represents the presence and a value of zero represents the absence of a small polar residue at the ith residue. The threshold for $\langle\langle SP(k)\rangle\rangle$ was 0.3.

$$\langle SP(j)\rangle = \left[\sum_{i=j-3}^{j+3} SP(i)\right]/7 \qquad (1)$$

$$\langle\langle SP(k)\rangle\rangle = \left[\sum_{j=k-3}^{k+3} \langle SP(j)\rangle\right]/7. \qquad (2)$$

The environmental parameters for discriminating the candidates in breaking regions were the average of the Kyte and Doolittle hydrophobicity index $\langle H(j)\rangle$ [5], the helical periodicity score $\langle HSP(j)\rangle$, the average amphiphilicity index of tryptophan and tyrosine $\langle A'(j)\rangle$ [6] and the density of glycine $\langle G(j)\rangle$.

The first three parameters were defined in the previous study [2]. $\langle G(j) \rangle$ was calculated by evaluating the existence of glycine using a window size of 7 as well as $\langle SP(i) \rangle$. We calculated the deviation of the parameters in a segment 15 residues long around the potential breakers from the corresponding values for the dataset of false data, namely the breaker candidates in the secondary structure core regions. The details of this calculation were described previously [2]. The discrimination function obtained from the principal component analysis for the deviation of the environmental parameters is defined by (3) for small polar residues:

$$\text{Score}(l) = a_0 + a_1 \Delta \langle H(l) \rangle + a_2 \Delta \langle \text{HSP}(l) \rangle + a_3 \Delta \langle A'(l) \rangle + a_4 \Delta \langle G(l) \rangle \quad (3)$$

in which l represents the position of a candidate secondary structure breaker. The parameters, $\Delta \langle H(j) \rangle$, $\Delta \langle \text{HSP}(j) \rangle$, $\Delta \langle A'(j) \rangle$, $\Delta \langle G(j) \rangle$ are the deviation from the average values for the false data set. The coefficients of the discrimination score were as follows: $a_1 = 2.433$, $a_2 = 1.306$, $a_3 = 1.026$, $a_4 = 178.064$, and $a_0 = -0.003$. Our current system for predicting secondary structure breakers was improved by combining this predicting method for small polar residue breakers with the previous method [2].

3 Results and Discussion

The clusters of small polar residues were considered to have the same effect as glycine because of the small size of their side chains, and we assumed that they could function as a fourth type of secondary structure breaker. We investigated the distribution of the clusters of small polar residues and found that they were located mainly in the breaking region, but they are also found in the secondary structure core region. This fact indicates that the fate of the clusters of small polar residues is determined by several physicochemical environmental factors, similar to that found in the previous study for the other three types of secondary structure breakers [2]. In order to reveal the difference between the small polar residue potential breakers in the breaking regions and those in the secondary structure core, a segment around a numerated potential breaker was divided into three regions with a fixed length of five residues each. Figure 1 shows the averages of four physicochemical properties for the three regions surrounding the clusters of small polar residues in the breaking region and secondary structure core, respectively. The properties are averaged over all data in each data set, and a general trend can be observed from the histograms. The hydrophobicity of the segments surrounding the clusters in the breaking region is lower than in the secondary structure core. Similarly, the helical periodicity in the breaking region is lower than that in the secondary structure core. Glycines are found with higher frequency in the breaking region than in the secondary structure core. Further, bulky amino acids with polar groups, such as tryptophan and tyrosine, are also more abundant in the

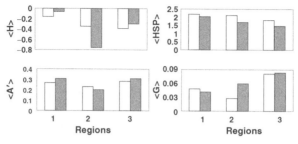

Fig. 1. Levels of averages of four kinds of properties, $\langle H(j) \rangle, \langle \mathrm{HSP}(j) \rangle, \langle A'(j) \rangle$, and $\langle G(j) \rangle$ in three regions were compared between potential breakers in secondary structure core and breaking regions. The *shaded bars* and *open bars* represent the properties in the breaking region and secondary structure core, respectively

breaking region. These differences seem reasonable: the hydrophobic core of a protein with high hydrophobicity is generally formed by secondary structures, and the helical structure at the interface region shows high periodicity. The bulkiness of the side chains in tryptophan and tyrosine may hinder ordering of the structure due to an excluded volume effect. Glycine residues give high flexibility to polypeptides, which tend to break the secondary structures. The differences between breaking region and secondary structure core for these four properties is not large, however, a combination of these effects may lead to accurate identification of secondary structure breakers. Principal component analysis of four environmental factors was conducted in order to discriminate the candidate breakers in the breaking regions. The relevant discrimination function (3) was also obtained. Figure 2 shows the plot of the ratio of the number of small polar residue predicted breakers to the total number of residues at each position vs. position of residue. As a control for secondary structure breakers, the same ratio for proline is plotted in the same diagram. The results shown in Fig. 2 indicate that almost all discriminated breaks by small polar residues are located in the loop core and the breaking regions and the shape of the distribution was very similar between the two types of breakers (small polar residues and proline). This observation strongly suggests that the effect of the environmental parameters is substantial for breaking secondary structure and the clusters of small polar residues are a fourth type of secondary structure breaker.

Table 1 shows the summary of the accuracy of the prediction by four mechanisms. In both this study and the previous study, we used only long secondary structures for the analysis [2], but secondary structure breakers at the termini of short secondary structures can also be discriminated by the same method. Thus, we evaluated the accuracy of our new system by taking short secondary structures into consideration. About 70% of predicted breakers (all four types) were actually located in the breaking region and 94% were located in the loop region that consists of both the breaking region and the loop core region. The predicted breakers covered 80% of loop regions.

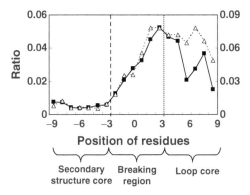

Fig. 2. Ratio of the number of predicted small polar residue secondary structure breakers to that of all amino acid residues after discrimination. The plot for proline is shown with an *open triangle* for comparison with small polar residues. The ordinate on the right-hand side indicates the ratio for proline. The position of 0 on the horizontal represents the end of secondary structure, and the negative side and the positive side represent secondary structure core and loop region, respectively

Table 1. (a) Number and accuracy of predicted breakers in loop region, breaking region, and secondary structure core; (b) number and ratios of loop regions and secondary structure cores that contain predicted secondary structure breakers. P, G, A, and SP represents the breakers proline, glycine, amphiphilic residues, and small polar residues, respectively

(a)

	P	G	A	SP	total	accuracy of prediction (%)
total predicted breakers	6,525	6,923	1,968	3,547	12,819	–
loop region	6,249	6,522	1,816	3,355	11,995	93.6
breaking region	4,533	4,719	1,352	2,512	8,972	70.0
loop core	1,716	1,803	464	843	3,023	23.6
secondary structure core	276	401	152	192	824	6.4

(b)

regions	total number	number of region with predicted breakers	ratio
loop region	11,127	8,861	0.80
breaking region	20,250	10,828	0.54
loop core	3,925	2,268	0.58
secondary structure core	5,657	690	0.12

About 20% of loop regions did not have any one of the four types of breakers. However, a fifth mechanism by which secondary structure can be broken was suggested from the analysis of charge symmetry in our current investigations [7]. We are analyzing the environmental conditions required for breaking by charge symmetry in order to further refine our method of secondary structure breaker prediction so that it includes a charge symmetry algorithm.

The current version of the system for predicting secondary structure breakers is available at the following URL: http://bp.nuap.nagoya-u.ac.jp/sosui/sosuibreaker/sosuibreaker_submit.html.

Acknowledgment

This work was partly supported by a Grant-in-Aid for the National Project on Protein Structural and Functional Analyses, and also by a Grant-in-Aid for the 21st Century COE "Frontiers of Computational Science" from the Ministry of Education, Culture, Sports, Science and Technology of Japan.

References

1. Rost B (2001) J. Struct. Biol. 134: 204–218
2. Imai K and Mitaku S (2005) Biophysics, 1: 55–65
3. Chou PY, Fasman GD (1978) Adv. Enzymol. Relat. Areas Mol. Biol. 47: 45–148
4. Levitt M (1978) Biochemistry 17: 4277–4285
5. Kyte J and Doolittle RF (1982) J. Mol. Biol. 157: 105–132
6. Mitaku S, Hirokawa T, Tuji T (2002) Bioinformatics 18: 608–616
7. Imai K and Mitaku S (2006) CBIJ 5: 65–77

Neighborhood Parallel Simulated Annealing

Keiko Ando, Mitsunori Miki, and Tomoyuki Hiroyasu

Summary. When applying SA to continuous optimization problems, the appropriate adjustment of the neighborhood ranges becomes necessary to obtain the good performance. In this paper, We propose a Neighborhood Parallel Simulated Annealing (NPSA) for continuous optimization problems, which provides global search using the periodic exchange of different neighborhood ranges with parallel computers. The proposed approach, NPSA, automatically determines the appropriate neighborhood range, and shows a good performance in solving typical test problems.

1 Introduction

Simulated annealing (SA) is a method for solving combinatorial optimization problems. SA was formulated by Kirkpatrick et al., and its algorithm imitates physical annealing. It is very useful for solving several types of optimization problems with nonlinear functions and multiple local optima [1–3].

The advantages and disadvantages of SA are well summarized in [4]. The most significant disadvantages are the difficulty in determining the appropriate parameters, and the long calculation time for determining the optimum solution. The neighborhood range and temperature are the important parameters in SA. In combinatorial optimization problems, the neighborhood range can be determined by replacing two adjoining elements in solutions. Thus, a neighborhood range is decided automatically for every problem; therefore, the temperature schedule becomes the most important factor. On the other hand, when SA is applied to continuous optimization problems, a neighborhood range can be freely decided in the range of the Eucledean space. However, the appropriate neighborhood range is greatly dependent on the landscape of the objective function. Therefore, the control of the neighborhood range becomes very important when applying SA to a continuous optimization problem, but it is difficult to determine a neighborhood range automatically in this case.

Some research has been carried out wherein the neighborhood range was adjusted adaptively according to the landscape of the objective function.

Corana's method [6] controls the neighborhood range according to a landscapes maintaining an acceptance ratio of 0.5. The authors proposed a new method called SA/AAN wherein the appropriate neighborhood range can be determined based on the arbitrary acceptance ratio. These methods do not require the tuning of their parameters for many problems; however, the target values of the acceptance ratio are necessary.

In this research, we propose a parameter-free SA method called neighborhood parallel simulated annealing (NPSA). This method determines the neighborhood range automatically and does not require the acceptance ratio in order to adjust a neighborhood range.

2 Neighborhood Parallel Simulated Annealing (NPSA)

2.1 The NPSA Concept

In NPSA, parallel processing is performed and neighborhood ranges are assigned during each process. The neighborhood ranges are self-adjusted through the exchange of neighborhood ranges between processors. Such an algorithm enables the exchange of the neighborhood ranges between processors based on the energy of the solutions. The processor with a low energy is assigned a small neighborhood range, so that the solution can search locally. On the other hand, the processor with a high energy is assigned a large neighborhood range, so that the solution can avoid the local minima. Thus, the processors cooperate and adjust neighborhood ranges as the search proceeds.

2.2 The NPSA Algorithm

The NPSA algorithm is pictorially depicted in Fig. 1. The process followed in NPSA is described as:

1. Parallel processing is performed.
2. A different neighborhood is assigned during in each process.

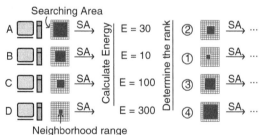

Fig. 1. NPSA algorithm

3. SA is executed in parallel in each processor.
4. At every cooling cycle, the energy values (evaluation values) are gathered in one process.
5. The energy values are sorted, and the rank is determined based on the energy.
6. The neighborhood range is assigned according to the rank as follows: Processor B is ranked 1; therefore, this processor's neighborhood is determined to be the smallest. Processor D is ranked 4; therefore, this processor's neighborhood is determined to be the biggest. The same logic is applied to the remaining cases.
7. Steps 3–6 are repeated.

3 Effectiveness Verification of the Proposed Method

3.1 Optimization Problems

In order to examine the search performance of the proposed method, two standard test functions are used, namely the Rastrigin and the Griewank functions. The optimum solutions are located at the origin for the Rastrigin and Griewank functions, and the function values are 0, and its function value is also 0. Two, three, and five dimensions of these functions are used as test functions.

The parameters are indicated in Table 1. The number of both cooling cycles and parallel processors were set to 32. Thirty-two parallel processors are used in both NPSA and PSA, and the annealing step was set to the same value as that in sequential methods by setting the cooling cycle to 1/32 times.

3.2 Results and Discussion

The minimum energy value obtained using the Rastrigin function is indicated in Fig. 2, while that using Griewank function is indicated in Fig. 3. The comparison between the Random Search, Conventional SA, Corana,

Table 1. Parmeters

Method	NPSA, PSA		Coran, SA/AAN, SA, Random	
Function	Rastrigin	Griewank	Rastrigin	Griewank
max. temperature	10	20	10	20
min. temperature	0.01	0.001	0.01	0.001
cooling cycles	320	960	320×32	960×32
cooling rate	0.8	0.726	0.8	0.726
number of processors			32	
min.neighborhood range		width of design space $\times 10^{-5}$		

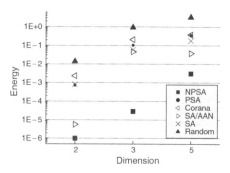

Fig. 2. Energy value (Rastrigin function)

SA/AAN, and NPSA methods in Figs. 2 and 3 clearly demonstrates the proposed method's (NPSA) exceedingly superior performance on all dimensions.

We now consider the effectiveness of NPSA by comparing the energy histories of PSA, which is the standard method of SA, and NPSA, which demonstrates a superior performance. It is clear from Figs. 4 and 5 that NPSA yields a better solution and arrives at the optimum solution faster than PSA. The neighborhood range selected can be considered as the reason for this.

This difference in performance stems from the variation of the neighborhood range. Figure 6 shows the history of the neighborhood ranges. PSA maintains range 1, which is the best range arrived at through extensive preliminary experimentation. On the other hand, NPSA utilizes the new adaptive mechanism for assigning the neighborhood range. In Fig. 6, the solutions trapped in a local minima in the first stage; therefore, the neighborhood range become quite large. In the middle stage, the solution has converged on the global minimum, and the neighborhood range has become increasingly smaller for local search. In the final stage, this mechanism can search the global optimum area. Thus, NPSA can adaptively vary the neighborhood range as search process proceeds, thereby facility effective searching.

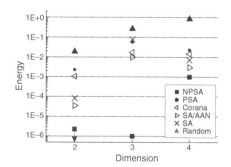

Fig. 3. Energy value (Griewank function)

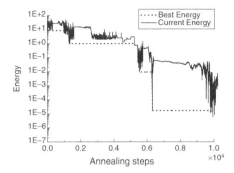

Fig. 4. Energy history of NPSA

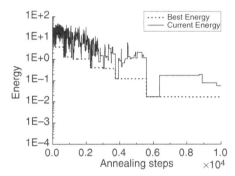

Fig. 5. Energy history of PSA

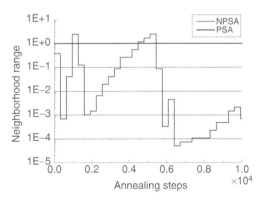

Fig. 6. History of neighborhood ranges

4 Conclusion

When simulated annealing is applied to a continuous optimized problem, neighborhood range adjustment is indispensable. This research proposed the NPSA method with a new neighborhood range generation mechanism that eliminates unnecessary parameters. The proposed method is found to be very effective for solving continuous optimization problems by SA.

References

1. Kita, H. (1997) Simulated Annealing. Japan Society for Fuzzy Theory and Intelligent Informatics, vol. 9, no. 6, pp. 870–875
2. Aarts, E. (1989) Simulated Annealing and Boltzmann Machines: A Stochastic Approach to Combinatorial Optimization and Neural Computing. Wiley, New York
3. Metropolis, N., Rosenbluth, A., Rosenbluth, M., Teller, A., and Teller, E. (1953) Equation of State Calculation by Fast Computing Machines. J Chem Phys, vol. 21, pp. 1087–1092
4. Ingber, L. (1993) Simulated Annealing: Practice versus theory. J Math Comput Model, vol.18, no.11, pp. 29–57
5. Kiku, M. (1996) Invitation to genome information. Kyoritsu shuppan, Japan
6. Corana, A. (1987) Minimizing Multimodal Functions of Continuous Variables with the "Simulated Annealing Algorithm". ACM Trans Mathl Software, vol. 13, no. 3, pp. 262–280

New Computing System Architecture for Simulation of Biological Signal Transduction Networks

Hitoshi Hemmi, Tetsuya Maeshiro, and Katsunori Shimohara

A new architecture has been developed for an ultra high-speed simulator that computes the evolution of biochemical signal transductions in cells. Unlike ordinary computers used for most scientific simulations, this architecture is not based on an ALU/FPU type sequential arithmetic process; instead, it directly implements the target phenomena into electronic circuits that operate massively in parallel. This methodology squeezes the potential computing power of the circuits on the silicon far more efficiently than CPU-oriented circuitry.

1 Introduction

A new computing system has been developed for an ultra high-speed simulator that computes the evolution of signal transductions in life forms, such as biochemical reactions. Unlike ordinary computers used for most scientific simulations, this computing system is not based on an ALU/FPU type sequential arithmetic process; instead, it directly implements the target phenomena into electronic circuits that operate massively in parallel. This methodology squeezes the potential computing power of the circuits on the silicon far more efficiently than CPU-oriented circuitry. As a result, the cost performance of the developed simulation system is much higher than software simulations operating on conventional computers. Actually, its performance is comparable to supercomputer systems, while such costs as electricity and placement area are only several times larger than PCs. This manuscript describes why such huge computation power is required, and the key technologies for realizing it.

2 Problem Scale

The signal transduction advanced research package (Starpack) was developed to simulate huge networks in the living cell. There exist several systems for the same purpose, like E-cell [1]. However, they all are the software-based system and unable to show satisfactory performance for practical use. This section calculates how large the required performance is.

From a microscopic viewpoint, the phenomena of life are the results of biochemical reactions. Inside a living cell, about 10 billion biochemical molecules react at a speed of 10,000 times per second. Molecules are classified into about 20,000 substances. A substance may be both substrates of reactions and products of other reactions. Accordingly, they form a huge network with the reactions. Knowing the amount of the evolution of substances, one can clarify the phenomena of growth and disease, which in turn are essential for personalized medical care or efficient food production, etc. However, the network is too complex to solve its behavior analytically; only simulation methodology is applicable. To do so, it is necessary to update the amount of substances 200 million times per second so that the simulation speed matches the real reactions. For more practical case of actual use, for example, 1,000 times faster than the actual speed, the number increases to 200 billion per second: a figure 200 times larger than one billion operations per second, which is the typical speed of the floating-point arithmetic of ordinary computers. In addition, in software programs, several dozens to several thousand machine clocks are usually needed to update one substance's amount. Therefore, the required computing power is four to six orders larger than PCs.

Now, it is apparent that the software-based systems are unable to achieve the requested performance unless using cost-consuming supercomputers.

3 Hardware System of the Starpack

3.1 Core Architecture

The main part of the Starpack architecture consists of a large number of processing units (PUs); a counter exists in that represents the amount of a substance (Fig. 1). It is decreased if representing a substrate of a reaction and decreased if representing a product. These calculations are operated in parallel, and the value of each counter evolves as the simulation progresses. Each reaction may occur randomly, but the average frequency depends on enzyme density and reaction characteristics. In the system, these reaction speed controls are achieved by combinations of random number generators and regulations of the degree of incrementing or decrementing. As mentioned earlier, some counters have to be incremented or decremented repeatedly if they are related to some reactions. In the system, these plural operations are computed simultaneously only in one time step. While providing such useful high-level operations, the circuit scale for a PU is still very small due to the simple architecture.

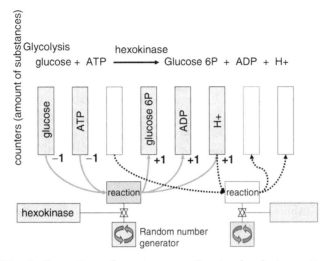

Fig. 1. Operations of counters according to glycolysis reaction

3.2 Prototype System Performance

Figure 2 shows the first Starpack prototype system. The system implements above architecture on the Xilinx XC6000 FPGA (field programmable gate array) chips. There are 64 FPGA chips in the system, and each chip includes 64 PUs. So there are 4,096 PUs in total. Basically each chip is supplied with its own clock, and can operate asynchronously with others. In the current system, however, the clock frequencies are same 100 MHz for all the chips. Each PU does one update per clock pulse, so the overall performance of the

Fig. 2. Starpack prototype system

prototype system is 409.6 billion update per second. This figure satisfies the requested condition at once.

However, there is still a problem. The prototype system implements 4,096 PUs and there are corresponding 4,096 substances: this figure is less than 20,000 substances in a real cell. Timesharing techniques, i.e., overloading plural substances to one PU, may overcome the problem, but there will be some overhead to do so. More efficient measure should be a straightforward one: increasing the PUs. Fortunately, by tuning the logical design of the FPGA circuitry, we can increase the number of PUs per FPGA chip, about 1.5 times larger. Also, the architecture has scalable characteristic, so by simply increasing FPGA chips in the system, full covering of the 20,000 substances in one cell can be easily achieved. Moreover, in the time, the simulation speed is about 10,000 times faster than the real reaction speed; it is one order faster than the assumption calculation.

4 Conclusion

We calculated performance requirements for the simulation of the biological signal transduction networks, and present necessity of the hardware simulation system. Next, the core architecture of the Starpack system was described. Calculation of the prototype system performance shows that this system almost satisfies the required condition, and therefore, is an essential tool to analyze such large-scale networks as signal transduction in the cell.

Acknowledgment

This research was conducted as part of the "Research on Human Communication" with funding from the National Institute of Information and Communications Technology.

Reference

1. Tomita M et al. (1999). Bioinformatics 15:72–84

Convergence Analysis of GMRES Methods for Least Squares Problems

Ken Hayami and Tokushi Ito

Summary. In previous papers, we proposed two methods for applying the GMRES method to over-determined and under-determined linear least squares problems. In this paper, we will give a preliminary convergence analysis of these methods as well as the CGLS method, which predicts that all the methods may show similar convergence behaviours, when the effect of rounding error is small.

1 Introduction

Consider the least squares problem

$$\min_{x \in \mathbf{R}^n} \|b - Ax\|_2, \tag{1}$$

where $A \in \mathbf{R}^{m \times n}$, and $m \geq n$ or $m < n$. We also allow the rank-deficient case when the equality in rank $A \leq \min(m, n)$ does not hold.

The least squares problem (1) is equivalent to the normal equation

$$A^\mathrm{T} A x = A^\mathrm{T} b. \tag{2}$$

The standard iterative method for least squares problems is the (preconditioned) conjugate gradient least squares (CGLS) method [1], which basically applies the conjugate gradient method to the normal equation.

Even with preconditioning, the convergence behaviour of the CGLS method may deteriorate for highly ill-conditioned problems due to rounding errors. Zhang and Oyanagi [2] proposed applying the Orthomin(k) method directly to the least squares problem (1), instead of treating the normal equation (2). This was done by introducing a mapping matrix $B \in \mathbf{R}^{n \times m}$ to transform the problem to a system with a square coefficient matrix $AB \in \mathbf{R}^{m \times m}$, and then applying the Krylov subspace method Orthomin(k) to this nonsymmetric system.

2 Solution of Least Squares Problems Using the GMRES Method

In [3–5], we further extended their method by applying the more robust (restarted) generalized minimal residual (GMRES(k)) method [6] instead of the Orthomin(k) method, and also introduced an alternative method of treating the system with a coefficient matrix $BA \in \mathbf{R}^{n \times n}$.

The first method is to use the Krylov subspace $\mathcal{K}_i(AB, r_0) := \langle r_0, ABr_0, \ldots, (AB)^{i-1}r_0 \rangle$ in $\mathbf{R}^m$ generated by $AB \in \mathbf{R}^{m \times m}$, as in [2], and to solve the least squares problem $\min_{z \in \mathbf{R}^m} \|b - ABz\|_2$ using the GMRES(k) method. This method is called the AB-GMRES method.

The other alternative is to use the same matrix $B \in \mathbf{R}^{n \times m}$ to map the initial residual vector $r_0 \in \mathbf{R}^m$ to $\tilde{r}_0 = Br_0 \in \mathbf{R}^n$, and then to construct the Krylov subspace $\mathcal{K}_i(BA, r_0) := \langle \tilde{r}_0, BA\tilde{r}_0, \ldots, (BA)^{i-1}\tilde{r}_0 \rangle$ in $\mathbf{R}^n$ and to solve the least squares problem $\min_{x \in \mathbf{R}^n} \|Bb - BAx\|_2$ using the GMRES(k) method. This method is called the BA-GMRES method.

In [5], we showed that if

$$\mathcal{R}(A) = \mathcal{R}(B^{\mathrm{T}}), \quad \mathcal{R}(A^{\mathrm{T}}) = \mathcal{R}(B) \tag{3}$$

holds, then the AB-GMRES and BA-GMRES methods (without restarts, i.e. $k = \infty$) determine the least squares solution of (1) without breakdown for all $b \in \mathbf{R}^m$ and for all $x_0 \in \mathbf{R}^n$. Here, exact arithmetic is assumed, and "breakdown" denotes the case when a divide by 0 occurs.

For the full rank overdetermined case: $m \geq n = \operatorname{rank} A$, let

$$B = CA^{\mathrm{T}}, \tag{4}$$

where $C \in \mathbf{R}^{n \times n}$ is an arbitrary nonsingular matrix. Then the sufficient condition (3) holds. A simple example for C is $C := \{\operatorname{diag}(A^{\mathrm{T}}A)\}^{-1}$.

For the full rank underdetermined case: $\operatorname{rank} A = m \leq n$, let

$$B = A^{\mathrm{T}}C, \tag{5}$$

where $C \in \mathbf{R}^{m \times m}$ is an arbitrary nonsingular matrix. Then the sufficient condition (3) holds. A simple example for C is $C := \{\operatorname{diag}(AA^{\mathrm{T}})\}^{-1}$.

Numerical experiments on over and underdetermined least squares problems showed that the two GMRES based methods using the matrix B can be more efficient compared to the (preconditioned) CGLS method for ill-conditioned problems [3, 5].

3 Convergence Analysis

In this paper, we will analyze the convergence of the AB-GMRES method and the BA-GMRES method.[1]

3.1 Overdetermined Case

First, we will consider the overdetermined case $m \geq n$ with $B := CA^T$ as in (4), where $C \in \mathbf{R}^{n \times n}$ is restricted to be symmetric and positive-definite, the following holds.

Theorem 1. *Let $A \in \mathbf{R}^{m \times n}$, $m \geq n$ and $B := CA^T$ where $C \in \mathbf{R}^{n \times n}$ is symmetric and positive-definite. Let the singular values of $\tilde{A} := AC^{1/2}$ be σ_i $(1 \leq i \leq n)$. Then, σ_i^2 $(1 \leq i \leq n)$ are eigenvalues of AB and BA. If $m > n$, all the other eigenvalues of AB are 0.*

Proof. Let $\tilde{A} := AC^{1/2} = U\Sigma V^T$ be the singular value decomposition of $\tilde{A}$. Here, $U \in \mathbf{R}^{m \times m}$, $V \in \mathbf{R}^{n \times n}$ are orthogonal matrices, and

$$\Sigma = \begin{bmatrix} \sigma_1 & & 0 \\ & \ddots & \\ & & \sigma_n \\ & 0 & \end{bmatrix} \in \mathbf{R}^{m \times n},$$

where $\sigma_1 \geq \cdots \geq \sigma_n \geq 0$ are the singular values of $\tilde{A}$. Then,

$$AB = ACA^T = \tilde{A}\tilde{A}^T = U\Sigma\Sigma^T U^T, \tag{6}$$

$$BA = CA^T A = C^{1/2}\tilde{A}^T \tilde{A} C^{-1/2} = C^{1/2}V\Sigma^T\Sigma(C^{1/2}V)^{-1}. \quad \square \tag{7}$$

Let $U = [\boldsymbol{u}_1, \ldots, \boldsymbol{u}_m]$, $V = [\boldsymbol{v}_1, \ldots, \boldsymbol{v}_n]$,

$$\boldsymbol{r}_0 = \boldsymbol{b} - A\boldsymbol{x}_0 = \sum_{i=1}^{m} \rho_i \boldsymbol{u}_i, \tag{8}$$

and $\boldsymbol{r}_k = \boldsymbol{b} - A\boldsymbol{x}_k$.

First, consider the AB-GMRES method. The kth residual vector $\boldsymbol{r}_k$ satisfies

$$\|\boldsymbol{r}_k\|_2 = \min_{\boldsymbol{\zeta} \in \mathcal{K}_k(AB, \boldsymbol{r}_0)} \|\boldsymbol{b} - AB(\boldsymbol{z}_0 + \boldsymbol{\zeta})\|_2 = \min_{p_k \in Q_k} \|p_k(AB)\boldsymbol{r}_0\|_2,$$

---

[1] We would like to thank Professor Michael Eiermann for discussions which lead to the following analysis.

where the minimum is attained by $p_k^{AB} \in Q_k$, and
$Q_k := \{p_k \mid p_k(x) : \text{polynomial of } x \text{ with degree} \leq k, \, p_k(0) = 1\}$.
(6) and (8) give

$$p_k(AB)r_0 = \sum_{i=1}^{n} \rho_i \, p_k(\sigma_i^2) u_i + p_k(0) \sum_{i=n+1}^{m} \rho_i u_i.$$

Thus,

$$\|r_k\|_2 = \min_{p_k \in Q_k} \left[\sum_{i=1}^{n} \{\rho_i \, p_k(\sigma_i^2)\}^2 + \sum_{i=n+1}^{m} \rho_i^2 \right]^{1/2},$$

and $p_k^{AB} \in Q_k$ minimizes $\sum_{i=1}^{n} \{\rho_i \, p_k(\sigma_i^2)\}^2$, and $\|r_k\|_2$ decreases monotonically with k.

Since

$$A^{\mathrm{T}} = C^{-1/2} \tilde{A}^{\mathrm{T}} = C^{-1/2} V \Sigma^{\mathrm{T}} U^{\mathrm{T}} = C^{-1/2} \sum_{i=1}^{n} \sigma_i v_i u_i^{\mathrm{T}},$$

$$A^{\mathrm{T}} r_k = \sum_{i=1}^{n} \rho_i \sigma_i p_k^{AB}(\sigma_i^2) C^{-1/2} v_i$$

and

$$A^{\mathrm{T}} r_0 = \sum_{i=1}^{n} \rho_i \sigma_i C^{-1/2} v_i.$$

Note that for the general inconsistent case ($b \notin \mathcal{R}(A)$), the convergence of the method can be monitored by $\|A^{\mathrm{T}} r_k\|_2 / \|A^{\mathrm{T}} r_0\|_2$.

Next, consider the BA-GMRES method, r_k satisfies

$$\|Br_k\|_2 = \min_{\xi \in \mathcal{K}_k(BA, Br_0)} \|Bb - BA(x_0 + \xi)\|_2 = \min_{p_k \in Q_k} \|p_k(BA)Br_0\|_2,$$

where the minimum is attained by $p_k^{BA} \in Q_k$. Note from (7) that

$$p_k(BA) = C^{1/2} V p_k(\Sigma^{\mathrm{T}} \Sigma)(C^{1/2} V)^{-1},$$

and from (8) that

$$Br_0 = CA^{\mathrm{T}} \sum_{i=1}^{m} \rho_i u_i = C^{1/2} V \Sigma^{\mathrm{T}} U^{\mathrm{T}} \sum_{i=1}^{m} \rho_i u_i = \sum_{i=1}^{n} \rho_i \sigma_i C^{1/2} v_i.$$

Hence,
$$Br_k = \sum_{i=1}^{n} \rho_i \sigma_i p_k^{BA}(\sigma_i^2) C^{1/2} v_i.$$

Since $B = CA^{\mathsf{T}}$,
$$A^{\mathsf{T}} r_k = C^{-1} B r_k = \sum_{i=1}^{n} \rho_i \sigma_i p_k^{BA}(\sigma_i^2) C^{-1/2} v_i$$

and
$$A^{\mathsf{T}} r_0 = \sum_{i=1}^{n} \rho_i \sigma_i C^{-1/2} v_i.$$

Note that for the overdetermined case $m \geq n$, the BA-GMRES method requires less computational work per iteration and memory compared to the AB-GMRES method, since $BA \in \mathbf{R}^{n \times n}$ and $AB \in \mathbf{R}^{m \times m}$.

The natural way to precondition the CGLS method using the same symmetric positive matrix $C \in \mathbf{R}^{n \times n}$ is to apply the CG method to
$$(C^{1/2} A^{\mathsf{T}} A C^{1/2})(C^{-1/2} x) = C^{1/2} A^{\mathsf{T}} b.$$

Note that $BA = CA^{\mathsf{T}} A$ and $C^{1/2} A^{\mathsf{T}} A C^{1/2}$ are similar.

3.2 Underdetermined Case

Similarly, consider the underdetermined case rank $A = m \leq n$ with $B = A^{\mathsf{T}} C$ as in (5), where $C \in \mathbf{R}^{m \times m}$ is restricted to be symmetric and positive-definite. The following holds.

Theorem 2. *Let $A \in \mathbf{R}^{m \times n}$, $m \leq n$, where $B := A^{\mathsf{T}} C$ and $C \in \mathbf{R}^{m \times m}$ is symmetric and positive-definite. Let the singular values of $\tilde{A} := C^{1/2} A$ be σ_i ($1 \leq i \leq m$). Then, σ_i^2 ($1 \leq i \leq m$) are the eigenvalues of AB and BA. If $m < n$, all the other eigenvalues of BA are 0.*

Proof. Let $\tilde{A} := C^{1/2} A = U \Sigma V^{\mathsf{T}}$ be the singular value decomposition of $\tilde{A}$. Here, $U \in \mathbf{R}^{m \times m}$, $V \in \mathbf{R}^{n \times n}$ are orthogonal matrices, and

$$\Sigma = \begin{bmatrix} \sigma_1 & & 0 \\ & \ddots & \\ & & \sigma_m \end{bmatrix} \in \mathbf{R}^{m \times n},$$

where $\sigma_1 \geq \cdots \geq \sigma_m \geq 0$ are the singular values of $\tilde{A}$. Then,
$$AB = AA^{\mathsf{T}} C = C^{-1/2} \tilde{A} \tilde{A}^{\mathsf{T}} C^{1/2} = C^{-1/2} U \Sigma \Sigma^{\mathsf{T}} (C^{-1/2} U)^{-1},$$
$$BA = A^{\mathsf{T}} CA = \tilde{A}^{\mathsf{T}} \tilde{A} = V \Sigma^{\mathsf{T}} \Sigma V^{\mathsf{T}}. \quad \square$$

Let $\tilde{A} := C^{1/2}A = U\Sigma V^T$ be the singular value decomposition of $\tilde{A}$, and $U = [\boldsymbol{u}_1, \ldots, \boldsymbol{u}_m]$, $V = [\boldsymbol{v}_1, \ldots, \boldsymbol{v}_n]$, similar to the overdetermined case.

First, consider the AB-GMRES method. $\boldsymbol{r}_k$ satisfies

$$\|\boldsymbol{r}_k\|_2 = \min_{p_k \in Q_k} \|p_k(AB)\boldsymbol{r}_0\|_2,$$

where the minimum is attained by $p_k^{AB} \in Q_k$. Hence,

$$\boldsymbol{r}_k = p_k^{AB}(AB)\boldsymbol{r}_0 = \sum_{i,j=1}^{m} \rho_j\, p_k^{AB}(\sigma_i^{\,2})(\boldsymbol{u}_i^T C^{1/2}\boldsymbol{u}_j)C^{-1/2}\boldsymbol{u}_i$$

and

$$A^T \boldsymbol{r}_k = \sum_{i,j=1}^{m} \rho_j\, p_k^{AB}(\sigma_i^{\,2})(\boldsymbol{u}_i^T C^{1/2}\boldsymbol{u}_j) \sum_{l=1}^{m} \sigma_l \boldsymbol{v}_l \boldsymbol{u}_l^T C^{-1/2}\boldsymbol{u}_i.$$

Note that for the underdetermined case $m \leq n$, the AB-GMRES method requires less computational work per iteration and memory compared to the BA-GMRES method, since $AB \in \mathbf{R}^{m \times m}$ and $BA \in \mathbf{R}^{n \times n}$.

The natural way to precondition the CGLS method using the same symmetric positive matrix $C \in \mathbf{R}^{n \times n}$ is to apply the CG method to

$$(C^{1/2}AA^T C^{1/2})(C^{1/2}\boldsymbol{z}) = C^{1/2}\boldsymbol{b}.$$

Note that $AB = AA^T C$ and $C^{1/2}AA^T C^{1/2}$ are similar.

Next, for the BA-GMRES method, we have

$$\|B\boldsymbol{r}_k\|_2 = \min_{p_k \in Q_k} \|p_k(BA)B\boldsymbol{r}_0\|_2,$$

where the minimum is attained by $p_k^{BA} \in Q_k$.

Hence,

$$B\boldsymbol{r}_k = p_k^{BA}(BA)B\boldsymbol{r}_0 = \sum_{i,j=1}^{m} \rho_j \sigma_i\, p_k^{BA}(\sigma_i^{\,2})(\boldsymbol{u}_i^T C^{1/2}\boldsymbol{u}_j)\boldsymbol{v}_i\,.$$

4 Conclusion

From the above analyses, we may expect that the AB-GMRES, BA-GMRES and the preconditioned CGLS methods exhibit similar convergence behaviours for over and underdetermined least squares problems, when the effect of rounding is small, as was shown in numerical experiments in [3, 5].

References

1. Björck, A. (1996). *Numerical Methods for Least Squares Problems*, SIAM, Philadelphia
2. Zhang, S.-L. and Oyanagi, Y. (1991). Orthomin(k) method for linear least squares problem, *J Inform Proc*, 14:121–125
3. Ito, T. and Hayami, K. (2004). Preconditioned GMRES methods for least squares problems, *NII Technical Report*, National Institute of Informatics, NII-2004-006E, 1–29
4. Hayami, K. and Ito, T. (2006). Application of the GMRES method to singular systems and least squares problems, *Proceedings of the 7th China-Japan Seminar on Numerical Mathematics*, August 2004, Zhangjiajie, Science Press, Beijing, 33–44
5. Hayami, K. and Ito, T. (2005). The solution of least squares problems using GMRES methods, *Proceedings of the Institute of Statistical Mathematics*, 53(2): 331–348, (in Japanese)
6. Saad, Y. and Schultz, M. H. (1986). GMRES: A generalized minimal residual method for solving nonsymmetric linear systems, *SIAM J Sci Stat Comput*, 7:856–869

A Numerical Method for Calculating the Green's Function Arising from Electronic Structure Theory

Tomohiro Sogabe, Takeo Hoshi, Shao-Liang Zhang, and Takeo Fujiwara

Summary. We developed a fast numerical method for complex symmetric shifted linear systems, which is motivated by the quantum-mechanical (electronic-structure) theory in nanoscale materials. The method is named shifted Conjugate Orthogonal Conjugate Gradient (shifted COCG) method. The formulation is given and several numerical aspects are discussed.

1 Introduction

The quantum-mechanical feature of electrons plays a crucial role in nanoscale materials and its mathematical foundation is reduced to linear-algebraic problems with given large matrices H, called Hamiltonian. The physical properties of electrons can be described by the Green's function G that is defined as inversed matrix $G(z) := (zI - H)^{-1}$ with a complex variable z whose real part corresponds to energy. Since the standard matrix-inversion procedure requires an impractical computational cost in case of large matrices or in nanoscale materials, there is a strong need for the fast solution of the Green's function, see [3].

Here we introduce a new method for calculating the Green's function [3]. Let H be an N-by-N real symmetric Hamiltonian matrix, then any element of the Green's function can be written as:

$$G_{ij}(z) = e_i^{\mathrm{T}}(zI - H)^{-1}e_j, \tag{1}$$

where e_i denotes the ith unit vector, and the complex energy $z = \sigma + i\delta \in C$. Note that the i,j entry of (1) can be obtained by two steps: first, compute $(zI - H)x = e_j$, and then compute $e_i^{\mathrm{T}}x$. Since an integral of $G_{ij}(z)$ with respect to σ is needed to obtain any physical quantity [3], the numerical integration requires $\{(\sigma_k + i\delta)I - H\}x^{(k)} = e_j$ for $k = 1,\ldots,m$. Hence, the problem reduces to solving the following shifted linear systems with complex symmetric matrices:

$$(A + \sigma_k I)x^{(k)} = b \quad \text{for} \quad k = 1,\ldots,m. \tag{2}$$

The paper is organized as follows: in Sect. 2, we describe the algorithm and the property of COCG for solving complex symmetric linear systems. In Sect. 3, to solve (2) efficiently, a numerical method named shifted COCG is proposed and seed switching technique is introduced. In Sect. 4, we report some numerical experiments. Finally, we make some concluding remarks in Sect. 5.

2 The COCG Method

Matrix A is called complex symmetric if A is not Hermitian but symmetric $A = A^T \neq A^H$. To solve the linear systems, the COCG method [4] has been proposed and is known as one of the most successful Krylov subspace methods.

Algorithm 1: COCG

x_0 is an initial guess,
$r_0 = b - Ax_0$, $p_{-1} = 0$, $\beta_{-1} = 0$,
for $n = 0, 1, \ldots$ until $\|r_n\| \leq \epsilon_1 \|b\|$ do:
$$p_n = r_n + \beta_{n-1} p_{n-1},$$
$$\alpha_n = \frac{r_n^T r_n}{p_n^T A p_n},$$
$$x_{n+1} = x_n + \alpha_n p_n,$$
$$r_{n+1} = r_n - \alpha_n A p_n,$$
$$\beta_n = \frac{r_{n+1}^T r_{n+1}}{r_n^T r_n}.$$
end

Observing Algorithm 1, we see that the nth residual can be written as $r_n (:= b - Ax_n) = R_n(A)r_0$, where $R_0(\lambda) = 1$, $R_1(\lambda) = (1 - \alpha_0 \lambda) R_0(\lambda)$, and

$$R_n(\lambda) = \left(1 + \frac{\beta_{n-2}}{\alpha_{n-2}} \alpha_{n-1} - \alpha_{n-1} \lambda \right) R_{n-1}(\lambda) - \frac{\beta_{n-2}}{\alpha_{n-2}} \alpha_{n-1} R_{n-2}(\lambda). \quad (3)$$

It is known that if breakdown does not occur, then the nth residual satisfies

$$r_n \perp \overline{K_n(A, r_0)} = \text{span}\{\bar{r}_0, \bar{A}\bar{r}_0, \ldots, \bar{A}^{n-1}\bar{r}_0\}, \quad (4)$$

which leads to conjugate orthogonality $r_i \perp \bar{r}_j$ for $i \neq j$.

3 A Shifted COCG Method

In this section, we develop the COCG method for solving complex symmetric shifted linear systems. We start with the following theorem:

Theorem 1 (Frommer [2]). *Let $W_1 \subseteq W_2 \subseteq \cdots \subseteq W_k$ be a sequence of nested subspaces of C^N such that W_n has dimension n and $W_n \cap (K_{n+1}(A,b))^\perp = \{0\}$, $n = 1, \ldots, k$. Let $r_n := R_n(A)b$, $r_n^\sigma := R_n^\sigma(A + \sigma I)b$ be residual vectors satisfying*

$$r_n, r_n^\sigma \perp W_n, \quad n = 1, \ldots, k. \tag{5}$$

Then r_n and r_n^σ are collinear.

Corollary 1. *Let r_n and r_n^σ be the residual vectors of COCG started with $x_0 = x_0^\sigma = 0$. Then r_n and r_n^σ are collinear, i.e., there exists $\pi_n^\sigma \in C$ such that $r_n = \pi_n^\sigma r_n^\sigma$.*

Proof. Since it follows from (4) that the COCG residuals satisfy (5) with $W_n = \overline{K_n(A,b)} = \overline{K_n(A + \sigma I, b)}$, this result follows from Theorem 1.

Here we give a formula for computing r_{n+1}^σ by using the information of r_{n+1}. It follows from the polynomial (3) that we have

$$r_{n+1} = \left(1 + \frac{\beta_{n-1}}{\alpha_{n-1}}\alpha_n - \alpha_n A\right)r_n - \frac{\beta_{n-1}}{\alpha_{n-1}}\alpha_n r_{n-1}, \tag{6}$$

$$r_{n+1}^\sigma = \left(1 + \frac{\beta_{n-1}^\sigma}{\alpha_{n-1}^\sigma}\alpha_n^\sigma - \alpha_n^\sigma(A + \sigma I)\right)r_n^\sigma - \frac{\beta_{n-1}^\sigma}{\alpha_{n-1}^\sigma}\alpha_n^\sigma r_{n-1}^\sigma. \tag{7}$$

Substituting the relation $r_n = \pi_n^\sigma r_n^\sigma$ into the previous recurrence (7), we have

$$r_{n+1} = \left(1 + \frac{\beta_{n-1}^\sigma}{\alpha_{n-1}^\sigma}\alpha_n^\sigma - \alpha_n^\sigma(A + \sigma I)\right)\frac{\pi_{n+1}^\sigma}{\pi_n^\sigma}r_n - \frac{\beta_{n-1}^\sigma \alpha_n^\sigma \pi_{n+1}^\sigma}{\alpha_{n-1}^\sigma \pi_{n-1}^\sigma}r_{n-1}. \tag{8}$$

To obtain the computational formula for r_{n+1}^σ, three parameters α_n^σ, β_{n-1}^σ, and π_{n+1}^σ are essentially required. Hence, we give the formulas for the three parameters. First, comparing the coefficients of Ar_n in (6) and (8), we find

$$\alpha_n^\sigma = \left(\pi_n^\sigma / \pi_{n+1}^\sigma\right)\alpha_n. \tag{9}$$

Second, comparing the coefficients of r_{n-1} leads to

$$\frac{\beta_{n-1}}{\alpha_{n-1}}\alpha_n = \frac{\beta_{n-1}^\sigma \alpha_n^\sigma \pi_{n+1}^\sigma}{\alpha_{n-1}^\sigma \pi_{n-1}^\sigma}.$$

Substituting the result of (9) into the previous equation, we have

$$\beta_{n-1}^\sigma = \left(\pi_{n-1}^\sigma / \pi_n^\sigma\right)^2 \beta_{n-1}. \tag{10}$$

Finally, comparing the coefficients of r_n, we find

$$\left(1 + \frac{\beta_{n-1}^\sigma}{\alpha_{n-1}^\sigma}\alpha_n^\sigma - \alpha_n^\sigma \sigma\right)\frac{\pi_{n+1}^\sigma}{\pi_n^\sigma} = 1 + \frac{\beta_{n-1}}{\alpha_{n-1}}\alpha_n.$$

Substituting (9) and (10) into the previous equation, we obtain

$$\pi_{n+1}^{\sigma} = \left(1 + \frac{\beta_{n-1}}{\alpha_{n-1}}\alpha_n + \alpha_n\sigma\right)\pi_n^{\sigma} - \frac{\beta_{n-1}}{\alpha_{n-1}}\alpha_n\pi_{n-1}^{\sigma} = R_{n+1}(-\sigma). \quad (11)$$

The earlier formulation is based on the assumption that the seed and shifted systems are $A\boldsymbol{x} = \boldsymbol{b}$ and $(A+\sigma I)\boldsymbol{x}^{\sigma} = \boldsymbol{b}$. Similarly, it can be readily generalized to solve m shifted linear systems $(A+\sigma_i I)\boldsymbol{x}^{(i)} = \boldsymbol{b}$ using the seed system $(A+\sigma_s I)\boldsymbol{x} = \boldsymbol{b}$. The resulting algorithm is given later.

Algorithm 2: Shifted COCG

$\boldsymbol{x}_0^{(i)} = \boldsymbol{p}_{-1}^{(i)} = \boldsymbol{0}$, $\boldsymbol{r}_0 = \boldsymbol{b}$,
$\beta_{-1} = 0$, $\pi_0^{(s,i)} = \pi_{-1}^{(s,i)} = \alpha_{-1} = 1$,
for $n = 0, 1, \ldots$ until $\frac{\|\boldsymbol{r}_n\|}{\|\boldsymbol{b}\|} \leq \epsilon_1$ do:
$\quad \boldsymbol{p}_n = \boldsymbol{r}_n + \beta_{n-1}\boldsymbol{p}_{n-1}$,
$\quad \alpha_n = \dfrac{\boldsymbol{r}_n^{\mathrm{T}}\boldsymbol{r}_n}{\boldsymbol{p}_n^{\mathrm{T}}(A+\sigma_s I)\boldsymbol{p}_n}$,
$\quad \boldsymbol{x}_{n+1} = \boldsymbol{x}_n + \alpha_n\boldsymbol{p}_n$,
$\quad${Begin shifted system}
$\quad$for $i(\neq s) = 1, \ldots, m$ do:
$\quad\quad$if $\|\boldsymbol{r}_n^{(i)}\| > \epsilon_2\|\boldsymbol{b}\|$ then
$\quad\quad\quad \pi_{n+1}^{(s,i)} = R_{n+1}^{(s)}(\sigma_s - \sigma_i), \leftarrow (11)$

$\quad\quad\quad \beta_{n-1}^{(i)} = \left(\pi_{n-1}^{(s,i)}/\pi_n^{(s,i)}\right)^2 \beta_{n-1}$,
$\quad\quad\quad \alpha_n^{(i)} = \dfrac{\pi_n^{(s,i)}}{\pi_{n+1}^{(s,i)}}\alpha_n$,
$\quad\quad\quad \boldsymbol{p}_n^{(i)} = \dfrac{1}{\pi_n^{(s,i)}}\boldsymbol{r}_n + \beta_{n-1}^{(i)}\boldsymbol{p}_{n-1}^{(i)}$,
$\quad\quad\quad \boldsymbol{x}_{n+1}^{(i)} = \boldsymbol{x}_n^{(i)} + \alpha_n^{(i)}\boldsymbol{p}_n^{(i)}$,
$\quad\quad$end if
$\quad$end
$\quad${End shifted system}
$\quad \boldsymbol{r}_{n+1} = \boldsymbol{r}_n - \alpha_n(A+\sigma_s I)\boldsymbol{p}_n$,
$\quad \beta_n = \dfrac{\boldsymbol{r}_{n+1}^{\mathrm{T}}\boldsymbol{r}_{n+1}}{\boldsymbol{r}_n^{\mathrm{T}}\boldsymbol{r}_n}$.
end

3.1 Shifted COCG with Seed Switching Technique

We can see from Algorithm 2 that if $|\pi_n^{(s,i)}| = |R_n^{(s)}(\sigma_s - \sigma_i)| \geq 1$, then $\|\boldsymbol{r}_n^{(i)}\| \leq \|\boldsymbol{r}_n\|$. Hence, if we could find a seed system such that $|R_n^{(s)}(\sigma_s - \sigma_i)| \geq 1$, then all shifted systems could be solved. However, it is extremely hard to find such a system in a priori except some special cases discussed in [2]. In this section, we will avoid such problem by using the following strategy:

1. Choose a seed system, and then start Algorithm 2.
2. If the seed system was solved at nth iteration, then find the new one.
3. Start Algorithm 2 from $(n+1)$th iteration using the new seed system.

In (b), as one of criteria for choosing the new seed system $\tilde{s}$, we adopt $\tilde{s} = \arg\max_{i \in I}\{\|\boldsymbol{r}_n^{(i)}\|\}$, where I denotes an index set of unsolved systems. In (c), we need two steps to switch the old seed system to the new one. First, compute

$$\pi_{n+1}^{(s,\tilde{s})} = R_{n+1}^{(s)}(\sigma_s - \sigma_{\tilde{s}}), \quad \beta_n^{(\tilde{s})} = \left(\frac{\pi_n^{(s,\tilde{s})}}{\pi_{n+1}^{(s,\tilde{s})}}\right)^2 \beta_n$$

for obtaining $r_{n+1}^{(\tilde{s})}$ and $\beta_n^{(\tilde{s})} p_n^{(\tilde{s})}$. Since it follows from $r_{n+1}^{(\tilde{s})} + \beta_n^{(\tilde{s})} p_n^{(\tilde{s})}$ that we have $p_{n+1}^{(\tilde{s})}$, we can start COCG solving the system $(A + \sigma_{\tilde{s}} I)x^{(\tilde{s})} = b$ from $(n+1)$th iteration step. Second, to solve remaining systems by using the new seed $\tilde{s}$, it requires generating $\alpha_{n+1}^{(i)}, \beta_n^{(i)}$ from the new seed. We see that they can be readily generated by the following polynomial

$$\pi_{n+1}^{(\tilde{s},i)} = R_{n+1}^{(\tilde{s})}(\sigma_{\tilde{s}} - \sigma_i) \quad \text{for all } i \in I.$$

To obtain the above polynomial, we need to compute

$$\alpha_i^{(\tilde{s})} = \left(\frac{\pi_i^{(s,\tilde{s})}}{\pi_{i+1}^{(s,\tilde{s})}}\right) \alpha_i, \quad \beta_j^{(\tilde{s})} = \left(\frac{\pi_j^{(s,\tilde{s})}}{\pi_{j+1}^{(s,\tilde{s})}}\right)^2 \beta_j$$

for $i = 0, \ldots, n$, $j = 0, \ldots, n-1$. Hence, the switching strategy requires only scalar operations, and moreover we can see that if breakdown does not occur, iterating the process from (b) to (c) enables us to keep solving the systems without losing the dimension of the Krylov subspace that has been generated until the last switching.

4 Numerical Examples

In this section, we report the results of numerical experiments. The problem originally comes from [3] and is written as follows:

$$(\sigma_k I - H)x^{(k)} = e_1, \quad k = 1, \ldots, m,$$

where $\sigma_k = 0.4 + (k - 1 + i)/1{,}000$, $H \in R^{2{,}048 \times 2{,}048}$ is a symmetric matrix, $e_1 = (1, 0, \ldots, 0)^T$, and $m = 1{,}001$. Since $(\sigma_k I - H)$ is complex symmetric, the iterative solvers we used are COCG and shifted COCG. We can also apply shifted Bi-CGSTAB(ℓ) [2] and GMRES [1] to the earlier problem since they can be used for general nonHermitian shifted linear systems. However, they do not exploit properties of complex symmetric matrices. This leads to high computational costs per iteration step.

All experiments were performed on an ALPHA work station with a 750 MHz processor using double precision arithmetic. Codes were written in Fortran 77 and compiled with the optimization option −O4. The switching criterion is $\tilde{s} = \arg\max_{i \in I}\{\|r_n^{(i)}\|\}$. The stopping criteria are $\epsilon_1, \epsilon_2 \leq 10^{-12}$. We report two examples for $k = 301$ and $k = 501$ as an initial seed system. True residual 2-norm histories are given in Fig. 1.

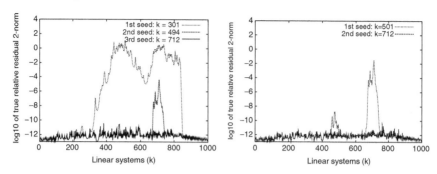

Fig. 1. The true relative residual 2-norm histories after each seed switching finished. The first seed is $k = 301$ on the left and $k = 501$ on the right

In Fig. 1 on the left, 518 systems remained unsolved when the first seed system $k = 301$ converged. Then, the next seed $k = 494$ solved 412 more systems. Finally, the third seed $k = 712$ solved all of the remaining systems. In Fig. 1 on the right, we chose $k = 501$ as an initial seed system. This choice led to 147 unsolved systems. Finally, the next seed $k = 712$ solved all of the remaining systems.

Numerical results are shown in Table 1. Total MVs denotes the total number of matrix–vector multiplications. We can see from Table 1 that shifted COCG required only about 0.27% of Total MVs of COCG.

5 Concluding Remarks

Since the original problem (1) is a fundamental quantum-mechanical equation, the present method is applicable, in principle, to various nanoscale materials, such as silicon, carbon, metals, polymers, and so on, if the Hamiltonian for electrons is given as an explicit matrix H.

The present paper gives an interdisciplinary research between mathematics and physics, which shows that the computational science can give an important contribution to nanoscience through the development of general numerical algorithms, when a fundamental equation is formulated in physics.

Table 1. Numerical results of COCG and shifted COCG

initial seed	switching	total MVs
$k = 301$	2	330
$k = 501$	1	328
COCG	–	124,396

References

1. Datta B and Saad Y (1991) Lin Alg Appl 154–156:225–244
2. Frommer A (2003) Computing 70:87–109
3. Takayama R, Hoshi T, Sogabe T, Zhang S-L and Fujiwara T (2006) Rhys Rev B 73 165108:1–9
4. van der Vorst HA and Melissen JBM (1990) IEEE Trans Mag 26(2):706–708

General Contribution

Direct Numerical Simulation of Turbulent Channel Flow Imposed with Spanwise Couette Flow

Yohji Seki, Kaoru Iwamoto, and Hiroshi Kawamura

Summary. Direct numerical simulation (DNS) of a turbulent channel flow with a three-dimensional mean velocity profile has been performed. In this study, a fully developed turbulent channel flow driven by a streamwise constant mean pressure gradient and a spanwise wall shear is assumed. The present results indicate that the energy redistribution of the Reynolds shear stresses depends upon the spanwise wall shear cases.

1 Introduction

Studies of three-dimensional turbulent flows (3DTFs) provide insight into understanding and prediction of complex flows encountered in engineering applications. For example, all the components of the Reynolds shear stresses take place in 3DTFs, although only some of them appear in a fully developed channel flow. This three-dimensional character is useful for assessment of turbulence models.

Most of the earlier studies in 3DTFs [1–3] focused on time-evolution of turbulence structures and statistics. These confirmed that the initial response of the flow to spanwise deformations was a drop in the turbulence intensity, which was caused by a partial breakdown of the near-wall streaky structures.

DNS of a fully developed turbulent channel flow with the spanwise Couette flow has been carried out. The generated database is valuable for quantitative and qualitative studies of wall-bounded 3DTFs and for design and testing of turbulence closure models. In this paper, the effect of the spanwise wall shear on statistical quantities are examined.

2 Numerical Procedures

A turbulent channel flow driven by a streamwise constant mean pressure gradient and a spanwise wall shear is assumed. A Reynolds number Re_τ, which is based on the streamwise friction velocity u_τ and the channel half width δ, is invariably 180. DNS was carried out with four different cases of the spanwise wall shear $W^r = 0, 0.6, 1.3, 2.0$, where W is the relative velocity between top and bottom walls and the superscript r indicates the ratio between W and a mean centreline velocity of a turbulent Poiseuille flow ($W = 0$). The case of $W^r = 0$ corresponds to the turbulent Poiseuille flow at $Re_\tau = 180$. The simulation has been made by the use of a finite difference method in which the following conditions are assumed: incompressible viscous flow, non-slip conditions on the walls, and the periodic boundary conditions in the streamwise (x) and spanwise (z) directions. For all the spanwise wall shear cases, $256 \times 128 \times 256$ grid points were used in the x, y (normal to the wall), and z directions, respectively. The spacings between collocation points in the streamwise and spanwise directions were $\Delta x^+ = 9.0$ and $\Delta z^+ = 4.5$, respectively. Nonuniform meshes were used in the wall-normal direction, with the minimum spacing of $\Delta y^+ = 0.20$ at the wall and the maximum spacing of $\Delta y^+ = 5.93$ at the centreline. The bottom wall and channel center are located at $y/\delta = 0$ and $y/\delta = 1$, respectively. The numerical algorithm used and other details of the simulation can be found in Abe et al. [4]. All the statistics shown later are obtained from the data accumulated over approximately 66 time units normalized by u_τ and δ. An overbar indicates an average over x, z and time, and a prime indicates perturbation from this average. A superscript + denotes normalization by the kinematic viscosity of the fluid ν and u_τ, for instance, $u^+ = u/u_\tau$ and $y^+ = yu_\tau/\nu$.

3 Results and Discussion

3.1 Turbulence Intensity

All the components of the root-mean-square of the velocity fluctuations for $W^r = 2.0$ are shown in Fig. 1 compared with those for $W^r = 0$. The spanwise component of the turbulence intensity increases with increase of W^r although

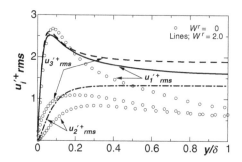

Fig. 1. Rms of velocity fluctuations

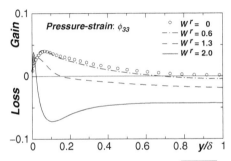

Fig. 2. Pressure-strain term of $\overline{u_3'^+ u_3'^+}$

the streamwise component is not different from that of $W^r = 0$ near the wall. In the case of $W^r = 2.0$, the streamwise and spanwise components become very close to each other for $y/\delta < 0.2$.

3.2 Redistribution of Energy

In incompressible turbulent flows, the pressure appears in the equation for the Reynolds stress through the velocity pressure-gradient term, Π_{ij}. This term Π_{ij} is split into a pressure-transport term and a redistributive one,

$$\Pi_{ij} = -\overline{\left(\frac{\partial (p' u_i')}{\partial x_j} + \frac{\partial (p' u_j')}{\partial x_i}\right)} + \overline{\left(p' \frac{\partial u_i'}{\partial x_j} + p' \frac{\partial u_j'}{\partial x_i}\right)}, \quad (1)$$

where indices $(1, 2, 3)$ are used to denote the x, y, and z directions, respectively. The second term on the right-hand side of 1 is known as the pressure-strain term, ϕ_{ij}. Figures 2 and 3 show ϕ_{ij} for $\overline{u_3'^+ u_3'^+}$ and $\overline{u_1'^+ u_1'^+}$, respectively. The negative level of ϕ_{33} increases with increase of W^r. On the other hand, ϕ_{11} becomes positive in the case of $W^r = 2.0$. These indicate that a portion of generated energy of $\overline{u_3'^+ u_3'^+}$ is transported to $\overline{u_1'^+ u_1'^+}$ except near the wall

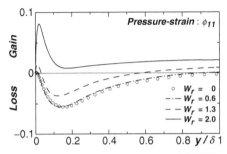

Fig. 3. Pressure-strain term of $\overline{u_1'^+ u_1'^+}$

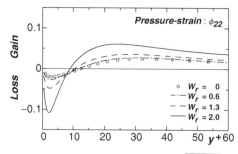

Fig. 4. Pressure-strain term of $\overline{u_2'^+ u_2'^+}$

although $(u_1')_{\text{rms}}$ and $(u_3')_{\text{rms}}$ become close to each other for $y/\delta < 0.2$ as seen in Fig. 1. The profile of ϕ_{22} is plotted in Fig. 4. The value of ϕ_{22} has been exploited to explain the splatting effect in the previous studies [5]. The negative level of ϕ_{22} increases with increasing W^r near the wall. It indicates that the splatting effect is enhanced by increasing W^r.

3.3 Instantaneous Velocity Field

Figure 5 shows instantaneous velocity vectors of $W^r = 2.0$ in a y–z crossstream plane. It is observed that weak flows ($u_2'^+ \sim 0.5$) in region (B) are ejected from the wall. On the other hand, strong sweep flows ($u_2'^+ \sim -6$) near the wall appear in region (A). Although the observed flow fields occasionally and partially contain this tendency, the values of strong sweep flows of $W^r = 2.0$ are dramatically stronger than those of $W^r = 0$ near the wall region (not shown here). In the case of $W^r = 0$, the sweep flows and the resultant impinging fluctuations induce the splatting effect [5]. In the presence of W^r, the strong sweep flows enhance the splatting effect.

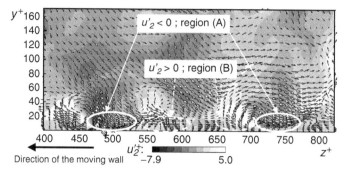

Fig. 5. Horizontal (y, z) section of instantaneous velocity field. The vector length shows magnitude of composition of u_2' and u_3'

4 Concluding Remarks

DNS of a turbulent channel flow with a spanwise Couette flow was performed to investigate the turbulence statistics and structures. The generated energy in the budget of $\overline{u_3'^+ u_3'^+}$ is transported to $\overline{u_1'^+ u_1'^+}$ and $\overline{u_2'^+ u_2'^+}$ in the presence of W^r except near the wall. The negative value of ϕ_{22} near the wall indicates that the splatting effect is enhanced by the spanwise wall shear.

Acknowledgments

We thank Professor M. A. Leschziner for invaluable discussions during the symposium. Part of the numerical results in this study were obtained using supercomputing resources at Information Synergy Center of Tohoku University. This work was supported by Research Center for the Holistic Computational Science (Holcs) at Tokyo University of Science.

References

1. Spalart P R (1989) J Fluid Mech 205:319–340
2. Moin P and Shih T-H (1990) Phys Fluids A 2:1846–1853
3. Coleman G N, Kim J, and Le A-T (1996) Int J Heat Fluid Flow 17:333–342
4. Abe H, Kawamura H, and Matsuo M (2001) ASME J Fluids Eng 123:382–393
5. Moin P and Kim J (1982) J Fluid Mech 118:341–377

Reynolds-Number Dependence of Transport Barriers in Chaotic Mixing

Yoshinori Mizuno and Mitsuaki Funakoshi

Summary. The dependence of the chaotic mixing of fluids on the Reynolds number Re in a partitioned-pipe mixer (PPM) is studied by numerical simulations. Some of tubular transport barriers expand in their cross-sectional area with increasing Re. The barriers exist within the regions where the flow is locally symmetric, and the expansion of the barriers occurs in accordance with the expansion of the regions where the flow is symmetric with increasing Re.

1 Introduction

Many studies on fluid mixing have been carried out from the viewpoint of the theory of dynamical systems [1]. A fluid can be mixed well within a region where the trajectories of fluid particles are chaotic because the fluid repeatedly undergoes stretching and folding by the flow. The mixing is called chaotic mixing. On the other hand, transport barriers associated with stable periodic and quasiperiodic orbits of fluid particles may exist and prevent mixing. Such transport barriers disappear when the periodic orbits become unstable.

An example of the systems for chaotic mixing due to three-dimensional steady flows is a partitioned-pipe mixer (PPM) [2]. A PPM consists of an infinitely long cylinder and plane plates fixed inside the cylinder alternately and orthogonally to each other. A schematic view of one periodic part of a PPM is shown in Fig. 1. A pressure gradient in the axial direction and rotation of the cylinder cause a flow of fluid within the cylinder.

We here examine the Reynolds-number dependence of transport barriers in the PPM.

2 Velocity Field and Transport Barriers in PPM

Let (r, θ) be the polar coordinates in the cross-sectional direction of the cylinder, and z be the coordinate in its axial direction. The wall of the cylinder,

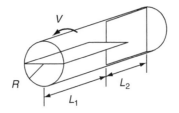

Fig. 1. Schematic view of one axial period of a PPM

expressed by $r = R$, is assumed to rotate at a constant velocity V. We consider only the case of $L_1 = L_2 = L$ and $R/L = 1/2$. Since periodicity of the velocity field in the axial direction is assumed, only one periodic part of the system, as shown in Fig. 1, is considered. The boundary condition for pressure p in the axial direction is given by $p(z = 2L) = p(z = 0) - p_d$, where p_d is a constant. We nondimensionalize the length in the cross-sectional directions with R, the length in the axial direction with L, the cross-sectional velocity components with V, the axial velocity component with VL/R, and the pressure with p_d. The governing equations and boundary conditions for the velocity field $\bm{v} = (v_r, v_\theta, v_z)$ of the fluid with constant density ρ and kinetic viscosity ν are given by follows:

$$\nabla \cdot \bm{v} = 0, \tag{1}$$

$$\frac{\partial \bm{v}}{\partial t} + (\bm{v} \cdot \nabla)\bm{v} = -\frac{1}{\beta Re}\nabla p + \frac{1}{Re}\nabla^2 \bm{v}, \tag{2}$$

where β and Re are defined by $\rho \nu V / p_d R$ and VR/ν, respectively,

$$\bm{v} = 0 \quad \text{on} \quad \begin{cases} \theta = 0 \text{ and } \pi, & \text{for } 0 \leq r < 1,\ 0 \leq z \leq 1, \\ \theta = \pi/2 \text{ and } 3\pi/2, & \text{for } 0 \leq r < 1,\ 1 \leq z \leq 2, \end{cases} \tag{3}$$

$$v_r = v_z = 0 \text{ and } v_\theta = 1 \quad \text{on} \quad r = 1,\ 0 \leq \theta \leq 2\pi,\ 0 \leq z \leq 2, \tag{4}$$

$$\left.\begin{array}{l} \bm{v}(r, \theta, z = 0) = \bm{v}(r, \theta, z = 2), \\ p(r, \theta, z = 0) = p(r, \theta, z = 2) + 1, \end{array}\right\} \text{ for } 0 \leq r \leq 1,\ 0 \leq \theta \leq 2\pi. \tag{5}$$

The parameter β is a measure of the ratio of the velocity of the flow due to the wall rotation to that due to the axial pressure gradient. We change the Reynolds number Re with β fixed to 0.01. The velocity field is numerically obtained. (See [3] for details.)

The trajectories of fluid particles are obtained by integrating the equation, $\dot{\bm{x}} = \bm{v}(\bm{x})$, where $\bm{x}$ is the position of a fluid particle and the dot means the differentiation with time nondimensionalized by R/V. This equation is numerically integrated by the fourth-order Runge–Kutta method.

Figure 2 shows the Poincaré sections taken in the middle of the horizontal plate for several values of Re. Only the upper halves of the Poincaré sections are shown. Closed curves seen in the Poincaré sections are the cross-sections of tubular transport barriers extending in the axial direction. The regions

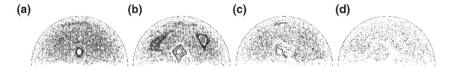

Fig. 2. Poincaré sections. (a) $Re = 10$, (b) 30, (c) 70, and (d) 150

enclosed by these curves are called islands. The region filled with dots is the chaotic region, where the fluid is expected to be well mixed.

There are four recognizable islands in Fig. 2a for $Re = 10$. The two triangle-shaped islands become larger with increasing Re up to 30 (Fig. 2b, and then disappear for $Re \simeq 50$. A similar nonmonotonic Re-dependence of transport barriers, that is, expansion (or appearance) and then shrinkage (or disappearance) of islands with increasing Re, is also observed in other studies on mixing due to three-dimensional steady flows [4,5]. The oval and crescent islands in Fig. 2a slightly expand(Fig. 2b) and then shrink and disappear(Fig. 2c,d).

3 Symmetry of Velocity Field

Chaotic behavior of fluid particles is impossible if the velocity field is geometrically one- or two-dimensional, or satisfies the Euler equation which is obtained by setting the term of $\nabla^2 \boldsymbol{v}/Re$ in (2) to zero. We here call these properties of the velocity field *geometrical* and *dynamical* symmetry, respectively. Although the velocity field in the PPM satisfies neither of these condition in the entire fluid region, we here assume that the velocity field is locally symmetric in the region where transport barriers exist [6]. On the basis of this assumption, the Re-dependence of transport barriers is revealed through the examination of the symmetry of the velocity field for different values of Re.

First, we focus on the geometrical symmetry. Since the barriers which have triangle-shaped cross-sections, seen in Fig. 2a,b, almost straightly extend in the axial direction, they are expected to be associated with the z-dependence of the velocity field which is evaluated by $|\partial \boldsymbol{v}/\partial z|$. Figure 3 shows the distribution of $S_g = \frac{1}{2} \int_0^2 |\partial \boldsymbol{v}/\partial z| \mathrm{d}z$. From the comparison between Figs. 2a,b and 3a,b, we find that S_g is small at the positions of the two triangle-shaped islands, and that the expansion of these islands is caused by the enhancement of the geometrical symmetry.

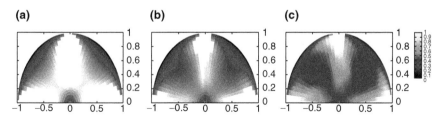

Fig. 3. Distribution of S_g. (a) $Re = 10$, (b) 30, and (c) 150

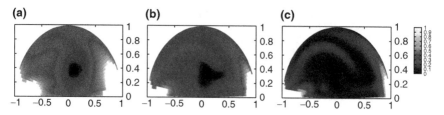

Fig. 4. Distribution of S_d. (a) $Re = 30$, (b), 70 and (c) 150

The other two transport barriers are associated with the dynamical symmetry. Figure 4 shows the distribution of $S_d = \frac{1}{Re}|\nabla^2 \boldsymbol{v}|$ on the same cross-section as those for the Poincaré sections in Fig. 2. From the comparison between Figs. 2b,c and 4a,b, the correspondence of the existence of the two islands to the dynamical symmetry of the velocity field is recognized.

For a large value of Re, there exists no recognizable islands, as shown in Fig. 2d for $Re = 150$, even though S_g and S_d are considerably small in some regions, as shown in Figs. 3c and 4c. The disappearance of the islands with the increase in Re is caused by the instability of the periodic orbits in the corresponding transport barriers.

4 Conclusion

We found that the expansion of transport barriers with the increase in Re is caused by the enhancement of the local symmetry of the velocity field, while the disappearance of the barriers with further increasing in Re is caused by the instability of the periodic orbits in the barriers.

The authors would like to thank Dr. G. P. King for valuable discussions. The present study was partly supported by the 21st Century COE Programs, "Research and Education on Complex Functional Mechanical Systems" and "Frontiers of Computational Science."

References

1. Aref H (1984) J. Fluid Mech. 143:1–21
2. Khakhar DV, et al. (1987) Chem. Eng. Sci. 42:2909–2926
3. Mizuno Y and Funakoshi M (2005) J. Phys. Soc. Jpn. 74:1479–1489
4. Hobbs DM and Muzzio FJ (1998) Chem. Eng. J. 70:93–104
5. Fountain GO, et al. (2000) J. Fluid Mech. 417:265–301
6. Yannacopoulos AN, et al. (1998) Phys. Rev. E 57:482–490

On the Treatment of Long-range Electrostatic Interactions in Biomolecular Simulations

Yoshiteru Yonetani

Summary. Cut-off treatment of electrostatic interactions in biomolecular simulations is discussed. Our test of cut-off length using a bulk water system shows that increasing cut-off length does not improve the simulation results. Moreover, the use of the long cut-off length can lead to a spurious artifact. A piece of evidence for this anomalous behaviour is also provided by evaluating distance dependent-Kirkwood factor $G_k(r)$.

1 Introduction

Molecular dynamics (MD) simulation has been widely used for studying molecular behaviour of various systems, liquids and solids of small molecules, and solutions of biological macromolecules such as proteins, nucleic acids, and membranes. In the current MD simulations, accurate estimation of the long-range Coulomb interactions remains to be an unsolved problem [1, 2]. The amount of the Coulomb force acting between two charges is proportional to the inverse of their square distance. Thus, the interaction force becomes smaller with increasing distance, but even at a long distance, it does not become negligibly small. This feature is the cause of difficulty in accurate estimation of the Coulomb interactions.

There are various treatments of Coulomb interactions implemented in MD simulations of biomolecular systems; the cut-off, the Ewald sum, the multipole expansion, and the continuum model. They are chosen depending on the purpose or the systems of simulations. The cut-off treatment is the simplest approach for evaluating intermolecular interactions, and it is widely used for conducting biomolecular simulations. It has been believed that increasing cut-off length makes simulation results better. In order to confirm whether the simulation results get better as the cut-off length increases, we perform a critical test of cut-off length. A bulk water system was chosen as a test system. This system is particularly preferred compared with solution systems of peptides and proteins, because results from such a homogeneous system are less influenced by initial conditions.

The result of our test was opposite to the belief: artificial property caused by cut-off was enhanced by increasing the cut-off length. Especially, in terms of the distance dependent Kirkwood factor $G_k(r)$, it was clearly shown that the orientational behaviour of water molecules becomes gradually worse as the cut-off length becomes longer. The artifact enhanced by the increased cut-off length led to a spurious artifact. Our results suggest that increasing cut-off should not be attempted.

2 Method

MD simulations were performed using a cubic MD cell containing 2201 water molecules. A periodic boundary condition was applied to the MD cell. The water model used was TIP3P [3], which was treated as a rigid body by the SHAKE constraint [4]. The equation of motion was numerically solved with the time-step 2 fs. The initial structure was prepared by the LEaP module of the AMBER7 software [5]. After minimizing the potential energy of the initial structure, MD simulations were started. The temperature was initially set to 10 K, which was raised to 100, 200, 300 and 350, and then lowered to 298 K every 10 ps. After that, the simulation at 298 K was continued for 2 ns. The pressure was kept at 1 atm throughout the simulation. The temperature and pressure were controlled by the Berendsen's weak coupling method [6] with the coupling constants of 0.1 ps and 0.2 ps, respectively. The group-based cut-off scheme was adopted to avoid the splitting of the water dipole [7]. Ten types of cut-off length (i.e., 9, 10, 11, 12, 13, 14, 15, 16, 17, 18 Å) were tested. In each case, the same length was used for both Coulomb and van der Waals interactions. For comparison, a simulation using the particle mesh Ewald (PME) method [8] with the real space cut-off of 9 Å was also performed. These MD simulations were carried out by the Gibbs module of the AMBER7.

3 Results and Discussion

The simulation with cut-off length 18 Å is shown in Fig. 1, where θ is the angle between the dipole vectors of two water molecules, and $\langle \ldots \rangle$ denotes the average over all molecular pairs. The order parameter $\langle |\cos\theta| \rangle$ shows the degrees of orientational order of water molecules. For example, this quantity becomes 0.50 when most of water molecules have random orientations. On the other hand, our $\langle |\cos\theta| \rangle$ calculated with cut-off length 18 Å largely deviates from the normal value 0.50, and it was about 0.51. This result suggests that a high degree of orientational order develops in our system. Simulations with other treatments (i.e., shorter cut-off 9 Å and PME) do not have such a large deviation. Therefore, the behaviour of $\langle |\cos\theta| \rangle$ we found in 18 Å cut-off calculation is an artifact due to the long cut-off length.

The structure obtained from the 18 Å cut-off simulation is shown in Fig. 2, where the gray sphere is an oxygen atom and white sphere is a hydrogen atom.

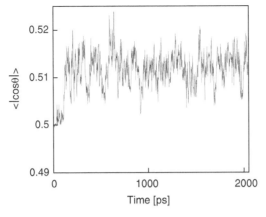

Fig. 1. Time evolution of $\langle |\cos\theta| \rangle$ in simulation with cut-off length 18 Å

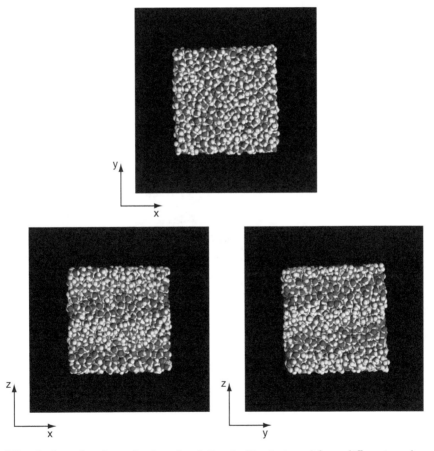

Fig. 2. Snapshot from the 2 ns simulation in Fig. 1 viewed from different angles

The structure is evidently far form natural liquid water. It appears to be a double layer structure in which two kinds of layers with different water orientations stack alternately, but the structure is not so simple. Further analysis revealed that the dipole moments have vortex distribution. The details of this structure will be discussed in [9].

In Fig. 3 we show the radial distribution function $g(r)$ and the distance dependent-Kirkwood factor $G_k(r)$, where r is the oxygen–oxygen distance between two water molecules. The $G_k(r)$ shows the orientational order as a

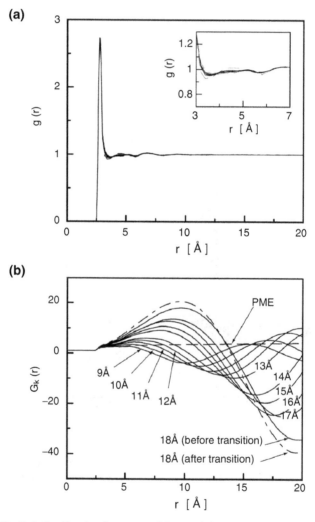

Fig. 3. (a) Radial distribution function $g(r)$ and (b) distance dependent-Kirkwood factor $G_k(r)$. r is the oxygen–oxygen distance between two water molecules

function of r, which is expressed as

$$G_\mathrm{k}(r) = \frac{1}{N}\left\langle \sum_i \left(\mathbf{u}_i \cdot \sum_{\substack{j \\ r_{ij} < r}} \mathbf{u}_j \right) \right\rangle_\mathrm{e},$$

where $\mathbf{u}_i$ is the normalized dipole vector of water molecule i, N is the number of molecules, and the $\langle\ldots\rangle_\mathrm{e}$ denotes the ensemble average. In Fig. 3, all calculations we performed here (i.e., 9–18 Å cut-off and PME) are given. These calculations resulted in giving almost the same $g(r)$ results, and there is no meaningful difference among them. On the other hand, $G_\mathrm{k}(r)$ function is much influenced by the cut-off length; $G_\mathrm{k}(r)$ with a longer cut-off length has a larger deviation from the PME result. That is, the orientational property of water molecules becomes worse as cut-off length increases.

So far, it has been expected that the use of a longer cut-off length (14–18 Å) makes simulation results better. Thus, many simulations of biomolecular systems including explicit water molecules have been performed by using such a long cut-off length. However, the present calculations clearly show that increasing cut-off length does not improve the simulation results. Moreover, the longer cut-off led to a spurious artifact.

More information about the cut-off artifact we introduced here is given in [9, 10]. This artifact is also discussed by van der Spoel and van Maaren [11], recently. These three articles describe various aspects of the layer structure artifact produced by a long cut-off length.

Acknowledgment

The author thanks Prof. Nobuhiro Go, Drs. Hidetoshi Kono, and Hisashi Ishida for valuable comments and discussions, and Dr. Kei Yura for encouraging this work.

References

1. Sagui C, Darden TA (1999) Annu Rev Biophys Biomol Struct 28:155–179
2. Norberg J, Nilsson L (2003) Q Rev Biophys 36: 257–306
3. Jorgensen WL, Chandrasekhar J, Madura JD, Impey RW, Klein ML (1983) J Chem Phys 79:926–935
4. Ryckaert J-P, Ciccotti G, Berendsen HJC (1977) J Comput Phys 23:327–341
5. Case DA, Pearlman DA, Caldwell JW, Cheahtam III TE, Wang J, Ross WS, Simmerling CL, Darden TA, Merz KM, Stanton RV, Cheng AL, Vincent JJ, Crowley M, Tsui V, Gohlke H, Radmer RJ, Duan Y, Pitera J, Massova I, Seibel GL, Singh UC, Weiner PK, Kollman PA (2002) AMBER7. University of California, San Francisco

6. Berendsen HJC, Postma JPM, van Gunsteren WF, DiNola A, Haak JR (1984) J Chem Phys 81:3684–3690
7. Leach AR (2001) Molecular Modeling: Principles and Applications. Pearson Education Limited, Harlow, England
8. Darden T, York D, Pedersen L (1993) J Chem Phys 98:10089–10092
9. Yonetani Y (2006) J Chem Phys 124:204501
10. Yonetani Y (2005) Chem Phys Lett 406:49–53
11. van der Spoel D, van Maaren PJ (2006) J Chem Theor Comp 2:1–11

DNS of Turbulent Channel Flow Obstructed by Rectangular Ribs

Fusao Kawamura, Yohji Seki, Kaoru Iwamoto, and Hiroshi Kawamura

Summary. Direct numerical simulation (DNS) of a turbulent channel flow obstructed by rectangular ribs with a passive temperature field has been performed for $Re_\tau = 80$ and $Pr = 0.71$, where Re_τ is based on the friction velocity, the channel half width and the kinematic viscosity. The flow remains turbulent at the very low Reynolds number owing to the ribs. Various time-averaged turbulence statistics such as the turbulent kinetic energy and the temperature variance are discussed.

1 Introduction

Separation and reattachment of turbulent flows occur in many flow systems of practical engineering applications such as diffusers, combustors and channels with sudden expansions. In these situations, convective heat transfer enhancement is expected. Therefore, separating and reattaching flows have been extensively studied with respect to the heat transfer enhancement. Numerical and experimental investigations for turbulent channel flows with transverse-rib roughness have been reported in numerous works (e.g., Nagano et al. [1]; Leonardi et al. [2]; Hanjalić et al. [3]).

In this study, DNS of turbulent heat transfer has been carried out for a channel flow obstructed periodically by rectangular ribs to examine their effect on the heat transfer. Various turbulence statistics such as the turbulent kinetic energy and the temperature variance are discussed.

2 Numerical Procedures

The configuration of the computational domain is shown in Fig. 1. The rectangular rib is located at the center of the computational domain and attaches directly to both the top and the bottom walls. The flow is assumed to be a fully developed turbulent flow of incompressible viscous fluid with a passive temperature field. It is driven by the streamwise mean pressure gradient.

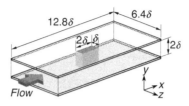

Fig. 1. Configuration of the computational domain. The bottom wall is hot, and the top one is cold

A constant temperature difference is applied in the thermal boundary condition. Hereafter, the superscript of $(^+)$ represents normalization by the kinematic viscosity ν and the friction velocity $u_{\tau 0}$ defined by the mean streamwise pressure gradient. Variables with the superscript of $(^*)$ are those normalized by the temperature difference ΔT.

A finite difference method is adopted for the spatial discretization. The periodic boundary condition is imposed in the horizontal directions. The non-slip conditions are adopted to the top and the bottom walls. The rib is computed by the immersed boundary method [4]. A linear temperature profile is applied in the rib since the heat conductivity of the rib is assumed to be much larger than that of the fluids. In this work, DNS has been performed for $Re_\tau = 80$ and $Pr = 0.71$. The numerical algorithm and other details of the simulations can be found in Abe et al. [5]. The computational conditions are summarized in Table 1, where L_x indicates the streamwise length of the computational domain.

3 Results and Discussion

3.1 Flow Field

The flow rate decreases owing to the additional pressure drag of the rib. The bulk Reynolds number $Re_m (= u_m \cdot 2\delta/\nu)$ drops down to 1,120, where u_m is the bulk mean velocity. Note that, in the case of the plane channel without the rib, an additional calculation at this Re_m has revealed that the flow tends to laminar. The instantaneous flow field is shown in Fig. 2. In an upstream region of the rib, a horseshoe vortex is formed near the wall. The flow separates at

Table 1. Computational conditions

grid number $(N_x \times N_y \times N_z)$	$256 \times 128 \times 256$
spatial resolution $(\Delta x^+, \Delta z^+)$	4.00, 2.00
spatial resolution (Δy^+)	$0.17 \sim 2.64$
time resolution (Δt^+)	0.0136
sampling time period $(t \cdot u_m / L_x)$	400 washout time

Fig. 2. Instantaneous flow field; low speed regions of u' and negative regions of the second invariant of deformation tensor II'^+ (dark gray, $u'^+ \leq -2.0$; light gray, $II'^+ \leq -0.01$). Lower half domain is visualized

the forward-edge of the rib, and many vortices are induced in this region. In the downstream, a three-dimensional wake is formed, which contains a lot of complex vortices. Figure 3 shows the vortex structures in the time-averaged flow field. The horseshoe vortices are clearly observed in front of the rib. In the downstream, four large-scale streamwize vortices exist in addition.

3.2 Nusselt Number

Figure 4 shows contours of the time-averaged local Nusselt number $Nu = \overline{q_w} \cdot 2\delta/\lambda(T_m - T_w)$, where $\overline{q_w}$ is the time-averaged heat flux at the wall, T_m the bulk mean temperature, T_w the temperature at the wall and λ the thermal conductivity. The heat transfer is significantly enhanced in front of the rib and the wake region. In the upstream region of the rib, Nu is remarkably large. This region corresponds to that of the horseshoe vortex. It indicates that the low temperature fluid in the channel central region is transported toward the

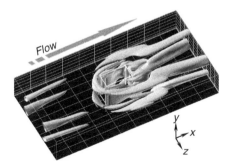

Fig. 3. Vortex structures in the time-averaged flow field visualized by the second invariant of deformation tensor ($\overline{II}^+ \leq -1.0 \times 10^{-5}$). Whole domain is visualized

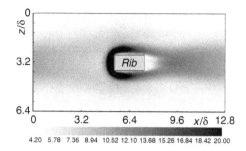

Fig. 4. Contours of Nu on the wall

wall by the flow impingement induced by the rib. In the wake region, Nu is also large as a consequence of effective mixing due to the large number of vortices seen in Fig. 2. Figure 5 shows the spatial-averaged Nusselt number Nu_{ave} on the wall in comparison with the result of DNS for a plane channel flow in the constant temperature difference condition by Tsukahara et al. [6]. In the present case, a remarkable enhancement of heat transfer is assessed quantitatively.

3.3 Turbulent Kinetic Energy and Temperature Variance

Figures 6 and 7 show contours of the turbulent kinetic energy $k^+ = \overline{u_i'^+ u_i'^+}/2$ and the temperature variance $k_\theta^* = \overline{\theta'^* \theta'^*}/2$ in the (x, z)-plane at the channel central height, respectively. The large turbulent kinetic energy is observed around the separation and the wake regions. On the other hand, the temperature variance does not increase in these regions. This is because the large-scale vortices strongly mix the temperature field and the temperature gradient is

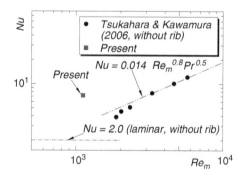

Fig. 5. Comparison of Nu_{ave} on the wall to that of plane channel flow

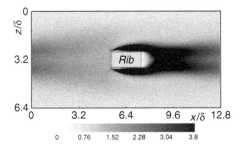

Fig. 6. Contours of k^+ in the (x,z)-plane at the channel central height

decreased. Thus the enhanced mixing is statistically confirmed. Note that turbulent heat flux $\overline{v'^+\theta'^*}$ is enhanced in the wake region (not shown here).

4 Conclusions

DNS of a turbulent heat transfer has been performed for a channel flow obstructed by rectangular ribs at $Re_\tau = 80$ and $Pr = 0.71$ in order to assess the effect of obstacles. The following conclusions are derived:

1. The flow remains turbulent at a very low Reynolds number of $Re_m = 1{,}120$. The heat transfer is significantly enhanced in front of the rib and in the wake region.
2. In the separation and the wake regions, the turbulent kinetic energy increases owing to complex vortices induced by the rib. The temperature variance, however, exhibits a small value in these regions since the large-scale vortices strongly mix the temperature field and thus the temperature gradient is decreased.

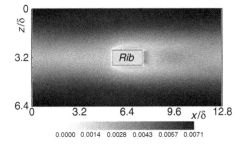

Fig. 7. Contours of k_θ^* in the (x,z)-plane at the channel central height

References

1. Nagano Y, Hattori H, Houra T (2004) Int. J. Heat Fluid Flow 25:393–403
2. Leonardi S, Orlandi P, Djenidi L, Antonia R (2004) Int. J. Heat Fluid Flow 25:384–392
3. Hanjalić K, Launder B (1972) J. Fluid Mech. 51:301–335
4. Fadlun E, Verzicco R, Orlandi P, Mohd-Yousof J (2000) J. Comput. Phys. 161: 35–60
5. Abe H, Kawamura H, Matsuo Y (2004) Int. J. Heat Fluid Flow 25:404–419
6. Tsukahara T, Iwamoto K, Kawamura H, Takeda T (2006) Proc. Turbulence, Heat and Mass Transfer 5 Croatia: 193–196

Source Models of Gravity Waves in an F-plane Shallow Water System

Norihiko Sugimoto, Keiichi Ishioka, and Katsuya Ishii

Summary. Gravity wave radiation from unsteady rotational flow is numerically investigated in an f-plane shallow water system. We propose three models for a source of gravity waves. The validity of these models are checked in the Ro–Fr plane, where Ro is the Rossby number and Fr is the Froude number.

1 Introduction

Gravity waves play a fundamental role on the atmosphere. Several observations suggest gravity waves are radiated from strongly rotational flow region [1]. However, the generation process of gravity waves is not studied in detail.

The first and second authors investigated gravity wave radiation from a rotational jet flow in an f-plane shallow water system with forcing, in which $Ro = 100$ and $Fr = 0.3$. They show the main source of gravity waves (SGW) can be expressed by three different forms [2]. Here, we propose three models for SGW and check their validity in a wide parameter range of Ro and Fr.

2 Experimental Configuration

We consider gravity wave radiation from unsteady jet flow in an f-plane shallow-water system. In the system, the Coriolis parameter and the gravitational acceleration are expressed by f and g, respectively. Cartesian coordinates (x, y) is chosen so that x, y are the longitudinal and latitudinal direction, respectively. The velocity (u_0, v_0) of the steady state, and the surface level h_0 in geostrophic balance with u_0 are $(u_0(y), v_0) = (U_0 \mathrm{sech}\{(y - y_0)/B\}, 0)$ and $h_0(y) = -(fBU_0/g) \arctan \{\exp(2y/B)\}$, where U_0, y_0, and B are parameters to determine the intensity, the position, and the width of the jet, respectively. We introduce the bottom topography h_b, so that the depth of the fluid $H_0 (= h_0 - h_\mathrm{b})$ is constant. This basic state of the zonal jet [3] is barotropically unstable. The equations for the disturbance (u, v, h) are

$$\frac{\partial u}{\partial t} + u\frac{\partial u}{\partial x} + v\frac{\partial u}{\partial y} - fv = -g\frac{\partial h}{\partial x} - \underbrace{\alpha(u - u_0)}_{\text{forcing}}, \qquad (1)$$

$$\frac{\partial v}{\partial t} + u\frac{\partial v}{\partial x} + v\frac{\partial v}{\partial y} + fu = -g\frac{\partial h}{\partial y} - \underbrace{\alpha(v - 0)}_{\text{forcing}}, \qquad (2)$$

$$\frac{\partial h}{\partial t} + u\frac{\partial (h - h_b)}{\partial x} + v\frac{\partial (h - h_b)}{\partial y} + (h - h_b)\left(\frac{\partial u}{\partial x} + \frac{\partial v}{\partial y}\right) = 0, \qquad (3)$$

where t is the time. The parameter α is zero except for the jet region.

The nondimensional parameters Ro (Rossby number) and Fr (Froude number) are determined by the basic state as $Ro \equiv U_0/fB$ and $Fr \equiv U_0/\sqrt{gH_0}$, respectively. Here, we choose $U_0 = 10\pi$ and $B = \pi/10$ for the numerical experiments. Then Ro and Fr are nondimensional reciprocals of the frequency f and the phase speed of gravity waves $\sqrt{gH_0}$, respectively.

The domain is set to be periodic in the x direction and infinite in the y direction, using a mapping method [4]. The mapped shallow water equations are computed using a spectral transform method [5]. Resolution is set to be $64(x) \times 1{,}024(y)$ grids. The fourth-order Runge–Kutta method is used for the time-integrations. In addition, we introduce a pseudoartificial viscosity term to absorb gravity waves near the boundary in the y direction. Details of experimental configuration are described in [6].

3 Source Models of Gravity Wave

Using the analogy with the aero-acoustic sound wave radiation theory, i.e., the Lighthill theory [7], Ford [8] showed the source field of gravity waves (near field) were obtained without a knowledge of the wave field (far field).

SGW is expressed by three different approximate forms in [2]. Details of the derivation are shown in [2]. Since these SGW have different characteristics and dependence of parameters Fr and Ro, we propose three source models for the zonal jet flow as follows:

- *Model 1:* When $Fr \ll 1$, the flow is nearly nondivergent in the jet region. We consider localized sources in the jet region as a simplified form

$$\text{SGW} \approx 2hu\frac{\partial}{\partial x}\left(\frac{\partial v}{\partial x}\frac{\partial u}{\partial y}\right). \qquad (4)$$

- *Model 2:* Far from the jet region, zonal gravity waves (no wavy structure in the x direction) is only propagated. To estimate these gravity waves, we introduce a simplified model of zonal SGW (averaged in the x direction,

$\overline{X}$) for $Fr \ll 1$, which is directly related to vortical motion [10]

$$\overline{\mathrm{SGW}} \approx h \overline{\frac{\partial}{\partial t} \nabla \cdot (\boldsymbol{\omega} \times \boldsymbol{u})}. \tag{5}$$

– **Model 3:** When Ro is high, we can simplify the model 2 further, using a scale law for balanced perturbations [9]. Then, we obtain

$$\overline{\mathrm{SGW}} \approx \frac{\partial^2}{\partial y^2} \overline{\left(-2hv\frac{\partial h}{\partial y} - \frac{\alpha}{f}v\right)}. \tag{6}$$

4 Discussion

To check the validity region of parameters for each model, we use the results of direct numerical simulation for a wide parameter range of Ro ($1 - 1000$, 10 cases) and Fr ($0.1 - 1$, 10 cases) in [6]. The model 1 shows the local distribution of SGW in the strong jet region for $Fr < 1$ case, and the models 2 and 3, which are related to the zonal SGW, estimate the time variations of $\partial \overline{h}/\partial t$ far from the jet region except for strong intermittent gravity waves, for $Fr < 1$ case (the model 2), and $Ro \geq 30$ and $Fr < 1$ case (the model 3). In Fig. 1, we compare the distribution of the total SGW and that of the model 1 for $Ro = 100$ and $Fr = 0.7$ at $t = 6.05$. The pattern of SGW localized in the strong jet region is expressed well by the model 1.

Figures 2 and 3 show the time variations of $\partial \overline{h}/\partial t$ (broken line) and the estimation (solid line) from the model 2 (left) and 3 (right) at eight y's far from the jet region. In Fig. 2, $Ro = 100$ and $Fr = 0.7$. The values of two estimation agree well with that of the simulation, except for around $t = 7.0, 9.8$. If we add the effect of forcing to the model 2 and 3, strong intermittent gravity waves (around $t = 7.0, 9.8$) are estimated well. In Fig. 3, $Ro = 10$ and $Fr = 0.1$.

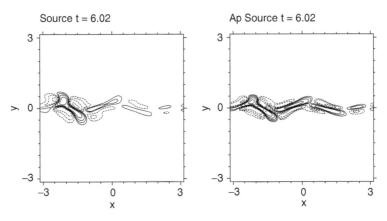

Fig. 1. SGW (left) and the model 1 (right) for $Ro = 100$ and $Fr = 0.7$ at $t = 6.05$.

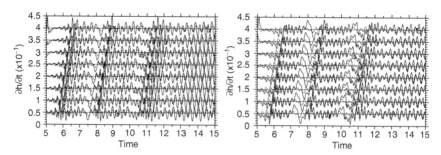

Fig. 2. $\partial \overline{h}/\partial t$ (*broken line*) and that calculated (*solid line*) from the model 2 (*left*) and the model 3 (*right*) at 8 values of y ($y = 5, 10, \ldots, 40$) for $Ro = 100, Fr = 0.7$

While the agreement of the model 2 is good (left), the model 3 is poor (right), since Ro is too small. When the effect of Colioris term in SGW is dominant for low Ro case [6], we should propose alternative model which include the effect of the Colioris term.

5 Concluding Remarks

We have proposed three models for source of gravity waves (SGW). The model 1 shows the local distribution of SGW in the strong jet region, and the models 2 and 3, which are related to the zonal SGW, estimate the time variations of $\partial \overline{h}/\partial t$ far from the jet region. We can apply three models to the jet flows with $Fr < 1$. In addition, for low Ro, in which the model 3 is not valid, we should propose alternative model which is related to the effect of Colioris term. It would be useful to propose more general model of SGW in future.

This study is supported by Grant-in-Aids for the 21st Century COE programs "Elucidation of the Active Geosphere" and "Frontiers of Computational Science." Authors thank Dr. K. Yoshimatsu for his useful comments.

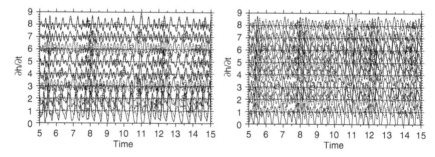

Fig. 3. $\partial \overline{h}/\partial t$ (*broken line*) and that calculated (*solid line*) from the model 2 (*left*) and the model 3 (*right*) at each y (same as Fig. 2) for $Ro = 10, Fr = 0.1$

References

1. Yoshiki M, Sato K (2000) J Geophys Res 105:17995–18011
2. Sugimoto N, Ishioka K, Yoden S (2006a) Fluid Dyn Res (in press)
3. Hartmann DL (1983) J Atmos Sci 40:817–835
4. Ishioka K (2006) Proc IUTAM Symposium 2006 Nagoya (submitted)
5. Ishioka K (2004) ispack-0.62, GFD Dennou Club.
6. Sugimoto N, Ishioka K, Yoden S (2005) Theor Appl Mech 54:299–305
7. Lighthill M J (1952) Proc R Soc London 211A:564–587
8. Ford R (1994) J Fluid Mech 281:81–118
9. Sugimoto N, Ishioka K, Yoden S (2006b) Fluid Dyn Res (in press)
10. Howe M S (1975) J Fluid Mech 71:625–673

Direct Numerical Simulation of Neutrally Stratified Ekman Boundary Layer

Katsuhiro Miyashita, Kaoru Iwamoto, and Hiroshi Kawamura

Summary. Direct numerical simulations (DNSs) are performed for the turbulent Ekman boundary layer over a smooth surface. Reynolds numbers Re_f are set to be $Re_f = 400$, 510, 600, 775, 1,140, and 1,393, where Re_f is based on the geostrophic wind G, the kinematic viscosity ν and Coriolis parameter f. A low-speed large-scale structure is found in the upper region ($6 < y/D < 15$) where y is the wall-normal direction and D the viscous Ekman layer depth, whereas the well-known streaky structure appears in the vicinity of the wall. The large-scale structure is also observed in a motion of a material line. The direction in which this structure is elongated does not coincide with the direction of the mean flow velocity. The reason is discussed based upon the motion of a material line whose initial position is horizontal in a vicinity of the wall.

1 Introduction

Rotation is one of the factors which affect the planetary boundary layer. The boundary layer under the effect of the system rotation is called the Ekman boundary layer (EBL). Among studies of EBL, researches using DNS have barely been performed. Coleman et al. [1] performed DNSs of EBL with low Reynolds numbers, and indicated that no horizontal roll vortices appeared under neutral stratification. Coleman et al. further carried out DNSs on buoyantly stable [2] and unstable [3, 4] cases, and obtained horizontal roll vortices only for moderately unstable cases. Later, Coleman [5] conducted DNS of EBL with a higher Reynolds number of $Re_f = 1{,}000$. Shingai et al. [6] employed larger computational domain and found that a large turbulent structure in the region where the mean velocity reaches its maximum. They also observed the inertial oscillation in the upper region. In the present study, we have calculated DNSs of the neutrally stratified EBL at the higher Reynolds numbers of $Re_f = 1{,}140$ and 1,393. Our objective is to investigate Reynolds number dependence and three-dimensional characteristics of EBL.

2 Simulated Flow and Numerical Methods

A calculated flow field is EBL of an incompressible viscous fluid over a smooth surface. The system rotates about a vertical axis. The flow is driven by the combination of the horizontal pressure gradient and the Coriolis force. The periodic boundary conditions are imposed in the streamwise (x) and spanwise (z) directions. The nonslip and the Neumann conditions are adopted for each component of the velocity on the lower and upper boundaries, respectively. The continuity and Navier-Stokes equations are solved. The fractional step method is used for coupling these equations. The second-order Crank-Nicolson and the Adams-Bashforth methods are employed as the time advance algorithms; the former for the vertical viscous term and the latter for the other terms. The finite difference method is applied for the spatial discretization. The fourth-order central difference scheme is adopted in the x- and z-directions, and the second-order central difference scheme in the wall-normal (y) direction. The Reynolds numbers Re_f are set to be $Re_f = 400$, 510, 600, 775, 1,140, and 1,393. Here Re_f is defined as

$$Re_f = \frac{GD}{\nu} = \frac{G}{\sqrt{\nu f/2}}, \tag{1}$$

where G is the geostrophic wind velocity, D the viscous Ekman layer depth ($= \sqrt{2\nu/f}$), ν the kinematic viscosity and f the Coriolis parameter. Hereafter, the superscripts of ($'$) and ($+$) represent fluctuating part from wall-parallel plane averaged one, and normalization by the friction velocity u_τ and ν, respectively.

3 Results

Figure 1 shows the spanwise wavelength λ_z of the maxima of the premultiplied spanwise energy spectra of the streamwise velocity fluctuation $k_z E_{u'u'}$ for the calculated Reynolds number, where k_z is the spanwise wavenumber. It is well known that the wavelength at the peak of the premultiplied energy spectrum indicates that of the mainly energy-containing scale (MES). In the vicinity of the wall, the spanwise wavelengths of MES is $\lambda_z^+ \sim 100$. This corresponds to the spanwise scale of the streaky structures near the wall. In the mid-height region ($y/D < 6$), the length scales increase linearly with increasing the height. In the upper region ($6 < y/D < 15$), however, the wavelengths of MES stay at a constant value of $\lambda_z/D \sim 15$–20. This suggests the existence of the large-scale structures with a spanwise scale which can be normalized by the viscous Ekman layer depth D. In the former study, Shingai et al. [6] suggested the spanwise wavelength of MES can be scaled by $\delta_{Q\max}$, which is the depth where the absolute value of the mean velocity Q becomes maximum in EBL. The present study, however, indicates the scaling by the viscous Ekman layer depth gives a better collapse.

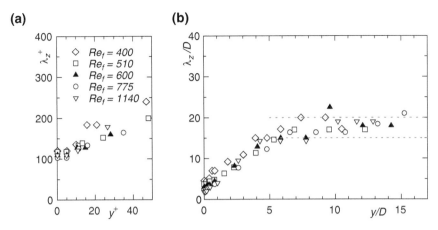

Fig. 1. Spanwise wavelength of the maxima of the premultiplied spectra nondimensionalized by (a) the wall-unit and (b) the viscous Ekman layer depth D

A bird's eye view of an instantaneous flow field for $Re_f = 1,393$ is displayed in Fig. 2a. Figure 2b shows an enlarged view of a part of Fig. 2a from a different view point. The streaky structures are observed near the wall and elongated along the direction of the mean velocity. On the other hand, large low-speed regions exist in the upper region. The large-scale structures rise up from the bottom because the low-speed fluid is conveyed from the bottom wall. The structures are not aligned in the direction of the mean velocity. They are

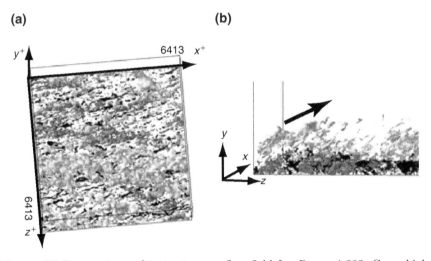

Fig. 2. Bird's eye views of instantaneous flow field for $Re_f = 1,393$. Gray: high-speed regions, $u'^+ > 3.0$; black: low-speed regions, $u'^+ < -3.0$; white: low-speed regions in the upper region, $u' < -1.2$ at $y/D > 5$

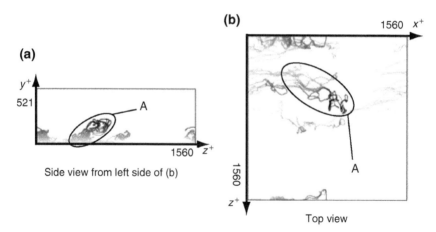

Fig. 3. A Motion of a material line after $t^+ = 162$ for $Re_f = 400$. The color indicates the height from the bottom wall, gray to black, $y^+ = 0$ to $y^+ = 182$ ($y/D = 7$)

elongated approximately in the direction of the geostrophic wind (in the x-direction in the present coordinate) and are also inclined slightly toward the higher pressure side (in the z-direction). The reason will be discussed later in conjunction with the movement of the material line released from a vicinity of the bottom wall.

To examine the large-scale structure, a motion of a passive material line in EBL is visualized in Fig. 3. The material line is initially released from a vicinity of the wall ($y^+ \sim 17$). Figure 3 shows the state of the material line at $t^+ = 162$ after the release. A strong ejection event is observed in the oval A in Fig. 3. If the advected spanwise distance is compared for the material line raised up to the outer region and the one staying in the vicinity of the wall, the distance of the former one is smaller than that of the latter. This is because the spanwise mean velocity $\overline{w}$ is larger in the vicinity of the wall than in the outer region. In fact, the spanwise movement of the material line near the wall in the oval A ($\Delta z^+ \approx 520$) is roughly equal to the product of the mean spanwise velocity ($\overline{w}^+ = 3.3$) and the elapsed time ($t^+ = 162$). Therefore, the inclination of the large-scale structure is caused by the combination of the strong ejection and the three-dimensional mean velocity profile.

References

1. Coleman G N, Ferziger J H and Spalart P R (1990) J Fluid Mech 213:313–348
2. Coleman G N, Ferziger J H and Spalart P R (1992) J Fluid Mech 244:677–712
3. Coleman G N, Ferziger J H and Spalart P R (1994) Boundary-Layer Meteor 70:247–272
4. Coleman G N and Ferziger J H (1996) Dyn Atmos Oceans 24:85–94
5. Coleman G N (1999) J Atmos Sci 56:891–900
6. Shingai K and Kawamura H (2004) J Turbulence 5:13

Application of the Renormalization Group Analysis to a Noisy Kuramoto–Sivashinsky Equation and its Numerical Simulation

Kazuto Ueno

Summary. The renormalization group analysis is applied to a noisy Kuramoto–Sivashinsky (KS) equation in d dimension. In the special case of d = 1, it is shown that the noisy KS equation belongs to the same universality class as the Kardar–Parisi–Zhang (KPZ) equation in the sense that they have the same scaling exponents in the long-wavelength limit.

1 Introduction

The most attractive continuous models for surface roughening are the Kardar–Parisi–Zhang (KPZ) equation $\partial h/\partial t = \nu \nabla^2 h + \lambda (\nabla h(\mathbf{x}))^2/2 + \eta(\mathbf{x}, t)$ and the Kuramoto–Sivashinsky (KS) equation $\partial h/\partial t = \nu \nabla^2 h - K(\nabla^2)^2 h + \lambda (\nabla h(\mathbf{x}))^2/2$. Here $h(\mathbf{x}, t)$ describes the height profile of a surface above a substrate point $\mathbf{x} = (x_1, \ldots, x_d)$ at time t. ν is the surface tension coefficient, K is a positive surface diffusion coefficient, and λ is the strength of the nonlinearity. The KPZ equation has a positive surface tension coefficient ($\nu > 0$) and is driven by a random forcing η, which is a Gaussian white noise with zero mean and correlation $\langle \eta(\mathbf{x}, t)\eta(\mathbf{x}', t') \rangle = 2D\delta^d(\mathbf{x} - \mathbf{x}')\delta(t - t')$, D being the strength of the noise. On the other hand, the KS equation is completely deterministic and is driven by inherent instabilities due to a negative surface tension coefficient ($\nu < 0$). In spite of such a difference between these equations, Yakhot conjectured that large distance and long-time behavior of the KS equation in one-dimension are equivalent to those of the KPZ equation (or a noisy Burgers equation) [1].

Here we summary some scaling exponents in surface growth [2]. The interface width is defined as $W(t) = \langle [h(\mathbf{x}, t) - \langle h(\mathbf{x}, t) \rangle]^2 \rangle^{1/2} \sim t^\beta$, where β is called the growth exponent. The other important scaling exponents are the roughness and the dynamic scaling exponents, α and z, respectively. These exponents are related as $\beta = \alpha/z$ by the scaling relation. In the KPZ equation in one-dimension, it is known that the values of α, β, and z are 1/2, and 1/3, and 3/2, respectively. Hereafter we refer to these as the KPZ scaling exponents. In order to clarify Yakhot's conjecture, a number of numerical investigations of

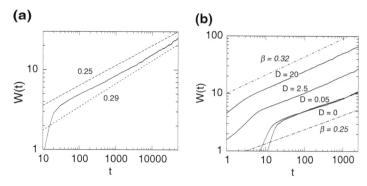

Fig. 1. (a) Time evolution of $W(t)$ for the deterministic KS equation in one-dimension for $L = 200{,}000$. Two straight *dashed lines* denote the line with exponent 0.25 and 0.29. (b) Time evolutions of $W(t)$ for the noisy KS equation in one-dimension at $D = 0$, 0.05, 2.5, and 20 for $L = 20{,}000$ [4]

the KS equation in one-dimension have been developed. However, much larger system size and longer time simulations were necessary to confirm the KPZ scaling exponents from the deterministic KS equation [3]. This is the case in our numerical simulation as shown in Fig. 1a. This displays the time evolution of $W(t)$ for the deterministic KS equation in one-dimension for system size $L = 200{,}000$. The exponent near $t \sim 50{,}000$ is about 0.29, which is still rather smaller than $1/3$ [4]. Therefore, we must find other way to circumvent the limitation of the system size and time in the numerical simulations to confirm the KPZ scaling from the deterministic KS equation.

2 Renormalization Group Analysis

Instead of considering the deterministic KS equation, we investigate the long-wavelength properties of a noisy KS equation in d dimension, $\partial h/\partial t = \nu \nabla^2 h - K(\nabla^2)^2 h + \lambda(\nabla h)^2/2 + \eta(\mathbf{x}, t)$, where $\eta(\mathbf{x}, t)$ is a Gaussian white noise with zero mean, and the correlation is given by $\langle \eta(\mathbf{x}, t) \eta(\mathbf{x}', t') \rangle = (2D - 2D_d \nabla^2) \delta^d(\mathbf{x} - \mathbf{x}') \delta(t - t')$. η is composed of the nonconserved and conserved noises whose strength is D and D_d, respectively.

There are five parameters ν, K, λ, D, and D_d in just above two equations. Applying a dynamic renormalization group (RG) analysis to this system, we obtain the RG flow equations describing the change in above five parameters under the RG transformation, which are consisted of coarse-graining and rescaling. From these five parameters we can define three dimensionless coupling constants by

$$(F(l), G(l), H(l)) = \left(\frac{\tilde{\nu}(l)}{\tilde{K}(l)\Lambda_0^2}, \frac{K_d}{8d(d+2)} \frac{\tilde{\lambda}(l)^2 \tilde{D}(l)}{\tilde{K}(l)^3 \Lambda_0^{8-d}}, \frac{K_d}{8d(d+2)} \frac{\tilde{\lambda}(l)^2 \tilde{D}_d(l)}{\tilde{K}(l)^3 \Lambda_0^{6-d}} \right)$$

with $K_d = S_d/(2\pi)^d$, $S_d = 2\pi^{d/2}/\Gamma(d/2)$ is the surface area of the d-dimensional unit sphere. $\tilde{\nu}(l)$, $\tilde{K}(l)$, $\tilde{\lambda}(l)$, $\tilde{D}(l)$, and $\tilde{D}_d(l)$ denote rescaled parameters in terms of a scaling parameter l, and Λ_0 is an upper cutoff in the Fourier space.

The RG flow equations for $(F(l), G(l), H(l))$ are $dF/dl = 2F - G/(1+F)^5\{a_0 + a_1 F + a_2 F^2 + a_3 F^3 + a_4 F^4 + (b_0 + b_1 F + b_2 F^2 + b_3 F^3 + b_4 F^4)(H/G)\}$, $dG/dl = (d+6)G - G^2/(1+F)^5\{c_0 + c_1 F + c_2 F^2 + c_3 F^3 + (d_0 + d_1 F + d_2 F^2 + d_3 F^3)(H/G) + (e_0 + e_1 F + e_2 F^2)(H/G)^2\}$, $dH/dl = (d+4)H - G^2/(1+F)^5\{f_0 + f_1 F + f_2 F^2 + (g_0 + g_1 F + g_2 F^2 + g_3 F^3)(H/G) + (h_0 + h_1 F + h_2 F^2 + h_3 F^3)(H/G)^2\}$, with the polynominals $a_0 = 2d^2 - 4d - 16$, $a_1 = 12d^2 - 84d + 108$, $a_2 = 21d^2 - 162d + 108$, $a_3 = 14d^2 - 114d + 100$, $a_4 = 3d^2 - 32d + 32$, $b_0 = 2d^2 - 8$, $b_1 = 10d^2 - 34d + 24$, $b_2 = 15d^2 - 60d - 12$, $b_3 = 8d^2 - 44d + 60$, $b_4 = d^2 - 18d + 20$, $c_0 = 16d^2 - 232d + 444$, $c_1 = 41d^2 - 482d + 420$, $c_2 = 34d^2 - 346d + 324$, $c_3 = 3(3d^2 - 32d + 32)$, $d_0 = 8d^2 - 122d + 120$, $d_1 = 19d^2 - 220d - 12$, $d_2 = 14d^2 - 152d + 180$, $d_3 = 3(d^2 - 18d + 20)$, $e_0 = -2d(d+2)$, $e_1 = -4d(d+2)$, $e_2 = -2d(d+2)$, $f_0 = 4d^2 - 12d - 40$, $f_1 = 5d^2 + 2d - 16$, $f_2 = d^2 - 4$, $g_0 = 24d^2 - 240d + 396$, $g_1 = 51d^2 - 458d + 428$, $g_2 = 3(12d^2 - 114d + 108)$, $g_3 = 3(3d^2 - 32d + 32)$, $h_0 = 14d^2 - 114d + 112$, $h_1 = 28d^2 - 190d + 12$, $h_2 = 17d^2 - 144d + 184$, and $h_3 = 3(d^2 - 18d + 20)$.

In $d = 1$ we recover the RG flow equations for $(F(l), G(l), H(l))$ in [4] by replacing $G \to G/6$ and $H \to H/6$. As shown in Fig. 2, except for the case of $\tilde{D}(0) = \tilde{D}_d(0) = 0$, as increasing l the remaining three trajectories reach the same fixed point $(F^*, G^*, H^*) = (10.7593, 113.442, 10.5436)$, which are determined from the equation $(dF/dl, dG/dl, dH/dl) = (0,0,0)$. Using these fixed point values, the scaling exponents are determined as follows $z = 2 - 6G^*/[F^*(1+F^*)^3]\{3 + F^* - (1 - F^*)H^*/G^*\} = 1.5$, $\alpha = 0.5\{z - 1 + 6G^*/(1+F^*)^3(1+H^*/G^*)^2\} = 0.5$, and from the scaling relation, we obtain $\beta = 1/3$, which are exactly the same values as the KPZ scaling exponents

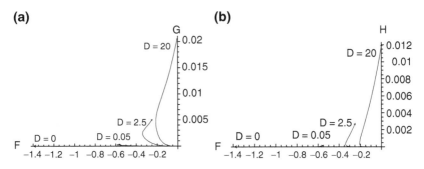

Fig. 2. The RG flow in the parameter space (F, G, H) in one-dimension for $0 \le l \le 2$, projected on (a) the (F, G) and (b) the (F, H) plane, for initial values $\tilde{\nu}(0) = -1$, $\tilde{K}(0) = 1$, $\tilde{\lambda} = 1$, $\tilde{D}(0) = D$ ($D = 0, 0.05, 2.5, 20$), $\tilde{D}_d(0) = 0$, and a cutoff wave number $\Lambda_0 = \pi/0.5$

mentioned earlier. Moreover, it is found that the RG flow for $(F(l), G(l), H(l))$ rapidly approaches the KPZ fixed point with increasing the strength of D. Once the value of D takes a different value from zero, the RG trajectory deviates from the F axis. For much smaller D it takes many RG transformations to reach the KPZ fixed point. The fluctuation-dissipation theorem (FD) must hold exactly in the KPZ equation in $d = 1$. We found that it is indispensable to introduce the D_d term to satisfy the FD theorem.

3 Confirmation of the RG Prediction by Numerical Simulation

Figure 1b displays the time evolution of $W(t)$ for the noisy KS equation in $d = 1$ with various $D = 0$, 0.05, 2.5, 20 for $L = 20,000$. As the noise strength D is increased, the time evolutions of $W(t)$ are shifted upward, since the fluctuations increase owing to the stochastic noises. For $D = 20$, the exponent of the dynamic scaling is evaluated as $\beta = 0.316$ by the method of least squares using the data between $50 < t < 2,000$, which is close to the exponent $1/3$ of the KPZ scaling. This result is consistent with the RG result in Fig. 2.

4 Conclusion

The RG analysis revealed that the RG flow for the parameters of the noisy KS equation rapidly approach the KPZ fixed point with increasing the strength of the noise. This was supplemented by the numerical simulations of the KS equation with a stochastic noise, in which the scaling behavior close to the KPZ scaling can be observed even in the moderate system size and time. It is still an open question whether the KS or noisy KS and the KPZ equations fall into the same universality class in the other dimensions. This is under investigation.

This work was supported by a Grant-in-Aid for the 21st Century COE "Frontiers of Computational Science."

References

1. V. Yakhot (1981) Phys. Rev. A **24**: 642–644
2. A.-L. Barabási and H. E. Stanley (1995) Fractal Concepts in Surface Growth. Cambridge University Press, Cambridge, England
3. K. Sneppen, J. Krug, M. H. Jensen, C. Jayaprakash, and T. Bohr (1992) Phys. Rev. A **46**: R7351–R7354
4. K. Ueno, H. Sakaguchi, and M. Okamura (2005) Phys. Rev. E **71**: 046138

Wavelet-Based Statistics in High-Resolution Direct Numerical Simulations of Three-Dimensional Homogeneous Isotropic Turbulence

Naoya Okamoto, Katsunori Yoshimatsu, and Yukio Kaneda

Summary. Statistics of kinetic energy and transfer localized in space and scale are examined by using the datasets of high-resolution direct numerical simulations of three-dimensional incompressible homogeneous isotropic turbulence with the Taylor micro-scale Reynolds number up to 732. We find that the standard deviation spectra of the local energy σ_e and that of energy transfer σ_t decrease with a representative wavenumber k_α approximately as $\sigma_e \propto k_\alpha^{-5/3}$ and $\sigma_t \propto k_\alpha^{-1/2}$ in the inertial subrange, respectively.

1 Introduction

In fully developed turbulent flows, eddies of a wide range of scales coexist and exhibit strong spatial intermittency. They interact with each other and the energy at large scales are transferred to small scales by the nonlinear interaction. Keeping track of both location and scale to represent the turbulent flows may play a key role in understanding of small scale statistics of turbulence and in exploration for universality. An orthonormal discrete wavelet expansion method is one of the most powerful tools to resolve information on both scale and location of turbulent flows [1,2]. Although wavelet analysis has been applied to direct numerical simulation (DNS) datasets with detailed information on turbulence that cannot be obtained experimentally and is free from experimental uncertainties, the corresponding Reynolds number would be too low to explore universality of turbulence. The reader may refer to the reviews on wavelet application to turbulence by Farge [3] and Addison [4].

In this paper, we examine wavelet-based statistics local both in physical and spectral spaces of turbulent energy and energy transfer, which are fundamental measures characterizing turbulence. A three-dimensional multiresolution analysis is made by using high resolution DNSs of incompressible homogeneous isotropic turbulence in a periodic box with resolution up to 2048^3 grid points and the Taylor microscale Reynolds number R_λ up to 732 [5,6]. The DNS at resolution 2048^3 exhibits the much wider inertial subrange than those in the DNSs at lower resolutions.

2 Three-Dimensional Orthonormal Wavelet Analysis of Incompressible Homogenous Isotropic Turbulence

2.1 DNS Data

Kaneda et al. [5] performed DNSs of incompressible turbulence obeying Navier-Stokes equation in a periodic box with sides 2π by the use of the Earth Simulator. We use the DNS data with resolution $256^3, 512^3, 1024^3$, and 2048^3 for $k_{max}\eta \sim 1$. Here, k_{max} is a maximum wavenumber in each of the DNSs and η is the Kolmogorov length scale. The Taylor microscale Reynolds numbers in the $256^3, 512^3, 1024^3$, and 2048^3 DNSs are 167, 257, 471, and 732, respectively. The reader may refer for the details of the DNSs to previous papers. R_λ up to 732 [5,6].

2.2 Multiresolution Analysis

By the use of three-dimensional multiresolution analysis based on orthonormal periodic wavelets in $\mathbf{T}^3 = [0, 2\pi]^3$, we expand scalar function $f(\mathbf{x})$ as

$$f(\mathbf{x}) = \bar{f} + \sum_{\alpha=1}^{\infty} \sum_{\iota_1,\iota_2,\iota_3=0}^{2^{\alpha-1}-1} \sum_{\mu=1}^{7} \mathcal{W}_{\iota}^{(\alpha,\mu)}[f] \Psi_{\mu,\iota}^{\alpha}(\mathbf{x}), \tag{1}$$

where $\bar{f} = \int_{\mathbf{T}} f(\mathbf{x})\, d\mathbf{x}/(2\pi)^3$, $\mathcal{W}_{\iota}^{(\alpha,\mu)}[f] = \int_{\mathbf{T}} f(\mathbf{x}) \Psi_{\mu,\iota}^{\alpha}(\mathbf{x})\, d\mathbf{x}/(2\pi)^3$, and $\iota = (\iota_1,\iota_2,\iota_3)$. $\Psi_{\mu,\iota}^{\alpha}(\mathbf{x}) = 2^{3\alpha/2} \Psi_\mu(2^\alpha \mathbf{x} - 2\pi\iota)$. Three-dimensional mother wavelets $\Psi_\mu(\mathbf{x})$ are constructed by tensor product of one-dimensional scaling function $\psi_0(x)$ and mother wavelet $\psi_1(x)$ as $\Psi_\mu(\mathbf{x}) = \psi_\xi(x_1)\psi_\eta(x_2)\psi_\zeta(x_3)$ ($\xi,\eta,\zeta = 0,1$, and $\mu = \xi + 2\eta + 4\zeta$). We use wavelets obtained by periodizing the Meyer wavelet [7].

2.3 Statistics of the Local Energy and the Net Energy Transfer in Wavelet Space

The local energy at level α ($\alpha = 1, 2, \ldots, \log_2 N$) and the net energy transfer to level α are defined by:

$$e_{\iota}^{(\alpha,\mu)} = \frac{1}{2}\{\mathcal{W}_{\iota}^{(\alpha,\mu)}[u_l]\}^2, \tag{2}$$

$$t_{\iota}^{(\alpha,\mu)} = -\mathcal{W}_{\iota}^{(\alpha,\mu)}[u_l]\mathcal{W}_{\iota}^{(\alpha,\mu)}[u_j\partial_j u_l + \partial_l p]. \tag{3}$$

Here, u_l ($l = 1, 2, 3$) is the l-th component of velocity, p is the pressure and $\partial_j = \partial/\partial x_j$, respectively. The summation convention is used for repeated indices but not for the Greek indices. $e_{\iota}^{(\alpha,\mu)}$ and $t_{\iota}^{(\alpha,\mu)}$ are counterparts of the local energy and the net energy transfer proposed by Meneveau [2]. We

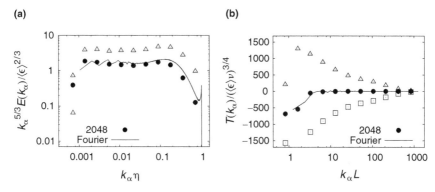

Fig. 1. Compensated dual spectra of (a) the local energy and (b) the net energy transfer for the 2048^3 DNS. ——, Fourier spectra; •, mean spectra; △, mean plus one standard deviation; □, mean minus one standard deviation. L: the integral length scale, $\langle\epsilon\rangle$: the mean energy dissipation rate per unit mass, ν: the kinematic viscosity

take no summation over μ. Mean wavelet spectra $E(k_\alpha)$ and $T(k_\alpha)$ are written as $E(k_\alpha) = N_\alpha \langle e_l^{(\alpha,\mu)} \rangle / \Delta k_\alpha$ and $T(k_\alpha) = N_\alpha \langle t_l^{(\alpha,\mu)} \rangle / \Delta k_\alpha$, respectively. $\langle \cdot \rangle$ denotes the mean value of $\cdot$ at each level. $N_\alpha = 7 \cdot 2^{3(\alpha-1)}$. The fluctuations of $e_l^{(\alpha,\mu)}$ and $t_l^{(\alpha,\mu)}$ are measured by their standard deviation spectra defined by:

$$\sigma_e(k_\alpha) = N_\alpha \sqrt{\left\langle \{e_l^{(\alpha,\mu)}\}^2 \right\rangle - \left\langle e_l^{(\alpha,\mu)} \right\rangle^2} / \Delta k_\alpha, \quad (4)$$

$$\sigma_t(k_\alpha) = N_\alpha \sqrt{\left\langle \{t_l^{(\alpha,\mu)}\}^2 \right\rangle - \left\langle t_l^{(\alpha,\mu)} \right\rangle^2} / \Delta k_\alpha, \quad (5)$$

respectively. Here, $\Delta k_\alpha = (k_{\alpha+1} - k_\alpha) \ln 2$ and $k_\alpha = 2^\alpha / 3$.

Figure 1 shows compensated dual spectra of the local energy and the net energy transfer as well as their Fourier spectra in the 2048^3 DNS. Dual spectra consist of wavelet mean spectrum and the mean plus and minus one standard deviation spectra at each level. The mean spectra $E(k_m)$ and $T(k_m)$ are in good agreement with the Fourier spectra, respectively. The compensated standard deviation spectra of the local energy and the net energy transfer are shown in Fig. 2. We find simple scalings such as $\sigma_e \propto k_\alpha^{-5/3}$ and $\sigma_t \propto k_\alpha^{-1/2}$ for the 1024^3 and 2048^3 DNSs in the inertial subrange, though the scalings are not observed for the 256^3 and 512^3 DNSs.

3 Conclusions

We studied wavelet statistics of the local energy at each level and the net energy transfer to each level through three-dimensional multiresolution analysis of the datasets of incompressible homogenous isotropic turbulence with

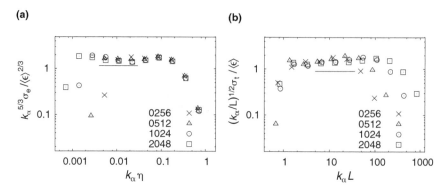

Fig. 2. Compensated standard deviation spectra for (a) the local energy and (b) the net energy transfer to level α for the 256^3, 512^3, 1024^3, and 2048^3 DNSs

$R_\lambda = 167, 257, 471$ and 732. The data for $R_\lambda = 471$ and 732 show that the standard deviation spectra of the local energy at each level σ_e and the net energy transfer to each level σ_t increase with k_α approximately as $\sigma_e \propto k_\alpha^{-5/3}$ and $\sigma_t \propto k_\alpha^{-1/2}$ in the inertial subrange, respectively.

Acknowledgment

The computations were carried out on HPC2500 system at the Information Technology Center of Nagoya University. The authors would like to express their thanks to Drs. T. Ishihara, M. Yokokawa, K. Itakura, and A. Uno for providing us with the DNS data. They are also grateful to Professors M. Farge and K. Schneider for valuable discussions on wavelets and their applications to turbulence and providing us with their wavelet source code for single processor. This work was supported by a Grant-in-Aid for the 21st Century COE "Frontiers of Computational Science" and also by Grant-in-Aids for Scientific Research (B)17340117 from the Japan Society for the Promotion of Science.

References

1. Yamada M, Ohkitani K (1991) Fluid Dyn Res 8:101–115
2. Meneveau C (1991) J Fluid Mech 232:469–520
3. Farge M (1992) Ann Rev Fluid Mech 24:395–457
4. Addison PS (2002) The illustrated wavelet transform handbook. Institute of Physics Publishing, Bristol and Philadelphia
5. Kaneda Y, Ishihara T, Yokokawa M, Itakura K, Uno A (2003) Phys Fluids 15:L21–L24
6. Yokokawa M, Itakura K, Uno A, Ishihara T, Kaneda Y (2002) Proc. IEEE/AVM SC2002 Conf., Baltimore, 2002; http://www.sc-2002.org/paperpdfs/pap.pap273.pdf
7. Daubchies I (1992) Ten lectuers on wavelets. SIAM, Philadelphia

Hydrophobic Hydration on Nanometer Length Scale

Takeshi Hotta and Masaki Sasai

Summary. Hydration structure around nanometer-size hydrophobic solutes is studied with molecular dynamics (MD) simulation by taking aqueous solution of single walled carbon nanotube (SWCN) as an example. We found a large amplitude density fluctuation in the narrow region between a pair of SWCNs. This density fluctuation strongly correlates with the mean force between SWCNs. For a triplet of SWCNs, water evaporates from the narrow region pinched by SWCNs and enhances the hydrophobic attraction among them.

1 Introduction

Hydrophobic effects have been intensively studied because of their central roles in chemistry and biology. Water molecules around a small hydrophobic solute strengthen the hydrogen bond (HB) network surrounding the solute, which has been called "iceberg-like cage" [1], and the gain of entropy in overlapping cages has been regarded as the origin of hydrophobic attraction between solutes [2]. On the other hand, it has been recently recognized that water molecules are repelled from the large hydrophobic object such as a graphite plate [3], leading to drying around the hydrophobic surface. Chandler and his colleagues suggested that the crossover from caging to drying should take place on the nanometer length scale [4]. In order to examine this crossover regime, we have studied hydration around nanometer spheres, C_{60} and $C_{60}H_{60}$, with MD simulation [5] and shown that there exist both fluctuating caging and drying hydration around these solutes. In this study we investigate hydration around another type of nanometer scale hydrophobic solute, SWCN: We may expect that the hydration structure around this cylindrical hydrophobic solute is different from that around spheres.

2 Method

2.1 Molecular Dynamics Simulation

Three systems were simulated; System I consisted of a single SWCN and 898 water molecules, System II consisted of a pair of SWCNs and 795 water molecules, and System III consisted of a triplet of SWCNs and 697 water molecules. All three systems were confined in a cubic box with the side length, $L_{\text{box}} = 31.064\,\text{Å}$, with the periodic boundary condition. We treated SWCNs as rigid bodies by fixing their positions. The orientation of each SWCN was fixed to the direction parallel to a ridge of the simulation box. For each of three systems, after short minimization of water configuration, we conducted MD simulations at 498 K for 1 ns duration. After equilibrating these configurations at 298 K, data were sampled from trajectories of 100 ps length. In this way we obtained 10 trajectories for each of the three systems.

3 Results

3.1 Hydrophobic Hydration Around a Single SWCN

We calculated the radial distribution function around a single SWCN in System I. The function has the first peak at $r = 6.6\,\text{Å}$ and the second peak at $r = 9.3\,\text{Å}$ (No Figure), where r is distance from the center of SWCN. The water density ρ_{hyd} in the hydration layer, which is in the range of $5.7\,\text{Å} < r < 11.6\,\text{Å}$, is $1.021\,\text{g/cm}^3$. This value is larger than the average density in this system, $\rho_{\text{av}} = 1.000\,\text{g/cm}^3$. Therefore, we should say there is no evidence of drying around a single SWCN.

3.2 Hydrophobic Interaction Between a Pair of SWCNs and Among a Triplet of SWCNs

We studied hydrophobic interaction in System II and System III by calculating potential of mean force (PMF). In System III three SWCNs were arranged on an isosceles triangle. Distance between centers of two SWCNs was fixed to be $10.50\,\text{Å}$ and the position of the third SWCN was varied to calculate PMF. PMFs, $V(R)$, are shown in Fig. 1, where R is distance between surfaces of SWCNs. $V_{\text{hyd}}(R)$ is small at $R \approx R_{\text{contact}} \approx 3\,\text{Å}$, so that $V(R_{\text{contact}}) \approx V_{\text{LJ}}(R_{\text{contact}})$. In the region $R_{\text{contact}} < R < R_{\text{separated}} \approx 7\,\text{Å}$, $V_{\text{hyd}}(R)$ is significant. For a pair of SWCNs, $V_{\text{hyd}}(R)$ brings about the free energy barrier which separates R_{contact} and $R_{\text{separated}}$. For a triplet of SWCNs, however, the global attractive gradient in $V_{\text{hyd}}(R)$ dominates and the free energy barrier between R_{contact} and $R_{\text{separated}}$ is small. We show in Fig. 2 the

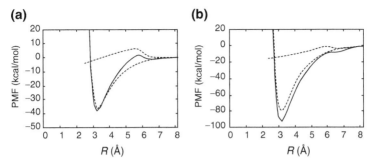

Fig. 1. PMFs. (**a**) PMF between a pair and (**b**) among a triplet of SWCNs are shown by *solid lines*. Contribution from the Lennard-Jones potential to PMF, $V_{\mathrm{LJ}}(R)$, is shown by *dotted lines*. The solvent contribution to PMF, $V_{\mathrm{hyd}}(R)$, is shown by *dashed lines*

correlation between $\mathbf{F}(R)$ and ρ_{int} obtained from 10 trajectories. Here, ρ_{int} is water density averaged over the interstitial region between solutes, and $\mathbf{F}(R)$ is the force averaged over 100 ps acting between solutes. $\mathbf{F}(R)$ is defined to be positive when attractive.

Widely scattered points in Fig. 2 show that there is large trajectory-to-trajectory fluctuation in density and force, implying that the free energy barrier between different hydration structures is too large to be surmounted within 100 ps. $\mathbf{F}(R)$ anticorrelates with ρ_{int}. Thus, density fluctuation between solutes critically affects the hydrophobic interaction. Profiles of density are shown in Fig. 3. For a triplet of SWCNs, when the third SWCN is moved toward a fixed pair of SWCNs, the interstitial region among three SWCNs is dried preceding the contact of SWCNs is made. This precedence of drying lowers the barrier of PMF which separates R_{contact} and $R_{\mathrm{separated}}$ as shown in Fig. 1b.

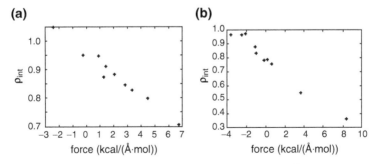

Fig. 2. Correlation between $\mathbf{F}(R)$ and ρ_{int} with $R = 5.7\,\text{Å}$. Each point is the average over each MD trajectory of 100 ps length. (**a**) A pair and (**b**) a triplet of SWCNs

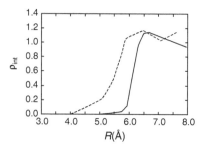

Fig. 3. Density profiles, ρ_{int}, (**a**) between a pair of SWCNs (*dashed line*), and (**b**) among a triplet of SWCNs (*solid line*)

4 Discussion

For a pair of SWCNs, the narrow interstitial region pinched by SWCNs has two sides open toward the bulk space. Water molecules in the interstitial region can make HBs with molecules outside the region through these open sides. In this way the HB network penetrates the region pinched by SWCNs and water molecules in the region are stabilized. Thus, the water density is not much reduced in average and the hydration gives a repulsive force between a pair of SWCNs when solutes approach each other from $R \approx R_{\text{separated}}$. The attractive force acts only when the region has a smaller number of water molecules in density fluctuation. For a triplet of SWCNs, water molecules among solutes are surrounded by three hydrophobic walls. The region does not have an open side, so that water molecules in the region can make at most a single HB with molecules outside the region. Water molecules in the region are not much stabilized and tend to be expelled from the region. This is drying expected around the large hydrophobic surface, which gives the attractive interaction among solutes. In the present simulation we showed that not only the solute size but also the solute configuration are critically important to determine whether the drying takes place around the nanometer size solutes. Cooperative formation and destruction of the HB network among water molecules dominate the cooperative change of hydration between wetting and drying, and thus decisively affect the hydrophobic interaction between nanometer scale solutes.

References

1. Frank HS, Evans MW (1945) J. Chem. Phys. 13: 507–532
2. Lazaridis T (2001) Account. Chem. Res. 34: 931–937
3. Stillinger FH (1973) J. Solution Chem. 2: 141–158
4. Lum K, Chandler D, Weeks JD (1999) J. Phys. Chem. B 103: 4570–4577
5. Hotta T, Kimura A, Sasai M (2005) J. Phys. Chem. B 109: 18600–18608

Ultra-Fast Genome Wide Simulation of Biological Signal Transduction Networks: Starpack

Tetsuya Maeshiro, Hitoshi Hemmi, and Katsunori Shimohara

Summary. We developed a computer system that simulates signal transduction networks in biological cells, for instance biochemical reactions and gene networks. The simulation speed is at least ten thousand times faster than software-based simulations executed on conventional computers. The system, named Starpack, simulates any model that can be described as a network (entities and relationships among them). Simulation of a simple cell model indicates clear superiority of Starpack over conventional systems.

1 Introduction

We have developed a dedicated computer system, named Starpack (signal transduction advanced research PACKage), to simulate and predict signal transduction networks, for instance biochemical reactions and gene networks, in biological cells.

Computer simulations of biological phenomena are valuable for system level understanding of biological systems described at molecular level. Drug screening is another application. Simulations of the whole network are important because interactions of individual components are ultimately responsible for an organism's form and functions. Several systems have been proposed for simulation, for instance E-Cell [1], but they lack the execution speed required to simulate real biological organisms, which has a large number of elements and complex interactions among elements. For instance, a human cell has about 30,000 types of genes and these genes produce substances that participate in a signal transduction network composed of approximately hundred thousand types of elements. Conventional simulation systems are software-based, and even running on cluster or grid systems consisting of one thousand processors, completion of simulations in a practical time is unfeasible.

Fig. 1. Pen display GUI showing microarray data window (left) and signal transduction network window (lower right)

2 Starpack System

Starpack has a novel architecture that is completely different from ordinary computers, designed specifically to simulate models described as large networks. The simulation model consists of quantities associated with elements, and interactions among elements that govern modifications of elements' quantities. Starpack has several thousands of processors, and is scalable. Each processor stores the quantity of assigned elements, and interactions among elements are simulated by modifying the values that correspond to elements' (substances') quantities.

Starpack provides a comprehensive visualization of signal transduction networks under simulation, and its pen display GUI allows intuitive and direct manipulation of simulated networks and simulation parameters (Fig. 1). To run a simulation, the user specifies substances, signal transductions (interactions among substances), and initial quantities of substances. Microarray experiment data can also be used as initial values, allowing direct input of microarray experiment results. Signal transduction network is visualized as a network whose topology is optimized according to a spring model based on Fruchterman-Reingold model [2]. During simulations, quantities of substances are displayed in real time as (1) time course quantity curve, and (2) microarray data. While the time course curve is useful to grasp quantity variations, the latter visualization method is useful for direct comparison with microarray experiment results. It is also possible to show multiple microarray displays in sequence. Moreover, microarray display is a collection of squares representing the substances under simulation, and by clicking a square in the microarray window, the system shows detailed information of the substance such as the substance name, biochemical properties, and participating signal transductions.

3 Simulation

Current configuration of Starpack is optimized to simulate two types of biological phenomena, both related to the inference of biological network structures [3] (1) detailed biochemical reactions, modeled from reaction kinematics; and (2) gene interaction networks. The first category represents biochemical reactions such as the citric acid cycle that is present in almost all organisms. Simulations of biochemical reactions focus mainly on variations of substances (elements) that participate in reactions. On the other hand, simulation of gene interaction networks, the second category, hides biochemical details and focus on how genes are functionally interconnected. For instance, some genes are periodically activated and suppressed following circadian rhythms [4], and one is interested in which genes trigger or silence other genes. The two categories request different types of computational speed. The first category, detailed biochemical simulations, requires speed to simulate reactions. However, the second category, simulations of gene interaction networks, requires the execution speed to explore large number of network configurations. Furthermore, while only the genes are treated in simulations of gene networks, simulations of biochemical reactions require additional substances that participate in reactions.

As a preliminary test of simulation of biochemical reactions, a simple model generated from human erythrocyte was used, where 142 substances participate in 45 reactions [1,5]. Although smaller than real biological reaction networks, this network was selected for benchmarking purposes. Same model was simulated on a software-based simulation system (E-Cell [1]) to compare the simulation speed and precision. Starpack was one million times faster than the software-based simulation for the same simulation precision (Table 1). Although E-Cell was run on a single PC, even a cluster system with thousand PCs cannot fill the gap of six orders of magnitude, albeit the problem of interprocess communication latency. An advantage of Starpack is the ability to simulate with the precision of real reactions, which is practically impossible with software-based simulation because of unrealistic computation time that is required.

Starpack is faster by orders of magnitude than software-based simulations, mainly due to a completely different simulation methodology. Conventional simulation systems model biochemical reactions as differential equations and solve the equations by numerical methods, which require enormous computational time. On the other hand, Starpack models directly the biochemical

Table 1. Performance of Starpack compared to software simulation (ecell) and real reaction. N.A. means not available, as E-Cell could not simulate with the precision of real reaction. E-Cell was run on a Windows PC with Pentium4HT 3GHz CPU

precision (time)	vs. E-Cell	vs. real reaction
0.1 sec	1 million	1 million
0.01 sec	1 million	0.1 million
same as real reaction	N.A.	1,000

reactions, analogous to the first principle modeling of physical phenomena. This results in high speed, further accelerated by hardware encoding implementation. So a plausible strategy for the prediction of signal transduction networks is (1) execute low precision but fast simulations to explore parameter spaces; then (2) execute high precision simulation to improve the accuracy.

Primary input data for the second category of simulations, the simulation of gene interaction networks, is the microarray experiment data. Microarray experiments provide analysis of the whole organism, suitable for the objective of Starpack, the system level simulation of organisms. Microarray experiments generate time course variation of gene quantities, denoted as expression patterns. Then activation relationships, describing which gene triggers or suppresses which genes, should be estimated. Each activation pattern results in a different network structure. The clustering algorithm is a widely used method to analyze microarray data, which groups genes based on similarity of expression patterns. However, the clustering algorithm clarifies only the binary relationships among genes. On the other hand, due to its high execution speed, Starpack explores possible activation combinations interactions among genes including quantitative relationships.

4 Conclusions

The preliminary simulation of biochemical reactions indicates clear advantage of Starpack over conventional systems, as Starpack is one million times faster than the software simulation for the same simulation precision. Improvement of execution speed by orders of magnitude considerably reduces the turn-around-time of simulation experiments, and opens possibilities for new research methodologies, as numerous parameter sets can be tested simultaneously. This advantage becomes larger for real networks found in biological organisms, which have more substances and more complex signal transduction mechanisms. Moreover, Starpack is also orders magnitudes faster than actual biochemical reactions, thus suitable to accomplish "virtual biological experiments." The speedup is significant for simulations of slow growing organisms such as plants.

References

1. Takahashi K et al. (2003) Bioinformatics, 19:1727–1729
2. Fruchterman TMJ, Reingold EM (1991) Software – Practice and Experience, 21:1129–1164
3. Carter GW (2005) Briefings in Bioinformatics, 6:380–389
4. Ueda H et al. (2002) Nature, 418:534–539
5. Joshi A, Palsson BO (1989) Journal of Theoretical Biology, 141:515–528

Molecular Dynamics Study on Interaction of C-Component Dislocation Loop and Edge Dislocation in α-Zirconium

Kenji Yamaura, Akiyuki Takahashi, and Masanori Kikuchi

Summary. Interactions between an edge dislocation and a c-component dislocation loop are investigated using the molecular dynamics method. The remote interactions work as an obstacle to the dislocation motion, and slightly depend on the position of the slip plane of the dislocation. On the other hand, in the direct interactions, they are combined to form a new dislocation structure.

1 Introduction

During operation of nuclear power plant, neutron irradiation gives a significant damage to the materials used in the reactor pressure vessels (RPV). Therefore, there is a strong demand for understanding the mechanism to ensure the lifetime and long-term strength of the material. The fuel cladding tube is a structure in RPV. The behavior of the material under neutron irradiation must be clarified for expanding the time for the use of the material. The material of the fuel cladding tube is zirconium alloy, and has hexagonal close packing (HCP) crystal structure. Experimental studies have proven that a vacancy type of dislocation loop, say c-component dislocation loop, is formed under long-term irradiation [1]. Thus, in order to understand the strength change in the material under long-term neutron irradiation, it is primarily of interest to understand the interaction between a c-component dislocation loop and a mobile dislocation.

In this study, the interaction between a c-component dislocation loop and an edge dislocation is simulated using the molecular dynamics (MD) method, and the mechanism in atomic scale is investigated.

2 Simulation Method

In the MD simulation, Finnis-Sinclair type of interatomic potential for α-zirconium [2,3] is used. Figure 1 shows the simulation volume used in this

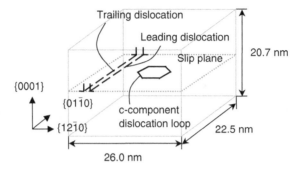

Fig. 1. Crystal model for MD simulation

study. The size of the volume is $26.0 \times 20.7 \times 22.5$ nm, and the total number of atoms is about 500,000. The xyz axes coincide with crystal orientation of $\{12\bar{1}0\}$, $\{01\bar{1}0\}$, and $\{0001\}$. A periodic boundary condition is applied to $\{12\bar{1}0\}$ and $\{01\bar{1}0\}$ direction, and the atoms near two edge surfaces in $\{0001\}$ direction can move only in (0001) plane, as well as acceleration equivalent to a shear stress of 200 MPa is given to the atoms. The temperature is controlled to 300 K by performing NVT ensemble MD before applying the shear stress. One edge dislocation is introduced into the volume by replacing an extra half atomic plane, and c-component dislocation loop is made by replacing atoms in platelet hexagonal shape.

3 Remote Interaction

First, we studied a remote interaction between a c-component dislocation loop and an edge dislocation. The snapshots of the MD simulation are shown in Figs. 2 and 3. At first, the edge dislocation spontaneously splits into two partial dislocations, leading and trailing partial dislocation. The partial dislocations have Burgers vectors of $a_0/3\ [10\bar{1}0]$ and $a_0/3\ [1\bar{1}00]$, respectively, where a_0 is the lattice constant. The Burgers vector of the dislocation loop is $a_0/6\ [\bar{2}023]$. When the edge dislocation is on the upper slip plane of the dislocation loop, the edge dislocation feels a strong resistance from the front of the dislocation loop. On the other hand, when the edge dislocation is on the lower slip plane

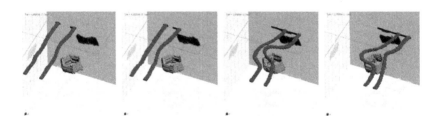

Fig. 2. The c-component dislocation loop is in tension side

Fig. 3. The c-component dislocation loop is in compression side

of the dislocation loop, the dislocation stops the motion at the center of the dislocation loop. In order to understand the interactions quantitatively, an elastic analysis of the interaction is performed. In the elastic analysis, the stress field of the dislocation loop is calculated with the Green function [4] with the elastic anisotropy, and then, the distribution of force acting on a straight edge dislocation in anywhere is calculated using the Peach–Koehler formula. The resutls are shown in Fig. 4. In the figure, the gray area in the figure is the position of the dislocation loop. In the figure, the edge dislocation on the upper slip plane to the dislocation loop is attractive to the front of the dislocation loop. As the edge dislocation glides further, the edge dislocation starts to feel a resistance from the dislocation loop. On the other hand, the edge dislocation on the lower slip plane to the dislocation loop, firstly, feels a resistance from the front of the dislocation loop. At the center of the dislocation loop, the sign of the force is turned to positive, which means that the edge dislocation prefers to glide further. These tendencies are qualitatively in good agreement with the results of the MD simulation.

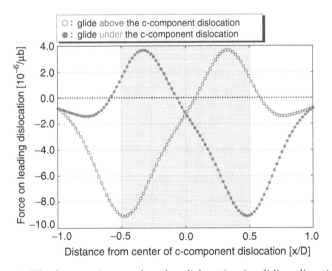

Fig. 4. The force acting on the edge dislocation in gliding direction

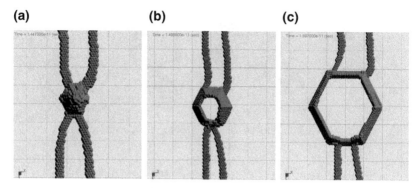

Fig. 5. Direct interaction of the dislocation loop and edge dislocation. The diameter of the loops is (**a**) 3.2 nm, (**b**) 6.5 nm, and (**c**) 13.0 nm

4 Direct Interaction

A direct interaction between a c-component dislocation loop and an edge dislocation on the same slip plane is also studied. As shown in Fig. 5, the edge dislocation is combined with the dislocation loop by the interaction. This interaction is analyzed by simple Burgers vector operations. By combining the leading partial dislocation and the dislocation loop, the Burgers vector of the combined part is turned into $a_0/6$ [0003]. The energy of the dislocation can be roughly estimated by assuming that the energy is proportional to square of the magnitude of Burgers vector. Due to the fact, the energy after the interaction is lowered than that before the interaction. Following the interaction, the trailing partial dislocation interacts with the dislocation loop, and the Burgers vector at combined part is again turned into $a_0/6$ [2$\bar{2}$03]. This interaction increases the total energy of the system. However, the energies before and after the interaction of leading and trailing parital dislocations are the same, and the trailing partial dislocation prefers to follow the leading partial dislocation to narrow the stacking fault between the partial dislocations.

5 Conclusions

Interactions between a c-component dislocation loop and an edge dislocation have been studied using MD simulations. If the dislocation loop and the edge dislocation are not on the same plane, the dislocation loop works as an obstacle to the edge dislocation motion. And the effect could be well explained with the elastic analysis. On the other hand, if the dislocation loop and the edge dislocation are on the same plane, the dislocation loop combines with the edge dislocation. The behavior could be explained with simple Burgers vector operations.

References

1. M.Griffiths, J. Nucl. Mater, 1988, 159, 190–218
2. G.J.Ackland, S.J.Wooding, D.J.Bacon, Phil. Mag. A, 1995, 71 (3), 553–565
3. U.Pinsook, G.J.Ackland, Phys. Rev. B, 1998, 58, 11252–11257
4. T.Mura, Micromechanics of defects in solids, 1987, Martinus Nijhoff

Theoretical Analysis of Intramolecular Interaction

Kenta Yamada and Nobuaki Koga

Summary. Ethane molecule, as one of the simplest molecules, interests us in intramolecular interaction in it. What does the energy difference about 3 kcal/mol between staggered and eclipsed conformations result from? In this paper, to answer this question, we applied the mBLW method, which we have developed, to an analysis of intramolecular interaction of this molecule, to find that the larger exchange repulsion in eclipsed conformer causes staggered conformer to be relatively stable.

1 Introduction

Conformations in ethane molecule that have often been argued are the staggered (SC) and eclipsed conformation (EC). While the dihedral angle between vicinal H atoms in SC is 60°, that in EC is 0°. It is well known that SC and EC are the equilibrium and transition structure, respectively, the former being about $3\,\mathrm{kcal\,mol^{-1}}$ more stable than the latter. Although ethane molecule is simple, the knowledge about the rotation around a C–C single bond of ethane may give us some fruitful insights about the structural characteristics of larger molecules such as proteins. Therefore we have an interest in what makes SC more stable than EC, and then we studied this in aspect with intramolecular interaction.

1.1 Intramolecular Interaction in Ethane

Intramolecular interaction in ethane that favors SC has been discussed many times, and two ways of explanation have so far been proposed [1–4]. One of them is the charge transfer (CT) interaction that could cause more stability in SC, and the other is the exchange repulsion (EX) interaction which would result in less stability in EC. The CT interaction is partial electron migration from occupied molecular orbitals (MOs) to vacant MOs, in this case from $\sigma_{\mathrm{C-H}}$ to its vicinal $\sigma^*_{\mathrm{C-H}}$. Due to the CT interaction, electron distribution in the region between donor and acceptor parts in one molecule (bonding region) increases. Consequently the CT interaction gives the system stability. The EX interaction arises from the Pauli exclusion principle. Because of this principle, electron distribution in the bonding region decreases and the distribution in the outside bonding region (antibonding region) increases. As a result, the EX interaction would destabilize the system. In order to understand which of the two interactions dominates, we analyzed intramolecular interaction in ethane using the modified Block-Locallized Wavefunction (mBLW) method we have developed.

2 Theoretical Concepts

Interaction energy (ΔE_{INT}) can be calculated theoretically as a difference in energy between systems with and without a specified interaction (1).

$$\Delta E_{INT} = E_{INT} - E_{noINT} \qquad (1)$$

The BLW and mBLW method give us the energy of a unique reference state without a specific interaction (E_{noINT}), whereas that with the interaction (E_{INT}) is obtained by a standard MO method [5]. Although the original BLW method can only be applied to an analysis of π-interaction [6], we have modified this method to analyze successfully σ-interaction [7]. The characteristics of the mBLW method are the following:

- It directly gives the energy of a reference state without an interaction, and the structure of a molecule in the reference state can be optimized.
- It is not necessary to break bonds in analyzing intramolecular interaction.
- Occupied MOs in a reference state (mBL-MOs) are obtained.
- Specifying vacant MOs with which occupied MOs interact is not needed.
- Any kinds of interaction may be analyzed.

3 Results and Discussions

3.1 Charge Transfer (CT) Interaction

Under the Hartree–Fock (HF) approximation, we calculated a reference state energy by the mBLW method to evaluate the interaction energy due to the CT interaction. The CT interaction energies thus obtained are shown in Table 1. Those by the natural-bond-orbital-with-deletion (NBO-d) method that has been well applied to an analysis of intramolecular interactions are also listed in it [8]. The NBO-d method often overestimates the interaction energy because the reference state wavefunction by this method is not relaxed. Accordingly, although the CT interaction energies for the conformers based on the mBLW calculation shows the same pattern as those evaluated by the NBO-d method, the latter values are about two times larger than the former values.

Table 1. CT interaction energies by the mBLW and NBO-d method (in kcal mol^{-1})

conformation	ΔE_{CT}	
	mBLW	NBO-d
SC	−12.34	−29.02
EC	−11.61	−23.60
EC′[a]	−12.08	−24.75

[a] EC′ is the conformation which rotates one methyl group by 60° around the C–C bond, fixing the other structural parameters of SC.

And this accounts for the larger difference in the NBO-d interaction energy between SC and EC (5.42 kcal mol^{-1}) than that in the mBLW interaction energy (0.73 kcal mol^{-1}). The same is true in the difference between SC and EC′ (0.26 vs. 4.27 kcal mol^{-1}). That is, the NBO-d study may lead to misdirected results in analysis because of the overestimation of interaction energy.

3.2 mBLW Optimized Structure

The structural parameters of ethane optimized by the mBLW and HF method are shown in Table 2. In the mBLW calculations, the CT interaction is forced not to occur.

While the C–C and C–H bond length in the mBLW structure are longer and shorter, respectively, than those in the HF structure, ∠ C–C–H increases a little upon deleting the CT interaction. These structural characteristics are as expected when the CT interaction is deactivated and rationalize the use of the mBLW method in analysis of intramolecular interaction.

3.3 Exchange Repulsion (EX) Interaction

Using this mBLW structure, the EX interaction was analyzed (Table 3). Since EC is different from SC in the C–C bond length and ∠ C–C–H as well as in

Table 2. Structural parameters optimized by the mBLW and HF method

conformation	C–C length[a]		C–H length[a]		∠ C–C–H[b]	
	mBLW	HF	mBLW	HF	mBLW	HF
SC	1.563	1.527	1.083	1.086	111.3	111.2
EC	1.574	1.541	1.082	1.085	111.7	111.7

[a] In Å.
[b] In degree.

Table 3. Energies of the states with and without the EX interaction and the EX interaction energies by the mBLW method (in kcal mol^{-1})

conformation	$E_{\text{noEX}}{}^a$	$E_{\text{EX}}{}^a$	ΔE_{EX}
SC	2.76	131.43	128.67
EC′	0.0	133.78	133.78

^{a}These energies are relative to E_{noEX} in EC′.

the conformation, we used EC′ instead of EC to analyze the difference only in the conformation.

As shown in Table 3, EC′ has larger ΔE_{EX} (133.78 kcal mol^{-1}) than SC in line with chemical intuition.

The HF calculations including all the interactions show that SC is 3.17 kcal/mol more stable than EC′, and the above analyses represent that the larger stability of SC is ascribed to the 5.11(= 133.78 − 128.67) kcal mol^{-1} smaller EX interaction as well as the 0.26 kcal mol^{-1} larger CT interaction. It follows from these results that the EX interaction is the main factor to determine the stability difference between SC and EC. The reason why EC′ is more stable than SC without the EX interaction would be because the distance in EC′ between the nearest-vicinal H atoms is about 0.2 Å smaller than that in SC and thereby the greater electron–nucleus attractive force exists between vicinal H atoms in EC′ without the EX interaction.

4 Conclusions

Our mBLW method can give a fictitious wavefunction by optimizing it with a specified interaction deleted. The state thus defined represents an artificial state without the interaction and therefore can be used as a reference state in evaluating an intramolecular interaction including σ-interaction. Accordingly, the analysis based on the mBLW method is expected to give reasonable results. Moreover, since the mBLW optimized structure well shows the structural behavior expected from the theoretical consideration, we believe that mBLW method is appropriate for an analysis of intramolecular interaction. And, in the analysis of intramolecular interaction in ethane, the results based on the methods led us to the conclusion that the EX interaction makes EC unstable and that thereby SC is relatively stable.

References

1. Weinhold F (2003) Angew Chem Int Ed 42:4188–4194
2. Pophristic V, Goodman L (2001) Nature 411:565–568

3. Mo Y, Wu W, Song L, Lin M, Zhang Q, Gao J (2004) Angew Chem Int Ed 43:1986–1990
4. Bickelhaupt FM, Baerends EJ (2003) Angew Chem Int Ed 42:4183–4188
5. Gaussian 98 Revision A.7, Frisch MJ et al. (1998) Gaussian, Pittsburgh PA
6. Mo Y, Peyerimhoff SD (1998) J Chem Phys 109:1687–1697
7. Yamada K, Koga N to be published
8. Foster JP, Weinhold F (1980) J Am Chem Soc 102:7211–7218

Effect of Nanoscale Structure of Solid Surface on Moving Droplet

Takuma Osawa and Ichiro Ueno

Summary. Effects of nanoscale roughness upon a moving droplet on a solid surface are investigated through classical molecular dynamics simulation. Droplet of argon liquid with 12-6 Lennard-Jones potential is driven by gravitational force applied to a direction parallel to the solid surface. Droplet behavior in traveling on the solid with single step molecule depression or double lines of solid molecule perpendicular to the direction of the gravity is examined.

1 Introduction

Wetting and dewetting of the solid material by the liquid can be observed in many natural phenomena and technological applications; such as falling rain droplet on the leaf, boiling phenomena, and thin film coating. These processes are accompanied with the movement of the boundary line of three phases; solid–liquid–gas interface, that is so-called "contact line (CL hereafter)." In the preceding theoretical and numerical work [1–4], assumptions in the region near CL have been commonly applied; there assumed to exist thin precursor film of constant thickness in front of the advancing CL, and the liquid would transverse above the thin film without changing its profile. Also, simulations of droplet that move in a variety of driving powers have been performed with molecular dynamics method [5, 6]. Recently an experimental work [7] indicated there exists the precursor film of nonconstant thickness ahead a droplet advancing on the solid surface. Insufficient knowledge, however, has been accumulated in the dynamics in this region. The authors focus upon the dynamics in the vicinity of moving CL by applying molecular dynamics simulation.

2 Numerical Procedure

Target geometries in the present study are shown in Fig. 1. Role-shaped droplet of liquid molecules is located on solid layer. The droplet is driven

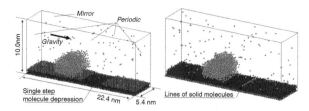

Fig. 1. Target geometries; single-molecule depression (*left*) and double lines

by gravitational force applied to a direction parallel to the solid surface. Various kinds of solid substrates are examined; a flat surface without any structure, and the surfaces with two different kinds of roughness patterns; (1) a fully span-wise single-step-molecule depression and (2) double line structure of solid molecules located perpendicular to the direction of the gravity are examined. Several kinds of gravitational forces in the range of 10^6–10^{11} m s^{-2} are applied in a preliminary calculations to check the droplet movement. Mirror boundary condition at the top and periodic boundary conditions in four sides are employed.

The Lennard-Jones molecules with potential cutoff at 3.5σ are employed. The potential function is described as $\phi(r) = 4\varepsilon\{(\sigma/r)^{12} - (\sigma/r)^6\}$, where σ is the length scale and ε energy scale. Set of parameters in the present study is listed in Table 1. We basically assume the liquid droplet as argon, and the solid as platinum for the sake of physical understanding.

3 Result and Discussion

When the droplet is advancing on the smooth solid surface, a frictional force occurs near the interface between the droplet and the solid. Averaged density profile shows layered structure formed in a range of the height upto 3.5σ from the solid surface. The layered structure was indicated in the preceding works without gravitational force such as [3]. Through the variation of the position of the droplet's center of mass driven by a variety of gravitational forces, it is found that the droplet moves at the constant speed on the smooth surface in the case of $g = 10^{11}$ m s^{-2}. The averaged value of the gravitational force is $m_{Ar} \times N_{Ar} \times g = 9.3$ pN, and the averaged frictional forces in the lateral direction is 10.7 pN, where m_{Ar} is the mass of the molecule consisting of solid

Table 1. Calculation conditions

	m (kg)	ε (J)	σ (m)
liquid	6.63×10^{-24}	1.67×10^{-21}	3.40×10^{-10}
solid	32.4×10^{-26}	62.84×10^{-21}	2.77×10^{-10}

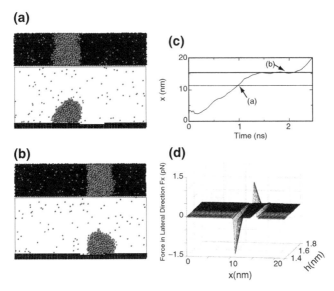

Fig. 2. Snapshots of the advancing droplet on the solid surface with depression (**a** and **b**), (**c**) time series of center-of-mass movement, and (**d**) spatial distribution of lateral force from solid surface with depression

base, N_{Ar} the number of the liquid molecule. This brings the droplet advancing on the solid surface at a constant velocity. We discuss the droplet behavior under this value of the gravity as the external body force in the latter cases with/without roughness.

Typical example of (a) top- and (b) side-view snapshots of the droplet, (c) time series of center-of-mass movement of the droplet in the case of single-step-molecule depression is shown in Fig. 2. And Fig. 2(d) indicates a spatial distribution of the lateral force attracting a liquid molecules at h in height from the solid surface. Positive value corresponds to the force acting toward the same direction as the gravity. When the droplet reaches the inlet of the depression (the point (a) as shown in Fig. 2), the bottom part of the droplet is exposed to lateral force in the negative direction of x from the solid surface. The whole droplet, at the same time, is driven forward by the gravitational force. The droplet thus forms a forward-bending shape; the advancing contact angle becomes larger. When the droplet is about to step over this depression structure at the point (b), on the other hand, the bottom part of the droplet is exposed to an additional positive force. It seems that these forces enable the droplet to step over the single-molecule step. The droplet struggles to step over the structure, however, because of the stiff layered structure in the bottom part of the droplet itself.

In the case of the double lines on the solid surface, the behavior of the droplet is significantly different; typical example of (a) top- and (b) side-view snapshots of the droplet and (c) a time series of the center-of-mass movement

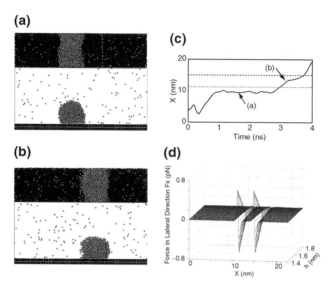

Fig. 3. Snapshots of the advancing droplet on the solid surface with two lines (**a** and **b**), (**c**) time series of center-of-mass movement, and (**d**) spatial distribution of lateral force from solid surface with two lines

is shown in Fig. 3. The droplet is stuck to the first single-line structure. After struggling to overcome the first line for a while, the droplet gets over it and travels toward the second line. As for the second-line structure, the droplet easily passes it almost without any stagnation. Spatial distribution of the lateral force to a single liquid molecule from the solid is illustrated in Fig. 3(d). The lateral force significantly affects the liquid molecules to form high-density layered structure to prevent passing the first line structure.

Difference between step molecule depression and double-line structure corresponds to difference of the spatial distribution of the lateral force from the solid surface including the nanoscale structure. These calculations indicate the significant effect of the structures of the scale of a single molecule upon the movement of the nanoscale droplet on the solid surface.

Acknowledgment

The authors gratefully acknowledge Dr. T Kimura at Kanagawa University, Dr. A Komiya at Tohoku University, Prof. S Maruyama and Dr. J Shiomi at The University of Tokyo, and Dr. S Matsumoto at JAXA for fruitful discussions and suggestions.

References

1. Troian SM, Herbolzheimer E, Safran SA (1990) Phys. Rev. Lett. 65: 333–336
2. De Coninck J, Hoorelbeke S, Valignat MP, Cazabat AM (1993) Phys. Rev. E 48: 4549–4555
3. Kataoka DE, Troian SM (1997) J. Colloid Interf. Sci. 192: 350–362
4. Matar OK, Troian SM (1999) Phys. Fluids 11: 3232–3246
5. Murata A, Mochizuki S (2003) Trans. JSME (the Japan Society of Mechanical Engineers) B 69: 1439–1444 (in Japanese)
6. Kimura T, Haramura Y (2004) Proceedings of 41st National Heat Transfer Symposium, Japan (Toyama, 2004) 1: 143–144 (in Japanese)
7. Ueno I, Komiya A, Watanabe T (2005) Proceedings 6th World Conference on Experimental Heat Transfer, Fluid Mechanics and Thermodynamics (ExHFT-6, Miyagi, 2005): CD-ROM (paper# 7-b-8(4 pages))

Proximity Effect in NS Junctions in the Presence of Anderson Impurities

Takehito Yokoyama, Yukio Tanaka, and Alexander A. Golubov

Summary. Proximity effect in the diffusive normal metal (DN)/insulator/superconductor junctions is studied in the presence of Anderson impurities in DN for various situations. The Kondo effect is approximately incorporated in the pair-breaking rate. It is shown that proximity effect is enhanced by the increase of Kondo temperature.

1 Introduction

Since the discovery of the BCS theory, proximity effect has been studied. Proximity effect is a phenomenon that Cooper pairs penetrate into the diffusive normal layer in diffusive normal metal/superconductor (DN/S) junctions, which strongly modifies the junction properties. To study the proximity effect, the quasiclassical Green's function theory [1] has been widely used because of its convenience and broad applicability. Based on this formalism, proximity effect in DN/S junctions with magnetic impurities is studied in [2]. It is shown that proximity effect is suppressed by magnetic impurities.

Although, in several works, proximity effect is studied in DN/S junctions with magnetic impurities in the quasiclassical scheme, most of them treat magnetic impurities following the Abrikosov and Gor'kov theory [3] within the Born approximation. However, beyond the Born approximation, magnetic impurities or Anderson impurities show the Kondo effect which gives the minimum of the resistance of a metal as a function of temperature [4]. So far detailed study of an interplay of Kondo effect and proximity effect has not been performed.

In the present paper we study proximity effect in DN/S junctions in the presence of Anderson impurities in DN for various situations, where we have used the Usadel equation. Density of states (DOS) is calculated by changing the magnitudes of the resistance in DN, Thouless energy and Kondo temperature. The Kondo effect is approximately incorporated in the pair-breaking rate. It is shown that the gap structure of the DOS in DN becomes deep as increasing Kondo temperature. This is because proximity effect is enhanced by the increase of Kondo temperature.

2 Formulation

We consider a junction consisting of normal and superconducting reservoirs (referred to N and S, respectively) connected by a quasi-one-dimensional DN with a length L much larger than the mean free path. The interface between the DN conductor and the S electrode has a resistance R_b while the DN/N interface has a resistance R'_b. The positions of the DN/N interface and the DN/S interface are denoted as $x = 0$ and $x = L$, respectively. We model infinitely narrow insulating barriers by the delta function $U(x) = H\Delta(x - L) + H'\Delta(x)$. We introduce dimensionless parameters as $Z = 2H/v_F$ and $Z' = 2H'/v_F$. Here v_F is Fermi velocity.

In order to study the proximity effect, we have to solve the Usadel equation [5]

$$D \frac{\partial}{\partial x}\left(\hat{R}_1 \frac{\partial}{\partial x}\hat{R}_1\right) + i\left[\hat{H} + i\frac{\alpha(\omega)}{2}\hat{\tau}_3 \hat{R}_1 \hat{\tau}_3, \hat{R}_1\right] = 0, \qquad (1)$$

$\hat{H} = i\omega\hat{\tau}_3 + i\Delta^u\hat{\tau}_2$. Here D, ω, $\hat{R}_1$, $\alpha(\omega)$ and Δ^u denote diffusion constant in the DN, Matsubara frequency, Green's function, pair breaking effect and pair weakening effect, respectively, where $\alpha(\omega)$ and Δ^u are given by [6]

$$\alpha(\omega) = \frac{n_i}{\pi N(0)}\left(\Theta(1-\tilde{\omega})\left(\tilde{\omega} - \frac{1}{2}\tilde{\omega}^2\right) + \Theta(1-\tilde{\omega})\frac{1}{2}\frac{\frac{3}{4}\pi^2}{(\log(\tilde{\omega}))^2 + \frac{3}{4}\pi^2}\right),$$

$$\Delta^u_\omega = -2\pi N(0) T \sum_{\omega' > 0} u(\omega, \omega') F_{\omega'}, \qquad (2)$$

$$u(\omega, \omega') = n_i V^2 |G_d(\omega)|^2 \Gamma_d |G_d(\omega')|^2 V^2. \qquad (3)$$

$G_d(\omega) = -i/\gamma \, \text{sign}(\omega)/1 + \frac{\pi|\omega|}{4T_K}$, $\Gamma_d = (\pi\gamma)^2/4T_K$, $\gamma = \pi N(0)V^2$, $\tilde{\omega} = \pi\omega/4T_K$ with anomalous Green's function $F_{\omega'}$, temperature T, DOS at the Fermi energy $N(0)$, d-electron propagator G_d, vertex function for the d-electron Γ_d, concentration of Anderson impurity n_i, mixing matrix of s-electron and d-electron V and Kondo temperature T_K. We use analytical continuation $\omega \to -i\varepsilon$ where ε denotes the quasiparticle energy.

For the convenience of the calculation, we introduce the θ parametrization: $\hat{R}_1(x) = \cos\theta(x)\tau_3 + \sin\theta(x)\tau_2$. We have solved the Usadel equation under the Nazarov's boundary conditions at the interfaces [7].

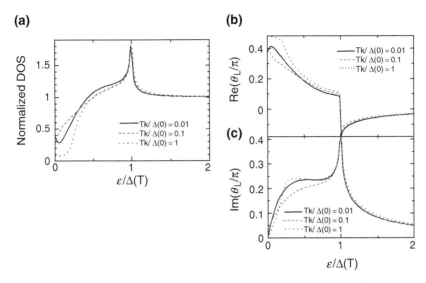

Fig. 1. DOS and real and imaginary parts of θ at $x = L$ for $R_\mathrm{d}/R_\mathrm{b} = 1$ and $E_\mathrm{Th}/\Delta(0) = 1$

Below we fix parameters as $T/T_\mathrm{C} = 0.01$, $Z = Z' = 3$, $R_\mathrm{d}/R'_\mathrm{b} = 0.1$ and $2n_i/(\pi N(0)\Delta(0)) = 1$. Here $\Delta(0)$ denotes the gap energy of the superconductor at $T = 0$.

3 Results

Here we focus on the DOS and proximity parameter θ at $x = L$. We show the dependence of the DOS on ε for $R_\mathrm{d}/R_\mathrm{b} = 1$ and $E_\mathrm{Th}/\Delta(0) = 1$ in Fig. 1a with various $T_\mathrm{K}/\Delta(0)$. As $T_\mathrm{K}/\Delta(0)$ increases, the gap structure of the DOS at $x = L$ becomes deep. This stems from the enhancement of both real and imaginary parts of θ as shown in Fig. 1b,c where θ_L denotes $\theta(L_-)$. This means that proximity effect is enhanced with the increase of Kondo temperature. A similar structure is also found for other parameters, e.g., $R_\mathrm{d}/R_\mathrm{b} = 5$ and $E_\mathrm{Th}/\Delta(0) = 0.1$ as shown in Fig. 2.

4 Conclusions

In the present paper, we have studied proximity effect in DN/S junctions in the presence of Anderson impurities in DN for various situations, where we have used the Usadel equation. DOS is calculated by changing the magnitudes of the resistance in DN, Thouless energy and Kondo temperature. The Kondo effect is approximately incorporated in the pair-breaking rate. It is shown

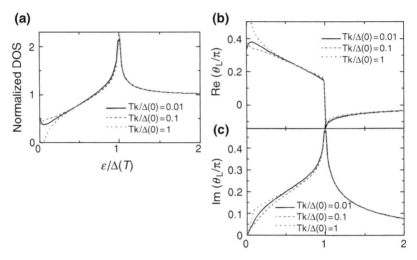

Fig. 2. DOS and real and imaginary parts of θ at $x = L$ for $R_d/R_b = 5$ and $E_{Th}/\Delta(0) = 0.1$

that the gap structure of the DOS in DN becomes deep as increasing Kondo temperature. This is because proximity effect is enhanced by the increase of Kondo temperature.

References

1. G. Eilenberger, Z. Phys. **214**, 195 (1968)
2. A.F. Volkov, A.V. Zaitsev, T.M. Klapwijk, Physica C **210** (1993) 21; S. Yip, Phys. Rev. B **52**, 15504 (1995); T. Yokoyama, Y. Tanaka, A.A. Golubov, J. Inoue, Y. Asano, Phys. Rev. B **71** (2005) 094506
3. A.A. Abrikosov, L.P. Gor'kov: Zh. Eksp. Teor. Fiz. **39** (1960) 1781 [Sov. Phys. JETP **12** (1961) 1243]
4. J. Kondo: Prog. Theor. Phys. **32** (1964) 37
5. K.D. Usadel Phys. Rev. Lett. **25** (1970) 507
6. T. Matsuura, S. Ichinose, Y. Nagaoka: Prog. Theor. Phys. **57** (1977) 713; O. Narikiyo and H. Fukuyama, J. Phys. Soc. Jpn. **58** (1989) 4557
7. Yu.V. Nazarov, Superlattices and Microstructuctures **25**, 1221 (1999); Y. Tanaka, A.A. Golubov, and S. Kashiwaya, Phys. Rev. B **68** (2003) 054513; T. Yokoyama, Y. Tanaka, A.A. Golubov, Phys. Rev. B **72**, 052512 (2005)

Wiedemann–Franz Law in Diffusive Normal Metal/p-Wave Superconductor Junctions

Takehito Yokoyama, Yukio Tanaka, Alexander A. Golubov, and Yasuhiro Asano

Summary. Wiedemann–Franz law in the diffusive normal (DN) metal/insulator/p-wave superconductor junctions is studied for various situations, where we have used the Usadel equation with extended Nazarov's boundary condition. Electrical and thermal conductance of the junction are calculated by changing the magnitudes of the resistance in DN, Thouless energy in DN, the transparency of the insulating barrier, and the angle between the normal to the interface and the lobe direction of p-wave pair potential. It is shown that the mid gap Andreev resonant states drastically enhance the electric conductance at low temperatures, whereas they suppress the thermal conductance. As a consequence, the Lorenz ratio deviates from its normal state value. We also find that the thermal conductance of the junctions reflects the existence of the line nodes of the pair potential.

1 Introduction

In diffusive normal metal/unconventional superconductor (DN/US) junctions, proximity effect and the formation of the midgap Andreev resonant states (MARS) at the interface of unconventional superconductors cause a number of unusual transport properties [1–6]. Recently, a theory reveals the anomalous charge transport in diffusive normal metal/p-wave superconductor (DN/P) junctions [6]. In particular, a giant peak in the density of states at zero energy is a remarkable feature in these junctions. Although tunneling conductance was studied for various types of junctions, the thermal conductance in unconventional superconductor (US) junctions has not been discussed in detail.

The Wiedemann–Franz (WF) law connects the electronic conductivity (σ) with the thermal conductivity (κ) of diffusive metals by the Lorenz number ($L = \kappa/\sigma T$). It is confirmed that WF law holds in a number of diffusive metals. In this paper, we discuss the influences of the proximity effect and the MARS on conductance, thermal conductance and the Lorenz number in superconducting junctions.

2 Formulation

We consider normal-reservoir (N)/DN/I/P junctions, where I denotes an insulator and the length of DN (L) is much larger than the mean free path. The x points the direction normal to the junction interface. The insulator is described by the Δ-function potential. The positions of the N/DN interface and the insulator are denoted as $x = 0$ and $x = L$ respectively. The interface between the DN and the P electrode has a resistance R_b while the interface between normal reservoir and DN has zero resistance. The DN has a resistance R_d. We apply the quasiclassical Keldysh formalism in the following calculation. The degree of the proximity effect is measured by $\theta(x)$ and the spatial dependence of $\theta(x)$ is determined by the following Usadel equation [7]:

$$D\frac{\partial^2}{\partial x^2}\theta(x) + 2i\epsilon \sin[\theta(x)] = 0, \qquad (1)$$

where D is a diffusion constant and the quasiparticle energy measured from the Fermi level is given by ϵ.

For the electrical conductance, we follow [6, 8] and obtain the following result at zero voltage

$$\sigma_S(T) = \frac{1}{2T}\int_0^\infty \frac{d\epsilon}{\cosh^2\left[\frac{\epsilon}{2T}\right]\left(\frac{R_b}{<I_{b4}>} + \frac{R_d}{L}\int_0^L \frac{dx}{\cosh^2\theta_{im}(x)}\right)},$$

where $\theta_{im}(x)$ denotes the imaginary part of $\theta(x)$. In the above, $<I_{b4}>$ is given in [6, 8].

With a similar calculation, we also obtain the thermal conductance at zero voltage represented by [8]

$$\kappa = \frac{1}{2e^2T^2}\int_0^\infty \frac{\varepsilon^2 d\varepsilon}{\cosh^2\left(\frac{\varepsilon}{2T}\right)\left(\frac{R_b}{\langle I_{b5}\rangle} + \frac{R_d}{L}\int_0^L \frac{dx}{\cos^2 Re\theta(x)}\right)}.$$

In the above, $<I_{b5}>$ is given in [8].

In the following, we will discuss the normalized conductance $\sigma_T(T) = \sigma_S(T)/\sigma_N$, the normalized thermal conductance $\kappa_T(T) = \kappa(T)/\kappa_N(T)$ and the normalized Lorenz ratio $L_T = \kappa_T(T)/\sigma_T(T)$. Here $\sigma_N = 1/(R_d + R_b)$ and $\kappa_N(T) = \pi^2 T/(3(R_b + R_d)e^2)$ are the electric conductance and the thermal conductance in the normal state, respectively.

3 Results

We show the conductance, the thermal conductance, and the Lorenz ratio normalized by their values in the normal state in Fig. 1. We choose $Z = 3$

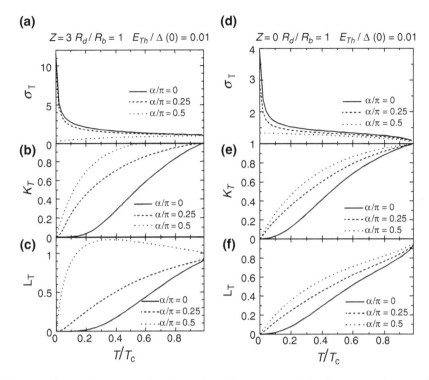

Fig. 1. The conductance, the thermal conductance and the Lorenz ratio are calculated for $Z = 3$ (left panels) and $Z = 0$ (right panels) where $R_d/R_b = 1$ and $E_{\mathrm{Th}}/\Delta(0) = 0.01$

(low transparent DN/P interface) and $Z = 0$ (no insulating barrier at DN/P interface), where $R_d/R_b = 1$ and $E_{\mathrm{Th}}/\Delta(0) = 0.01$ are fixed.

For $Z = 3$, σ_T shows a wide variety of temperature and α dependence because of the formation of the MARS (Fig. 1a). The result for $\alpha = \pi/2$ increases almost linearly with T. On the other hand, σ_T for $\alpha = \pi/4$ or $\alpha = 0$ drastically increases with decreasing temperature. This is because the resonant transmission via the MARS enhances the charge transport in the superconducting junctions.

Concerning the thermal conductance, κ_T is less sensitive to the change of α than σ_T (Fig. 1b). As α increases, the thermal conductance is suppressed. This indicates that the MARS suppress the thermal transport. κ_T is a monotonically increasing function of T/T_C and linear in T/T_C for small T/T_C except for $\alpha \sim 0$. The thermal conductance of the junctions reflects the existence of the line node of the pair potential. Although line node of the pair potential exists, κ_T has an exponential dependence on T for $\alpha = 0$ as in the s-wave case because the direction of the line node is perpendicular to that of the temperature difference.

The resulting Lorenz ratio L_T is shown in Fig. 1c. The WF law is valid for a wide temperature region as $\alpha \to \pi/2$. For $Z = 0$ (Fig. 1d–f), α dependences of the conductance, the thermal conductance and the Lorenz ratio are similar but weak compared to those in Fig. 1a–c.

4 Conclusions

In the present paper, we have studied the electric conductance and the thermal conductance in DN/p-wave superconductor junctions by using the Keldysh Green function method and the quasiclassical approximation. The main results are summarized as follows:

1. The MARS drastically enhance the electric conductance at low temperatures, whereas they suppress the thermal transport. As a consequence, the Lorenz ratio deviates from its normal state value.
2. The thermal conductance of the junctions reflects the existence of the line nodes of the pair potential: it has a linear dependence on temperature at low temperatures.

References

1. A.F. Volkov, A.V. Zaitsev, and T.M. Klapwijk, Physica C **210** (1993) 21
2. Y. Tanaka, A.A. Golubov, and S. Kashiwaya, Phys. Rev. B **68** (2003) 054513
3. T. Yokoyama, Y. Tanaka, A.A. Golubov, J. Inoue, and Y. Asano, Phys. Rev. B **71** (2005) 094506
4. C.R. Hu, Phys. Rev. Lett. **72** (1994) 1526; Y. Tanaka, and S. Kashiwaya, Phys. Rev. Lett. **74** (1995) 3451; S. Kashiwaya, Y. Tanaka, M. Koyanagi, and K. Kajimura, Phys. Rev. B **53** (1996) 2667; Y. Tanuma, Y. Tanaka, and S. Kashiwaya Phys. Rev. B **64** (2001) 214519; S. Kashiwaya and Y. Tanaka, Rep. Prog. Phys. **63** (2000) 1641
5. Y. Tanaka, Yu.V. Nazarov, and S. Kashiwaya, Phys. Rev. Lett. **90** (2003) 167003; Y. Tanaka, Yu.V. Nazarov, A.A. Golubov, and S. Kashiwaya, Phys. Rev. B **69** (2004) 144519
6. Y. Tanaka and S. Kashiwaya, Phys. Rev. B **70** (2004) 012507; Y. Tanaka, S. Kashiwaya, and T. Yokoyama, Phys. Rev. B **71** (2005) 094513
7. K.D. Usadel Phys. Rev. Lett. **25** (1970) 507
8. T. Yokoyama, Y. Tanaka, A. A. Golubov, and Y. Asano, Phys. Rev. B **72**, 214513 (2005)

Nodal Structures of Wave Functions in Chaotic Billiards

Hiromu Ishio

Summary. Gaussian random waves are found good models for description of nodal structures of waves in chaotic billiards.

1 Introduction

The study of nodes (zeros) in waves is a subject which has been attracting attention of theoretical and experimental scientists in various areas of physics and mathematics. This includes a wide range of potential applications such as electromagnetic waves, acoustic waves, water waves, vibrating elastic plates, quantum waves etc. The theoretical studies of wave morphology and statistics of nodal structures in chaotic systems suggest that random waves may typically be a good model to describe universality in such systems under general boundary conditions [1]. This supports a previous conjecture that there might be a relation between complicated nodal structures in wave functions and the underlying classical chaotic dynamics [2].

In this presentation, the theoretical analyses adopting Gaussian-random-wave model with mixed boundary conditions are compared with numerical calculations for eigenfunctions of planar chaotic billiards. The contents presented in the following are based on the paper [3].

2 Theory

In billiards consisting of a planar domain $D(\ni \mathbf{r} = \{x, y\})$ with mixed boundary ∂D, eigenstates $u_n(\mathbf{r}, \alpha)$ with wavenumber k_n satisfy $\nabla^2 u_n + k_n^2 u_n = 0$ in D and $k_n u_n \cos\alpha + \mathbf{n} \cdot \nabla u_n \sin\alpha = 0$ in ∂D. Here $\mathbf{n}$ is the inward normal, and $\alpha(-\pi/2 < \alpha \leq \pi/2)$ parametrizes the boundary condition. Dirichlet and Neumann conditions correspond to $\alpha = 0$ and $\alpha = \pi/2$, respectively.

In general, monochromatic waves in chaotic billiards may be considered as superpositions of plane waves travelling in different directions but with the same wavelength $\lambda = 2\pi/k$: $\psi(\mathbf{r}) = \sum_j \exp(i\mathbf{k}_j \cdot \mathbf{r} + \phi_j)$. Berry conjectured that they are typically modelled by using Gaussian random waves [2]. In the boundary-adapted Gaussian-random-wave model [1], using scaled coordinates $\mathbf{R} = \{X, Y\} = \{kx, ky\}$, the domain is approximated as the half-space $Y \geq 0$, so the boundary is the straight line $Y = 0$. Members of the ensemble of random waves, whose statistics are to be compared with those of the eigenstates in the billiards, are [1]

$$u(\mathbf{r}, \alpha) = \frac{2}{\sqrt{J}} \sum_{j=1}^{J} \frac{\sin(Y \sin \theta_j) - \tan \alpha \sin \theta_j \cos(Y \sin \theta_j)}{\sqrt{1 + \tan^2 \alpha \sin^2 \theta_j}} \cos(X \cos \theta_j + \phi_j). \tag{1}$$

This is a superposition of $J(\gg 1)$ plane waves satisfying the Helmholtz equation and boundary condition for $Y \gg 0$, travelling in directions θ_j equidistributed on $[0, \pi]$ with phases ϕ_j equidistributed on $[0, 2\pi]$.

The nodal line density $\rho_L(Y, \alpha)$ to be calculated is defined in such a way that the mean nodal line length per unit area equals $k/(2\sqrt{2})\rho_L(Y, \alpha)$ so that $\rho_L(Y, \alpha) \to 1$ for $Y \to \infty$ [4,5]. The average here is over the ensemble parameters θ_j and ϕ_j, or equivalently (since the ensemble is ergodic) over long thin strips between Y and $Y + dY$. The theory [1] involves the following averages derived by using more general techniques [5]: $B(Y, \alpha) \equiv \langle u^2 \rangle = 1 - (2/\pi) \int_0^{\pi/2} d\theta \, \mathrm{Re} f(\theta, Y, \alpha)$, $D_X(Y, \alpha) \equiv \langle (\partial u/\partial X)^2 \rangle = (1/2) - (2/\pi) \int_0^{\pi/2} d\theta \, (\cos \theta)^2 \mathrm{Re} f(\theta, Y, \alpha)$, $K(Y, \alpha) \equiv \langle u (\partial u/\partial Y) \rangle = (2/\pi) \int_0^{\pi/2} d\theta \sin \theta \mathrm{Im} f(\theta, Y, \alpha)$, where $f(\theta, Y, \alpha) = \exp(2iY \sin \theta)\{(1 - i \tan \alpha \sin \theta)/(1 + i \tan \alpha \sin \theta)\}$. The result is that the nodal line density is

$$\rho_L(Y, \alpha) = \frac{2}{\pi} \sqrt{\frac{2D_X}{B}} E\left(\frac{B(B-1) + K^2}{BD_X}\right), \tag{2}$$

where E denotes the complete elliptic integral whose definition is given in [6].

The other unanticipated feature of the theory is sharp peaks in $\rho_L(Y, \alpha)$ near $Y = \alpha$ for $(0 <)\alpha \ll \pi/2$. These were interpreted as "ghosts" of the Dirichlet boundary nodal line, which migrates from the boundary as the boundary condition changes from the Dirichlet to Neumann type, and found to be described analytically [1].

3 Stadium Billiard

The analytical results presented in the previous section are now compared with numerical calculations for eigenfunctions of chaotic billiards. For D we chose the chaotic quarter stadium [7] consisting of the unit square augmented by a quarter circle of radius 1. For boundary-condition parameters $\alpha = -\pi/4, 0, \pi/4$ and $\pi/2$, we calculated eigenstates in the range

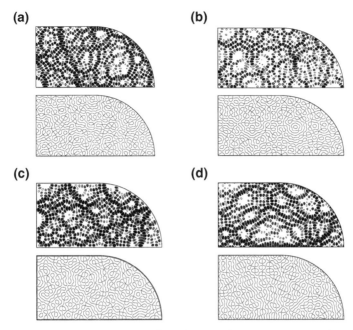

Fig. 1. Contour map of intensity $u^2(x,y)$ (*upper*) and nodal lines (*lower*) of the states **(a)** $k = 83.99$ for $\alpha = -\pi/4$, **(b)** $k = 84.08$ for $\alpha = 0$, **(c)** $k = 83.89$ for $\alpha = \pi/4$ and **(d)** $k = 83.91$ for $\alpha = \pi/2$.

$83 < k < 85$. According to the leading-order Weyl rule, this interval includes approximately 47 states between the 978th and the 1,026th. We used a mixed-boundary-condition adaptation of the numerical wave-expansion technique [8], employing combinations of (1). In this way, we captured about 74% of the states (about 35 states) in the chosen k interval for each value of α except $\alpha = 0$, for which we captured about 36% of the states (17 states).

Figure 1 shows a typical such state for each α. In general, both wave intensity and nodal line plots show complicated patterns typical of chaotic systems. The case for $\alpha = \pi/4$ is chosen to show the ghost nodal lines and to illustrate how this feature is hard to discern in the intensity plot.

Figure 2 shows how accurately the billiard eigenstates fit the Gaussian random theory for $\rho_L(Y, \alpha)$, even deep inside D and for the ghost nodal lines (inset in Fig. 2c).

4 Conclusion

For nodal statistics, the analytical results obtained before were compared with numerical calculations for eigenfunctions of chaotic billiards with mixed boundary conditions. We have found very good agreement between the analytical and numerical results, confirming that Gaussian random waves

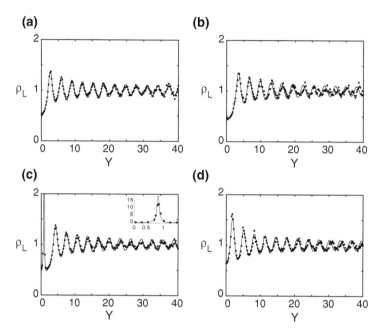

Fig. 2. Nodal line density $\rho_L(Y,\alpha)$ for **(a)** $\alpha = -\pi/4$, **(b)** $\alpha = 0$, **(c)** $\alpha = \pi/4$ and **(d)** $\alpha = \pi/2$. *Full curves*: (2) of Sect. (2); *bold dots*: billiard eigenstates. The inset in **(c)** is a magnification of the peak near $Y = \pi/4$, arising from the *ghost lines*. The figure is taken from [3].

are good mathematical models for statistical description of waves in chaotic systems.

Acknowledgments

The work is supported by a Grant-in-Aid for the 21st Century COE 'Frontiers of Computational Science' in Japan and the Royal Society in UK.

References

1. Berry MV, Ishio H (2002) J Phys A 35:5961–5972
2. Berry MV (1977) J Phys A 10:2083–2091
3. Berry MV, Ishio H (2005) J Phys A 38:L513–L518
4. Berry MV (2002) J Phys A 35:3025–3038
5. Berry MV, Dennis MR (2000) Proc R Soc A 456:2059–2079
6. Wolfram S (1996) The Mathematica Book, Cambridge U. P., Cambridge
7. Bunimovich LA (1974) Funct Anal Appl 8:254–255
8. Heller EJ (1991) in Les Houches Lecture Series, Vol. 52, edited by Giannoni M-J, Voros A, Zinn-Justin J, North-Holland, Amsterdam

Elongation of Water Residence Time at the Protein Interior in Aqueous Solution with Ectoine

Isseki Yu and Masataka Nagaoka

Summary. Ectoine is one of the most common compatible solutes found in halophilic bacteria. We investigated the effect of ectoine addition on the dynamic properties of water molecules in each area of a protein structure. The results indicated a significant elongation of water residence times, in the interior of the protein structure not easily accessed by bulk solvent molecules.

1 Introduction

Compatible solutes are compounds with low molecular weights that are produced by the cells of organisms and accumulate in them in response to such high environmental stresses as high salt concentration or high temperature. Although many experiments have been performed to examine how their addition to a protein solution influences its global thermodynamic characteristics, the mechanism of this influence has still not been fully investigated from a microscopic viewpoint.

Under the current circumstances, we have been trying to analyze their effects by using the molecular dynamics (MD) simulation method. Throughout our studies, ectoine (2-methyl-4-carbonyl-1,4,5,6-tetrahydropyrimidine) [1, 2] which is one of the most typical compatible solutes found in halophilic bacteria, has been used as a target compatible solute [3, 4]. So far, we have found that ectoine does not change the hydration structure of a target protein, chymotrypsin inhibitor 2 (CI2), essentially but significantly slows down the diffusion of water molecules near the protein surface in addition to in the bulk phase at high temperature (370 K) [4]. This finding provided a new microscopic basis for the current phenomenological explanation that *compatible solutes modify the stability of proteins not by interacting strongly with the macromolecules but by acting to alter the solvent properties*. In this work, we analyze quantitatively the magnitude of the slowdown of water diffusion in each residual area of CI2. As an index of the slowdown, we utilize average residence time (ART) of atoms in water molecules around each CI2 atom.

2 Methods

2.1 Molecular Dynamics Simulations of CI2

Molecular mechanics (MM) and molecular dynamics (MD) calculations were all performed using the AMBER7 program suite [5]. The force field parameter set, parm99, was used for all the molecules in the present systems (partial atomic charges for ectoine atoms were obtained in our previous work [3]). One MD simulation for a system of CI2 in water–ectoine mixture was performed in the following way. First, a CI2 unit was set, under the periodic boundary condition, at the center of a cubic box ($64 \times 64 \times 64$ Å^3) filled with 6,903 TIP3P [6] water molecules and 227 ectoine molecules (the simulation for this system is named MD$_m$). Second, a reference MD simulation was separately performed for a system of one CI2 in pure 8,374 TIP3P water molecules, i.e., MD$_p$. In this work, subscripts "m" and "p" represent the ectoine–water mixture solution and the pure water, respectively. Both systems were initially equilibrated each for 300 ps duration at 300 K under the NPT condition. Then, for each system, a further 2,000 ps (= 2.0 ns) simulation was performed at 300 K under the NVE and the neutral pH condition. Trajectories were generated by numerically solving equations of motions by the Verlet method with a time step of 2.0 fs. The electrostatic interaction was calculated using the particle-mesh Ewald (PME) method [7]. All the bonds involving hydrogen atoms were constrained by the SHAKE method [8].

2.2 Average Residence Time

We defined the ART $T_{W\alpha}$ of atoms in the hydration shell of a peptide atom α, i.e., $S_\alpha^{r_0}$. In this analysis, the solvation shell $S_\alpha^{r_0}$ is defined as a sphere of radius $r_0 (= 3.0$ Å$)$ of a peptide atom α in CI2. Then, $T_{W\alpha}$ is defined precisely by

$$T_{W\alpha} = \frac{1}{N_{W\alpha}} \sum_{i \in \text{Water}} \tau_{i\alpha}, \tag{1}$$

where $\tau_{i\alpha}$ is the total time that an atom i in water molecules spends inside $S_\alpha^{r_0}$ during the entire time $T_{\text{run}}(= 2.0$ ns$)$ of MD simulation, and $N_{W\alpha}$ is the total number of these water atoms that visit inside $S_\alpha^{r_0}$, defined by

$$N_{W\alpha} = N_{W\alpha}(r_0) = \sum_{i \in \text{Water}} h(r_0 - r_{i\alpha}^{\min}), \tag{2}$$

where i specifies a water atom which exists somewhere in the whole solution and

$$r_{i\alpha}^{\min} = \min_{0 < t < T_{\text{run}}} |\mathbf{r}_i(t) - \mathbf{r}_\alpha(t)| \tag{3}$$

and $h(x)$ is a step function as follows,

$$h(x) = \begin{cases} 1 & (x \geq 0) \\ 0 & (x < 0) \end{cases}.$$ (4)

Then, $h(r_0 - r_{i\alpha}^{\min})$ takes 1 if a water atom i exists within r_0 from a CI2 atom, i.e., if $r_0 - r_{i\alpha}^{\min} > 0$, where $r_{i\alpha}^{\min}$ is defined as the minimum distance between the water atom i and a CI2 atom α throughout the entire time duration.

3 Results and Discussions

3.1 Hydration Structure Near the Protein Surface

Because we could not observe any significant difference in C_α root mean square distances (RMSDs) and solvent accessible surface areas (SASAs) of CI2 between both solvent environments (data is not shown), it is understood that the CI2 conformation in each solvent is essentially the same. This indicates that during our simulation, the property difference of water molecules around CI2 is originating not from the difference in the CI2 conformation but from the change of solvent properties. On this standpoint, we have investigated the effect of ectoine addition on the hydration structure by the integrated coordination number (ICN) (cf. (1)–(4) in [4]) of solvent atoms near the protein surface.

As shown in Fig. 1, ICN $\langle N_{\text{CI2}}(r) \rangle$ shows the average number of solvent atoms, which exist within a distance r Å from the closest CI2 atoms. In the

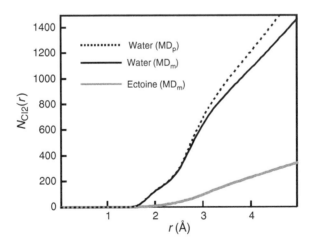

Fig. 1. Integrated coordination numbers (ICNs) of atoms in water molecules within the distance r (Å) from the closest atom in CI2 in MD_m (*black line*) and MD_p (*dot line*), and that in ectoine molecules in MD_m (*gray line*)

analysis, without distinction of atom types in each solvent molecule, they were counted to contribute to $\langle N_{CI2}(r)\rangle$s. Although the simulation temperature was 70 K lower than that in our previous system [4], we have observed the same tendency as that in the previous one; (1) near the surface of CI2 ($r < 2$ Å), those atoms in water molecules precede those in ectoine ones in the coordination, by the latter being excluded more from the surface, and (2) there exist little ectoine atoms within the solvation layer near the surface of CI2. Thus, one could conclude that the ectoine molecules do not strongly coordinate to the protein atoms also in the solvent environment, in which the temperature is closer to the actual intracellular one.

3.2 Residence Times of Water Atoms

We analyzed ARTs of water atoms to investigate how the ectoine coordination shown in Fig. 1 influences on the dynamic property of water molecules near the protein. ARTs of water atoms $T_{W\alpha}$s in both solvent environments are shown in Fig. 2. In both systems, water atoms show short-time residences in the areas that are exposed to the bulk solvent phase, such as active-site loop and N terminus strand, while long time residences ($T_{W\alpha} > 100$ ps) are observed mainly in the interior areas of CI2, such as β-sheet and C-terminus strand.

Although $T_{W\alpha}$s are elongated in many areas of CI2 in MD_p, it is interesting to indicate that the residence times have significantly elongated also

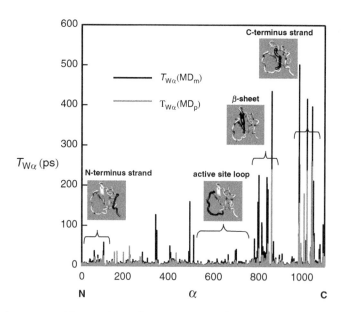

Fig. 2. Average residence time of water atoms that visited a sphere of radius $r (= 3.0$ Å$)$ of a peptide atom α (gray: in pure water, black: in ectoine solution)

in the areas that the bulk solvent cannot easily access to (e.g., β-sheet and C-terminus strand). It is still outstanding issue to evaluate quantitatively the influence of such water molecules that are strongly associated with these protein areas on the thermodynamic and kinetic stability of protein structure. However, the results suggest that ectoine stabilizes the protein structure not only by slowing down the water diffusion near the "outer" surface of protein but also by reducing the mobility of those water molecules that are strongly associated to the "interior" surface of protein that the bulk solvent cannot easily access to.

Acknowledgements

This work was supported partly by a Grant-in-Aid for Science Research from the Ministry of Education, Culture, Sport, Science and Technology in Japan and also by a Grant-in-Aid for the 21st Century COE program "Frontiers in Computational Science" at Nagoya University, and the Core Research for Evolutional Science and Technology (CREST) 'High Performance Computing for Multi-scale and Multi-physics Phenomena' from the Japan Science and Technology Agency.

References

1. Galinski, E. A. Experientia (1993) 49; 487–496
2. Galinski, E. A. Adv. Microb. Physiol. (1995) 37; 273–328
3. Suenobu, K.; Nagaoka, M.; Yamabe, T.; Nagata, S. J. Phys. Chem. A (1998) 102; 7505–7511
4. Yu, I.; Nagaoka, M. Chem. Phys. Lett. (2004) 388; 316–321
5. Case, D. A. et al. 2002, AMBER7, University of California, San Francisco
6. Jorgensen, W. L.; Chandrasekhar, J.; Madura, J. D.; Impey, R. W.; Klein, M. L. J. Chem. Phys. (1983) 79; 926–935
7. Darden, T.; York, D.; Pedersen, L. J. Chem. Phys. (1993) 98; 10089–10092
8. Ryckaert, J. P.; Ciccotti, G.; Berendsen, H. J. C. J. Comput. Phys. (1977) 23; 327–341

Analytical Model for Protein Folding

Yoko Suzuki and José N. Onuchic

Summary. A theoretical framework is constructed with the aid of a free-energy functional method that is capable of describing the interplay between geometrical and energetic effects on protein folding. We generalize a free-energy functional model proposed by Plotkin, which is based on polymer theory. This generalization is made by introducing cooperativity into the configurational entropy and the internal energy. Modifications to the configurational entropy enable the model to account for the loop–loop interactions. Modifications to the internal energy introduce many-body corrections. To demonstrate the efficiency of the modified analytical model, we compare our results with C_α structure-based (Gō) model simulations of chymotrypsin inhibitor II and the SH3 domain of src. This theoretical framework provides the understanding of physics of protein folding.

1 Introduction

The current understanding of the physical mechanisms controlling protein folding is strongly based on the energy landscape theory and the funnel concept. This theoretical framework provides the basis for understanding an entire new family of experiments that have being designed to provide a quantitative understanding of these mechanisms. These studies have been enormously aided by a series of numerical simulations. Still, without analytical models, it is impossible to obtain a full physical understanding of both simulation and experimental results. In this study, we generalize a free-energy functional model based on polymer theory, the Plotkin model [1–3], to make it more appropriate for comparison with protein folding simulations and experiments [4].

2 Theory

The Plotkin model for free energy F at temperature T is given by [1–3]

$$F = E - T(Ns_0 + S_{\text{bond}} + S_{\text{route}}). \tag{1}$$

The internal energy E is given by; $E = -\sum_{i=1}^{M} \epsilon_i Q_i$, where ϵ_i and Q_i are the contact energy and the formation probability for the ith contact, respectively. M is the number of residue pair contacts in the native structure [3]. The first term inside the parentheses in (1), Ns_0, accounts for the entropy loss by the change from the unfolded state to the folded one, where N is the number of residues in a protein. The second and third terms, S_{bond} and S_{route}, measure the bond entropy and route entropy, respectively. The bond entropy, S_{bond}, is the entropy loss due to contact formation.

The generalization has been made for S_{route} and E as follows [4].

The route entropy, S_{route}, is given by

$$S_{\text{route}} = k_B \lambda(Q) \sum_{i=1}^{M} [-Q_i \log Q_i - (1 - Q_i) \log(1 - Q_i)], \qquad (2)$$

where Q is the fraction of native contacts with $0 < Q < 1$. Equation 2 corresponds to the density of states obtained by counting the number of ways in which we can get a partially folded protein with MQ contacts. The constant k_B is the Boltzmann constant. The function $\lambda(Q)$ accounts for the entropy reduction due to the coupling to chain connectivity. The function $\lambda(Q)$ is generalized from the Plotkin's expression, $\lambda(Q) = 1 - Q^{\alpha_1}$, to a new expression;

$$\lambda(Q) = (1 - Q^{\alpha_1})^{\alpha_2} + \xi Q^{\alpha_3}, \qquad (3)$$

where $\alpha_1, \alpha_2, \alpha_3$, and ξ are determined phenomenologically by the best fit to numerical simulation. Exponents α_1 and α_2 in (3) account for the entropy reduction due to the formation of short-range contacts and that of long-range contacts. The second term of $\lambda(Q)$ in (3) is needed to account for loop–loop interactions, mostly via very long range contacts. The physical meaning of this term is a compensation for the overestimated entropy reduction arising from the mean-field approximation in the bond entropy, where parameters ξ and α_3 control the strength and range of the overestimated degeneracy.

In the following, the model with the new expression of $\lambda(Q)$ is called the generalized model.

The second modification is the introduction of many-body interactions to the internal energy E. These interactions appear to be necessary from the Gō model simulation as well as experimental observations. In order to introduce cooperativity due to residue–residue interaction along the polypeptide chain, an additional interaction term, $-(1/2) \sum_{i \neq j} e_{i,j}^{(2)} Q_i Q_j$, is introduced, where $e_{i,j}^{(2)}$ accounts for the interaction between contacts i and j.

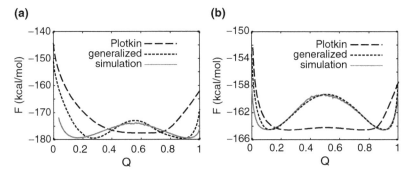

Fig. 1. Free energy $F(Q)$ at the folding temperature T_F for **(a)**CI2 and **(b)**SH3 in the Plotkin model (*dashed line*), the generalized model (*dotted line*), and the simulation (*solid line*)

3 Results

To quantitatively test the applicability of our model, we compare the theoretical free energy profiles at the folding temperature, T_F, with those obtained from Gō model simulations. For this comparison, the contact energy ϵ_i is fixed to a constant ϵ exactly as in the standard Gō models [5].

We perform these calculations for the chymotrypsin inhibitor II (CI2) and for the src SH3 domain. The parameters for both CI2 and SH3 are summarized in Table 1 of [4].

Figure 1a shows free energy profile for CI2 in the Plotkin model (dashed line), the generalized model (dotted line), and the simulation (solid line).

To better understand the cooperativity and origin of barriers in the folding mechanism, we investigate the individual contact formation probability, Q_i^*, as a function of the global Q. The results are shown in Fig. 2 for (a)the Plotkin and (b)generalized models and for (c)the simulation.

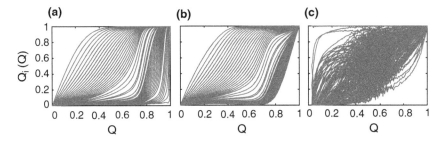

Fig. 2. Probability of each contact i, $Q_i^*(Q)$: **(a)** the plotkin model; **(b)** the generalized model; **(c)** the simulation

In the simulation, the free energy profile (solid line in Fig. 1a), has the two minima at $Q \sim 0.15$ and $Q \sim 0.95$. From Fig. 2c, the $Q \sim 0.15$ minimum corresponds to the unfolded basin where only local structure is present (only the short-range loops are formed). The folded basin corresponds to the other minimum at $Q \sim 0.95$.

In the Plotkin model, the free energy profile (dashed line in Fig. 1a) shows no barrier and has one minimum at $Q \sim 0.6$, where only the short-range loops are formed as shown in Fig. 2a and the long-range loops have difficulty to coalesce. Therefore, the protein is only partially folded.

In the generalized model, the free energy profile (dotted line in Fig. 1a) has a barrier and the line shape is closer to solid line than dashed line. It has two minima $Q \sim 0.3$ and $Q \sim 0.9$, which correspond to the unfolded state with local structure and the folded state, respectively, as shown in Fig. 2b. The energy barrier accounts for the transition between the two states. Such behavior is similar to the simulation. We can see that the newly introduced terms in the route entropy and the internal energy induce cooperativity and therefore make the results closer to the simulation ones.

The similar results can be seen in the case of SH3 domain, as in Fig. 1b, where dashed, dotted, and solid lines represent the results for the Plotkin model, the generalized model, and the simulation, respectively. In the Plotkin model, the free energy profile (dashed line in Fig. 1b) has a much smaller energy barrier than that obtained from the simulation (solid line in Fig. 1b). In the generalized model, the barrier height obtained from the simulation is reproduced, as in Fig. 1b.

4 Conclusion

In the current study, we have constructed an analytical model that generalizes Plotkin's model by incorporating geometrical and energetic cooperativity into the folding process. Comparison of the two models with the simulations of CI2 and src SH3 domain demonstrates the need of cooperativity for more realistic descriptions of the protein energy landscape.

By integrating simulation and theory, we can build a quantitative framework capable of describing the interplay between geometrical and energetic contributions to protein folding. Our work represents an initial step towards more complete energy functionals that are needed to deal with real experimental data.

Acknowledgments

We wish to thank Leslie Chavez for all the simulation data.

References

1. S. S. Plotkin and J. N. Onuchic (2002) Q. Rev. Biophys. 35:111–167
2. S. S. Plotkin and J. N. Onuchic (2002) Q. Rev. Biophys. 35:205–286
3. S. S. Plotkin and J. N. Onuchic (2000) Proc. Natl. Acad. Sci. USA 97:6509–6514; (2002) J. Chem. Phys. 116:5263–5283
4. Y. Suzuki and J. N. Onuchic (2005) J. Phys. Chem. B 109:16503–16510
5. H. Taketomi, Y. Ueda, and N. Go (1975) Int. J. Pept. Res. 7:445–459

Theoretical Research on Dynamics of the Genetic Toggle Switch

Tomohiro Ushikubo, Wataru Inoue, Mitsumasa Yoda, and Masaki Sasai

Summary. Genetic toggle switch is a network of two genes which repress each other. A synthetic network of this type has been embedded in *E. coli* to demonstrate its switching behavior [1]. Here, we numerically investigate this network in the wide range of parameters. When the DNA promoter change is slow as was expected in eukaryotic chromosome, the network shows a variety of different behaviors with three or four stable states. A transition state theory is applied to see whether the free energy landscape picture is consistent to describe the network.

1 Introduction

The genetic toggle switch was designed and embedded into *E. coli*, and the results were compared with theories [1, 2]. Most of those theoretical studies, however, assumed that the binding rate of repressor to promoter and the dissociation rate of repressor from the promoter are large, so that the promoter status quickly reaches the equilibrium and only the numbers of proteins remain to be dynamical stochastic variables. This assumption of rapid changing of the promoter status seems to hold in the light of *in vitro* data of rate constants, but the assumption is not confirmed in prokaryotic cells and is certainly invalid in eukaryotic cells. In this study we depart from this assumption and simulate the network in the wider parameter range.

2 Model

As shown in Fig. 1a, the network has two genes which repress each other. Chemical reactions of the gene expression are modeled by four processes (Fig. 1b). First process is the protein production: The rate is $g > 0$ in the repressor-unbound state, and 0 in the repressor-bound state. The synthesized protein acts as a repressor for another gene. Second is the protein degradation with the rate k. Third is the protein-binding to DNA with the rate $hn_i(n_i-1)$, where n_i is the number of the repressor proteins synthesized from the ith gene

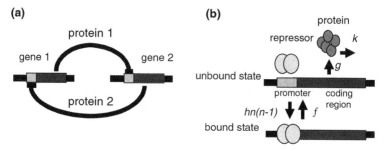

Fig. 1. Model of the genetic toggle switch. (**a**) Two genes suppress each other. (**b**) Scheme of gene expression. Repressor is assumed to work as a dimmer

with $i = 1$ or 2. Fourth is the protein dissociation from DNA with the rate f. In this study we use the value $X^{\mathrm{eq}} = f/h = 1,000$. Frequency of the DNA status change is measured by $\omega = f/k$ and the representative number of proteins is $X_{\mathrm{ad}} = g/2k$. Stochastic noise should be large when ω is small or X_{ad} is small. The master equation is written down using these parameters [3]. We numerically integrate the master equation with the Gillespie algorithm [4].

3 Number of Stable Point

In Fig. 2 the number distribution, $P(n_1, n_2) = \langle P(n_1, n_2, t) \rangle$, is plotted on the n_1–n_2 plane at various values of X_{ad} and ω, where $\langle \cdots \rangle$ is the average

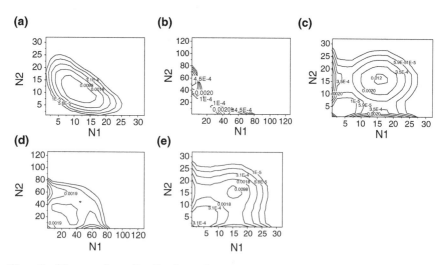

Fig. 2. The number distribution of protein 1 and protein 2. The parameter representing the relative speed of the change in the promoter status, X_{ad}, and the representative number of proteins, ω, are (**a**) $(X_{\mathrm{ad}}, \omega) = (10^{1.2}, 10^2)$, (**b**) $(X_{\mathrm{ad}}, \omega) = (10^{1.8}, 10^2)$, (**c**) $(X_{\mathrm{ad}}, \omega) = (10^{1.2}, 10^{-2})$, (**d**) $(X_{\mathrm{ad}}, \omega) = (10^{1.8}, 10^{-1})$, and (**e**) $(X_{\mathrm{ad}}, \omega) = (10^{1.2}, 10^{-1})$

along the simulated trajectory. When the representative number of proteins is small and the DNA status change is fast, the system becomes monostable as shown in Fig. 2a. In this case the switching behavior is masked by the large fluctuation of proteins. When the representative number of proteins is large and the DNA status change is fast, the fluctuation is suppressed and the system becomes bistable as shown in Fig. 2b. The maxima of $P(n_1, n_2)$ are found at $n_1 = 0$ and at $n_2 = 0$, and the transition between them can be utilized as the switching process. In Fig. 2c–e with the slow change in the DNA promoter status, we can find three or four stable points. These multistable features have not been studied in previous studies. In Fig. 2c, the maximal point of $P(n_1, n_2)$ is on the $n_1 = 0$ axis or on the $n_2 = 0$ axis and the third maximal point is found at $n_1 = n_2 \neq 0$. In Fig. 2d the third maximal point is at $n_1 = n_2 \approx 0$. In Fig. 2e the third maximal point at $n_1 = n_2 \neq 0$ is more evident than in Fig. 2c.

4 Transition State Theory

A transition path is defined on the n_1–n_2 plane. Using the steepest descent method, the path is searched from the local minimum of $P(n_1, n_2)$ toward the local maximum of $P(n_1, n_2)$. Free energy like quantity, $F = -\ln P(n_1, n_2)$, is defined. Using the analogy to transition state theory, the transition rate from local minimum point A to another local minimum point C through transition state B should be written as $k'_{AC} = k^{\ddagger}_{BC} \exp(-(F_B - F_A))$, where F_B and F_A are free energies averaged over the region of width $1/e$ in the number distribution around point B and point A, respectively. $k^{\ddagger}_{BC}$ is the transition rate from B to C. The transition rate k_{AC} from A to C can be directly calculated from the numerical data to be compared with the prediction of the transition state theory. The ratio between k'_{AC} and k_{AC} measures the degree that the transition state theory is broken and is called the transmission coefficient (TC). TCs are summarized in Table 1. Upper part of the table is for transitions involving the local minimum in F with $n_1 = n_2 \neq 0$. In this case, TC for transition from A to C and TC for the transition from C to A are almost same. On the other hand, the lower part of the table is for transitions involving the local minimum with $n_1 = n_2 \approx 0$. In this case, TCs for the forward and backward transitions are very deferent.

Table 1. Transmission coefficients at different X_{ad} and ω

(X_{ad}, ω)	k'_{AC}/k_{AC}	k'_{CA}/k_{CA}
$(10^{1.2}, 10^{-1})$	0.234	0.230
$(10^{2.0}, 10^{-3})$	0.012	0.020
$(10^{1.2}, 10^{-1})$	0.561	0.017
$(10^{2.0}, 10^{-1})$	0.060	4.8×10^{-4}

5 Discussion

The behavior of the two gene network switch was numerically investigated in the wide parameter region. We found rich behaviors of the network especially in the case that the rate of the status change of the promoter is slow. Such behaviors can be expected especially in eukaryotic cells. Since this network motif is abundant in both prokaryotic and eukaryotic cells, it is interesting to ask whether the change in the switching scheme found in the present study should be used in living cells through the possible modulation of reaction rates.

We compared the prediction from the transition state theory and the direct numerical data for the transition rate. For transitions involving the local minimum in F with $n_1 = n_2 \neq 0$, the transmission coefficient of the forward transition and that of the backward transition almost coincide, implying the transition state theory may be applicable with the correction of transmission coefficients. On the other hand, for transitions involving the local minimum with $n_1 = n_2 \approx 0$, correction with the transmission coefficient does not give a consistent result. The large fluctuation in the case of $n_1 = n_2 \approx 0$ should prevent the transition state theoretical interpretation, hence the free energy landscape picture seems not directly applicable in the case of $n_1 = n_2 \approx 0$.

References

1. Gardner TS, Cantor CR, Collins JJ (2000) Nature 403:339–342
2. Kepler TB, Elston TC (2001) Biophys J 81:3116–3136
3. Sasai M, Wolynes G (2003) Proc Natl Acad Sci USA 100:2374–2379
4. Gillespie DT (1977) J Phys Chem 81:2340–2361

Non-Condon Theory for the Energy Gap Dependence of Electron Transfer Rate

Hirotaka Nishioka, Akihiro Kimura, Takahisa Yamato, and Toshiaki Kakitani

Summary. By properly taking into account the effect of the conformation fluctuation of the molecular environment on the electron tunneling matrix element, we derived a new formula for the electron transfer (ET) rate as a function of the energy gap of the ET reaction. The ET rate is written by the sum of two terms due to the elastic electron tunneling mechanism and due to the inelastic electron tunneling mechanism, respectively. The term due to the inelastic electron tunneling mechanism contributes to enhance the ET rate at the very large energy gap, causing a deviation from the Marcus energy gap law. The theory is applied to the ET in the photosynthetic bacterial reaction center of *Rhodobacter sphaeroides*.

1 Introduction

Electron transfer (ET) is an intriguing reaction in chemistry and biology. In particular, the long-range ET reactions in protein environment play important roles in biological functions such as photosynthesis and respiration. The rate of such a ET was usually expressed by Marcus theory as follows:

$$k_{DA}(-\Delta G) = \frac{2\pi}{\hbar}|T_{DA}|^2 \frac{1}{\sqrt{4\pi\lambda k_B T}} \exp\left[-\frac{(-\Delta G - \lambda)^2}{\sqrt{4\lambda k_B T}}\right], \quad (1)$$

where T_{DA} is the electron tunneling matrix element, $-\Delta G$ is the free energy gap, and λ is the standard reorganization energy. In (1), the Condon approximation is used. Namely, T_{DA} is assumed to be independent of the nuclear dynamics. However, our computational studies of the molecular dynamics (MD) simulations and the quantum chemical calculations have revealed that the T_{DA} is strongly influenced by the thermal fluctuation of protein conformation [1, 2]. This result indicates that the Condon approximation is not valid and as a consequence, we cannot use (1) to evaluate the rate of the long-range ET reactions in protein.

Under this situation, we produced a new non-Condon theory of the rate of ET which happens through thermally fluctuating protein media [3]. The ET rate formula is divided into two terms of elastic and inelastic tunneling mechanisms. We applied this theory to the ET from the bacteriopheophytin anion (Bph$^-$) to the primary quinone (Q$_A$) in the bacterial photosynthetic reaction center of *Rhodobacter sphaerodies*. We analized the calculated energy gap dependence of the ET rates. In this paper, we explain origin of the "anomalous" inverted region caused owing to the inelastic tunneling mechanism.

2 Theory and Method

2.1 Non-Condon Theory

Developing the quantum transition rate theory of Prezhdo and Rossky [4], we produced a new non-Condon theory of the rate of ET which happens through a protein medium with conformation fluctuation [3]. The new theory is written as:

$$k_{DA}(-\Delta G) = \frac{2\pi}{\hbar} \frac{1}{\sqrt{4\pi\lambda k_B T}} \int_{-\infty}^{\infty} d\epsilon P(\epsilon) \exp\left[-\frac{(-\Delta G - \lambda - \epsilon)^2}{4\lambda k_B T}\right], \quad (2)$$

where $P(\epsilon)$ is the power spectrum for the autocorrelation function of $T_{DA}(t)$ with quantum correction as follows [3]:

$$P(\epsilon) = \frac{2}{1+\exp(-\epsilon/k_B T)} \frac{1}{2\pi\hbar} \int_{-\infty}^{\infty} dt \langle T_{DA}(t) T_{DA}(0) \rangle_T \exp(i\epsilon t/\hbar). \quad (3)$$

Here, $T_{DA}(t)$ is evaluated along the classical nuclear trajectory propagated on the initial electronic surface. The $\langle\ \rangle_T$ indicates the thermal average over the initial vibrational state.

We can divide the ET rate formula (2) into two terms of elastic and inelastic tunneling mechanisms as follows [3]:

$$k_{DA}(-\Delta G) = k_{DA}^{el}(-\Delta G) + k_{DA}^{inel}(-\Delta G), \quad (4)$$

where

$$k_{DA}^{el}(-\Delta G) = \frac{2\pi}{\hbar} \langle T_{DA}^2 \rangle_T \frac{1}{\sqrt{4\pi\lambda k_B T}} \exp\left[-\frac{(-\Delta G - \lambda)^2}{4\lambda k_B T}\right], \quad (5)$$

$$k_{DA}^{inel}(-\Delta G) = \frac{1}{\hbar^2} \frac{\langle T_{DA}^2 \rangle_T - \langle T_{DA} \rangle_T^2}{\sqrt{4\pi\lambda k_B T}} \int_{-\infty}^{\infty} d\epsilon \int_{-\infty}^{\infty} dt (A(t) - 1) \quad (6)$$

$$\times \exp(i\epsilon t/\hbar) \frac{2}{1+\exp(-\epsilon/k_B T)} \exp\left[-\frac{(-\Delta G - \lambda - \epsilon)^2}{4\lambda k_B T}\right]. \quad (7)$$

Here, $A(t)$ is the normalized autocorrelation function defined as:

$$A(t) = \frac{\langle T_{DA}(t) T_{DA}(0) \rangle_T - \langle T_{DA} \rangle_T^2}{\langle T_{DA}^2 \rangle_T - \langle T_{DA} \rangle_T^2}. \quad (8)$$

The inelastic term k_{DA}^{inel} becomes zero when the thermal fluctuation of T_{DA} does not happen and $A(t) = 1$ holds. The elastic term k_{DA}^{el} remains to be a finite value even when the thermal fluctuation of T_{DA} does not happen. And k_{DA}^{el} corresponds to the Marcus energy gap law.

2.2 Simulations Method

The Molecular structures used in our MD simulations consists of the whole protein of the reaction center in *Rhodobacter sphaeroides*. The initial configuration was obtained from the Protein Data Bank, entry code 1AIJ. The system was kept at 300 K. After 160 ps of equilibration, we generated a trajectory for 515 ps. We collected the conformation at energy 1 fs, which was used to calculate T_{DA}. In the calculation of T_{DA} for each protein conformation, we adopt a pruned system which consists of Bph, Q_A, and the three amino acids Trp^{M252}, Met^{M218}, and His^{M219}. The electronic states of the pruned protein are solved by the extended Hückel method. The method for computation of T_{DA} for the long-range ET is presented in [1]. More detailed procedures of our simulations are described in [3].

3 Result

On the basis of the calculated time course of $T_{DA}(t)$ for 515 ps, we calculated $P(\epsilon)$ defined in (3). The calculated $P(\epsilon)$ has an asymmetrical shape with respect to $\epsilon = 0$ due to quantum correction factor in (3). We observed the global form of $P(\epsilon)$ is a Lorentzian type at the large value of ϵ and keeps a considerable amplitude even at $\epsilon > 3,000$ cm^{-1}. (See the right side of Fig. 2.) Origin of this long tail in $P(\epsilon)$ is described in [3].

In Fig. 1, we show the logarithm of the calculated k_{DA} with use of (2) as a function of $-\Delta G$ by the solid line. In this calculation, we chose 0.60 eV for λ.

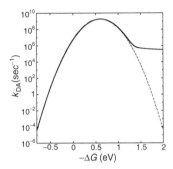

Fig. 1. The energy gap dependence of the calculated ET rate

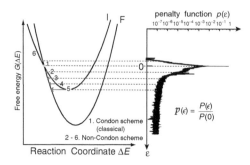

Fig. 2. Schematic diagram of Condon (1) and non-Condon (2–6) schemes

In Fig. 1, we also show the calculated k_{DA}^{el} having a shape of Marcus' parabola by the broken line. Figure 1 shows that the calculated energy gap dependence of k_{DA} is nearly Marcus parabola in most of the normal region $(-\Delta G < \lambda)$ and around maximum region $(-\Delta G \sim \lambda)$, but it does not decay substantially at large energy gap $(-\Delta G > 1.3\,\text{eV})$ in the inverted region.

4 Discussion

In Fig. 2, we explain the origin of this anomalous inverted region schematically. The elastic tunneling proceeds only through the crossing point between free energy curves of the initial and final states (arrow 1). In contrast to this, the inelastic tunneling can proceed through any noncrossing region (arrows 2–6) with a weight of penalty function defined as $p(\epsilon) = P(\epsilon)/P(0)$. The $p(\epsilon)$ obtained by simulations data does not decay rapidly after $3{,}000\,\text{cm}^{-1}$. Then, the ET eventually proceeds from around the bottom of the initial state (as arrow 5), achieving a large value of the ET rate in the case of large energy gap.

Acknowledgments

This work was supported by the Grant-in-Aid on Scientific Research (C) to T.K. from the Ministry of Education, Culture, Sports, Science and Technology of Japan. This work was also supported by Grants-in-Aid for the 21st Century COE Program "Frontiers of Computational Science," and for the Scientific Research on the priority area "Genome Information Science" from the Ministry of Education, Culture, Sports, Science and Technology of Japan, and a Grants-in-Aid from the Hori Information Science Promotion Foundation to T.Y.

References

1. Kawatsu T, Kakitani T, Yamato T (2002) J Phys Chem B 106:11356–11366
2. Nishioka H, Kimura A, Yamato T, Kawatsu T, Kakitani T (2005) J Phys Chem B 109:1978–1987
3. Nishioka H, Kimura A, Yamato T, Kawatsu T, Kakitani T (2005) J Phys Chem B 109:15621–15635
4. Prezhdo OV, Rossky PJ (1997) J Chem Phys 107:5863–5878

Density Functional Molecular Orbital Calculations on Longer DNA–DNA and PNA–DNA Double Strands

Takayuki Natsume, Yasuyuki Ishikawa, Kenichi Dedachi, and Noriyuki Kurita

Summary. Stable structures and electronic properties of hybridized DNA–DNA and PNA–DNA double strands with common base sequences were theoretically investigated by molecular orbital calculations based on the density functional theory. The computed hybridization energy in PNA–DNA is greater than that in the DNA–DNA double strand. The origin of the larger stability of PNA–DNA double strand is ascribed to the presence of greater positive charge on the hydrogen atoms involved in H-bonding, leading to stronger hybridization. This result suggests a possibility that PNA–DNA is more suitable for highly sensitive sensor than DNA–DNA.

1 Introduction

DNA array technology provides a method for rapid genotyping, facilitating the diagnosis of diseases for which gene mutations have been identified. High-throughput DNA sensors capable of detecting single-base mismatches are required for routine screening of genetic mutation and disease. Recently, a peptide nucleic acid (PNA) was employed to develop electrochemical sensors [1]. The PNA, an uncharged, hydrophobic peptide chain with tethered purine/pyrimidine bases, lacks sugar/phosphate components, and displays high affinity toward a complementary nucleic acid to form a double-stranded assembly. PNA–DNA hybridization would facilitate enhanced sensitivity in detection and ultimately could lead to highly sensitive high-throughput DNA sensors. The underlying hybridization mechanism and the interaction energy of PNA–DNA in the double-stranded assembly has never been studied theoretically or computationally. In our previous study [2], to elucidate the difference in electronic properties between DNA–DNA and PNA–DNA double strands, we theoretically investigated the binding energies between strands with three base-pairs by the first-principle molecular orbital (MO) method based on density functional theory (DFT). In the present study, we performed the same MO calculation for the four base-pairs double-strands to elucidate the dependence of binding energy on the number of base-pairs.

2 Computational Methods

Initially, we constructed the ideal B-form DNA–DNA double strand with four base-pairs of the 5'-CAGT-3' sequence by using the molecular modeling software HyperChem [3]. The 3' and 5' ends of the double strand were terminated by hydrogen atoms. To neutralize the negative charges of the PO_4 parts of DNA backbones, counter ions (Na^+) were added to each PO_4 part of the DNA backbones. The positions of these added hydrogen atoms and Na^+ ions were optimized by classical molecular mechanics (MM) calculations based on AMBER [4] force field. The remaining parts of the DNA–DNA double strand were fixed to the ideal B-form because the MM optimization often causes a distorted structure for stacked base-pairs in DNA double strand.

In constructing the PNA–DNA structure, one of the backbones in the DNA duplex was replaced by the corresponding PNA backbone, the structure of which was subsequently optimized by AMBER force field. In the present study, we adopted the antiparallel complex, in which the N-terminus of the PNA strand is oriented toward the 3' end of the complementary DNA strand. For these structures, the binding energies between single-strands were calculated by the Perdew-Wang91 functionals (PW91) [5, 6] and 6-31G** basis-sets. The basis set superposition error (BSSE) was corrected by the counterpoise calculations [7, 8]. All MO calculations were performed by the Gaussian03 program [9].

3 Results and Discussions

Figure 1 shows the constructed DNA–DNA and PNA–DNA structures with four base-pairs. It is noted that these structures were constructed based on

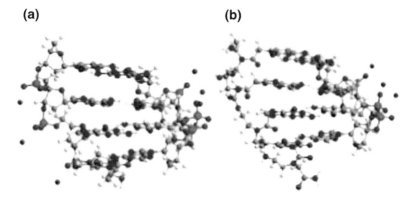

Fig. 1. Optimized structures of (a) DNA–DNA and (b) PNA–DNA with the base sequence (5'-CAGT-3') obtained by classical molecular mechanics calculations based on AMBER force field

Table 1. Binding energies (kcal mol^{-1}) between strands for DNA–DNA and PNA–DNA double strands obtained by the PW91PW91/6-31G** method. The values in parentheses are the discrepancies from those of the DNA–DNA

base-pairs	binding energy	
	DNA–DNA	PNA–DNA
3	−65.7	−71.5 (5.8)
4	−80.2	−85.2 (5.0)

the ideal B-form DNA–DNA structure having parallel stacked base pairs and optimized by fixing the structure of the base pairs, so that all base pairs are parallel to each other. In the previous molecular dynamics (MD) study [10], it was found that the geometries and dynamics of the H-bonds within base pairs are not significantly dependent on the detail nature of the backbone. Hence, the above process for constructing structures seems to be reasonable.

To check the reliability of the constructed structures, we compared the torsional angles of PNA backbones in PNA–DNA with those obtained by the previous MD simulation [10] as well as the experimental study [11, 12]. The result indicates that the values obtained by the present study are comparable to the previous results.

The binding energies for DNA–DNA and PNA–DNA having three and four base-pairs were obtained by the PW91PW91/6-31G** method. As shown in Table 1, the obtained binding energies for PNA–DNAs with three and four base-pairs are about 6 and 5 kcal mol^{-1} larger than those for DNA–DNA, respectively. This result indicates that the binding energy of PNA–DNA is significantly larger than that of DNA–DNA for both three and four base-pairs double strands. This finding is comparable to experimental results [12] showing the high affinity of PNA for a complementary nucleic acid, resulting in the formation of a double-stranded assembly.

To elucidate the origin of this large binding energy for PNA–DNA, we investigated the charge distributions around the hydrogen bonds between the two strands. Table 2 shows the Merz-Kollman's [13] atomic charges obtained by the PW91PW91/6-31G** method. The obtained atomic charges indicate that the hydrogen atoms contributing the PNA–DNA hydrogen bonds have larger positive charges than those for DNA–DNA, resulting in the larger binding energy between PNA and DNA shown in Table 1. In particular, the hydrogen atoms in the N(A)...H–N(T) and N–H(G)...N(C) bonds, which are located around the center of double strand, have the largest charge. These atomic charges are largely increased by the alternation of DNA backbone with PNA one. It should be noted that the local structures around the hydrogen bonds in the DNA–DNA and PNA–DNA are kept the same in the present calculation. Therefore, it can be concluded that the difference in binding energy between the strands for DNA–DNA and PNA–DNA originates from the

Table 2. Merz-Kollman's atomic charges around the hydrogen bonds between the two strands of DNA–DNA and PNA–DNA obtained by the PW91PW91/6-31G** method

concerned atoms	DNA–DNA	PNA–DNA
O(C)...H-N(G)	−0.56, 0.43, −0.91	−0.61, 0.49, −1.00
N(C)...H-N(G)	−0.65, 0.45, −0.80	−0.65, 0.42, −0.81
N-H(C)...O(G)	−0.88, 0.47, −0.58	−0.97, 0.52, −0.58
N(A)...H-N(T)	−0.76, 0.58, −1.08	−1.05, 0.54, −0.57
N-H(A)...O(T)	−0.57, 0.28, −0.54	−0.50, 0.30, −0.50
N-H(G)...O(C)	−0.54, 0.29, −0.55	−0.94, 0.45, −0.62
N-H(G)...N(C)	−0.59, 0.48, −0.80	−1.11, 0.99, −1.55
O(G)...H-N(C)	−0.58, 0.44, −0.83	−0.62, 0.52, −0.90
N-H(T)...N(A)	−0.59, 0.35, −0.56	−0.58, 0.32, −0.52
O(T)...H-N(A)	−0.52, 0.42, −0.84	−0.50, 0.42, −0.80
average	−0.62, 0.42, −0.75	−0.75, 0.50, −0.79

variation in backbone structure between DNA and PNA. This variation seems to cause the difference in electronic states of base-pairs in DNA–DNA and PNA–DNA, resulting in the remarkable difference in their binding energies as well as charge distributions.

In a previous experimental study [11], it was concluded that the negative charges in DNA backbones have a larger effect on the DNA–DNA double strand compared with PNA–DNA, in which the PNA backbone has no charge. In the present calculations, the negative charges of DNA backbones were neutralized by the Na^+ counter ions, so that the effect of backbone charges on the binding energy is excluded. Therefore, we concluded from Table 1 that PNA–DNA is more stable than DNA–DNA, even though the DNA backbone charges are neutralized by counter ions. This result suggests higher potential sensitivity in a PNA–DNA array sensor than in its DNA–DNA counterpart.

To elucidate the chemical reactivity of DNA–DNA and PNA–DNA, the energy levels of the highest occupied MO (HOMO) and lowest unoccupied MO (LUMO) were investigated (Table 3). The energy gaps between HOMO and LUMO for the double strands with three and four base-pairs are 0.30,

Table 3. Energy levels (eV) of HOMO and LUMO for DNA–DNA and PNA–DNA double strands obtained by the PW91PW91/6-31G** methods

base-pairs	DNA–DNA			PNA–DNA		
	HOMO	LUMO	H-L gap	HOMO	LUMO	H-L gap
3	−3.41	−3.11	0.30	−3.67	−2.86	0.81
4	−3.27	−2.99	0.28	−3.85	−3.33	0.52

0.28 eV (DNA–DNA), and 0.81, 0.52 eV (PNA–DNA), respectively. These results suggest that the chemical reactivity of PNA–DNA is more dependent on the number of base-pairs than that of DNA–DNA.

Therefore, we concluded from the DFT calculations for three and four base-pairs DNA–DNA and PNA–DNA double strands that PNA–DNA double strand is more stable than DNA–DNA. This result suggests a possibility that PNA–DNA is more suitable for highly sensitive sensor than DNA–DNA.

References

1. Wang J et al. (1997) Anal Chem 69:5200–5202
2. Natsume T et al. (2005) Chem Phys Lett 418:239–244
3. HyperChem 6.03, Hyper cube Inc., Florida, USA (2000)
4. Cornell WD et al. (1995) J Am Chem Soc 117:5179–5197
5. Perdew JP et al. (1992) Phys Rev B 46:6671–6687
6. Perdew JP et al. (1993) Phys Rev B 48:4978
7. Duijneveldt FB, Duijneveldt JGCM, Lenthe J (1994) Chem Rev 94:1873
8. Chalasinsky G, Szczesniak MM (1994) Chem Rev 94:1723–1765
9. Frisch MJ et al. (2003) Gaussian 03 (Revision B.04), Gaussian, Inc., Pittsburgh PA
10. Sen S, Nilsson L (1998) J Am Chem Soc 120:619–631
11. Eriksson M, Nielsen PE (1996) Nature Struct Biol 3:410–413
12. Rasmussen H et al. (1997) Nature Struct Biol 4:98–101
13. Besler BH, Merz KM, Kollman PA (1990) J Comp Chem 11:431–439

A Parallel Computing Framework for Nonexperts of Computers: Easy Installation, Programming and Execution of Master–Worker Applications Using Spare Computing Power of PCs

Takashi Noda, Hisaya Mine, Noriyuki Fujimoto, and Kenichi Hagihara

Summary. The parallel computing using a PC cluster or a grid is useful to solve large-scale problems in various scientific fields. However, common researchers will suffer from technical hurdles to enter their research with parallel computing unless they are computer experts. To relieve this situation, we developed a novel parallel computing framework, which supports master-worker paradigm and utilizes spare computing power of existing PCs in laboratories, offices and so on, and which is easy to install and use. This paper describes the design and implementation of our framework.

1 Introduction

Building a PC cluster and writing a parallel program may be financial and technical problems when a scientist, who is not a computer expert, needs a parallel computing in his/her study. For example, in parallel computing with a PC cluster, the purchase of many PCs and network equipments is a financial problem and installation and configurations of a parallel middleware such as MPICH, Condor, Globus Toolkit and so on are technical problems when building a parallel environment. In addition, to learn usage of parallel computing tools such as MPI libraries and concepts of communication, synchronization and load-balancing are technical problems when writing a parallel program.

To relieve these problems, we have developed a master–worker parallel computing framework in this research. An installation of the framework is easy. With the framework, a user can easily write a parallel program that utilizes spare computing power of PCs under operation in a laboratory and so on. Parallel programs over the framework are written in Java.

2 Design of the Framework

2.1 Platform Independency: Java and SOAP

The framework itself is written from scratch in Java with SOAP (thanks to Apache Tomcat and Apache Axis). Therefore, it is platform independent and independent of any other middleware such as Globus Toolkit, which is difficult to install and configure. This implies that the framework can utilize spare computing power of all the PCs in a laboratory (and so on) even if they are heterogeneous. We expect that those who gave up the introduction of such an existing middleware can also use the framework. Programming with the framework effectively hides all complexity of a master–worker program (i.e., communication, scheduling, and so on) except the problem inherent in the user application so that a user can concentrate on the implementation of the problem.

2.2 Efficient Parallel Computing: Scheduling Algorithm RR

We have adopted the master–worker paradigm, which is simple and applicable to a wide range of applications, and the scheduling algorithm RR [1,2], we had developed, in our framework. RR has a theoretically proved property such that spare computing power consumed by an application is almost optimal regardless how the spare computing power of each computer varies over time. This property is very suitable for our purpose. On the other hand, we decided not to use the MPI paradigm because an efficient MPI program is difficult to write without a lot of skills [3] especially if the computers are heterogeneous.

However, the functions such as Java, SOAP, and RR may decrease the performance of the framework. In particular, it is only in coarse-grained case (i.e., when communication latency is negligible) that the performance of RR is theoretically guaranteed. Therefore, experiments are required to verify how much the functions decrease performance of the framework. Moreover, for RR, we have proposed novel technique of simultaneous allocation and interruption of a task so that overhead of termination is reduced.

2.3 Usage of the Framework

All a user has to do for running his/her parallel program with the framework is the three steps below:

1. *Installation*: To install Java 2 SDK on a master PC, and to run a single command on each worker PC after extracting the archive of the framework.
2. *Programming*: To write three points in Java (1) preparing the arguments of tasks and (2) summing up the results of tasks in a master program, and (3) executing a task in a worker program (Figs. 1 and 2).
3. *Execution*: To run the master program with "java" command.

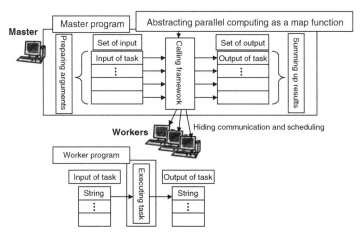

Fig. 1. Overview of the framework

3 Experiments

In the exploratory experiment, the following things turn out. Java cannot decrease the performance of the framework when each task runs over a few seconds. The latency of SOAP is a few dozens of milliseconds longer than that of TCP (Table 1). The bandwidth of SOAP is 1/7 times lower than that of MPI in the LAN environment because of computation overhead. On the other hand, communication overhead of SOAP is relatively low because increase of data size by SOAP header is less than 1 KB. However, in a master–worker program, low bandwidth is not a problem.

```
// Master program
// to call framework in arbitrary program
import hagi.scheduling.rr.master.RRMaster;

public class LeibnizPi {
  private final static int nTerms = 1000000;
  private final static int nTasks = 100;

  public static void main(String[] args) {
    String[][] argss = prepareArgss();

    String[] urls = new String[] {
      "worker01", "worker02", ...};
    String[][] retss =
      new RRMaster(urls, argss).start();

    double pi = sumupRetss(retss);
    System.out.println(pi);
  }
}
```

```
// Worker program
// to modify the template
import hagi.scheduling.rr.worker.Task;

public class TaskImpl implements Task {
  public String[] run(String[] args) {
    int k = Integer.parseInt(args[0]);
    int n = Integer.parseInt(args[1]);
    int begin = 1 + (n * 4) * k;
    int end = begin + (n * 4);

    double sum = 0.0;
    for ( int i = begin; i < end; i += 4 )
    {
      sum += 1.0 / i;
      sum -= 1.0 / (i+2);
    }

    return new String[] {
      Double.toString(sum) };
  }
}
```

Fig. 2. An example of a master and worker program

Table 1. Communication performance of SOAP

	latency (ms)	bandwidth (Mbps)
LAN (100 Mbps)		
SOAP	38	11
MPI	0.08	72
WAN (ADSL)		
SOAP	70	0.92
TCP	50	0.96

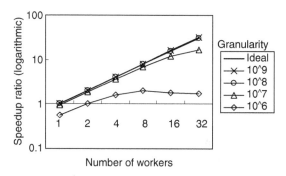

Fig. 3. Task granularity and speedup ratio (number of tasks is 1,000)

The performance evaluation shows that the speedup ratio of the framework is ideal when each task runs over a few seconds in the LAN environment up to 32 workers (Fig. 3). In this figure, the granularity (from 10^6 to 10^9) stands for computation amount of a task. A task of 10^8 runs on a few seconds. Execution time of 10^9 is 10 times longer than that of 10^8.

In addition, it was estimated by the communication latency and the task allocation time that ideal speedup is achieved in the WAN environments of 1,024 workers when each task runs over 4 min on the fastest worker in the environment.

References

1. Fujimoto N, Hagihara K (2003). Near-optimal dynamic task scheduling of independent coarse-grained tasks onto a computational grid, Proc. ICPP, 391–398
2. Fujimoto N, Tanaka K, Hagihara K (2005). Prediction-aware experimental evaluation of dynamic task scheduling algorithms for parametric study on a desktop grid, Proc. PDPTA, 475–480
3. Gropp W (2001). Learning from the Success of MPI, Proc. HiPC, 81–94

Derivation of TIS for Search Performance Quantification on HC Method

Naotoshi Hoshiya and Tomohiro Yoshikawa

Summary. The probabilistic behavior of EAs makes the theoretical investigation difficult. Then the search performances of EAs have been tested through actual simulations in most of the studies. This study tries to mathematically quantify the search performance of EAs. This paper defines "TIS" to indicate search performance of EAs and attempts to formulate it.

1 Introduction

Recently, evolutionary algorithms (EAs), which are based on the biological evolutionary process, are widely studied in a lot of fields. Genetic algorithm (GA) [1] is one of the most popular methods in EAs. It is employed for various combinational optimization problem such as Travelling Salesman Problem (TSP), Napsack Problem, and so on. However, we have to take the coding suitable for the problem to each solution, and there is no general way. Moreover, GA has various parameters such as number of chromosomes, that of generations, crossover rate, mutation rate, and so on, and the adjustment of these parameters greatly influences on the search performance. Therefore, when we apply GA, it is necessary to adjust the parameters. However, the optimal values of the parameters depend on each problem as well as the coding. This is because the search of GA progresses with probabilistic behavior, and it makes the theoretical investigation difficult. As a result, the search performances of GA have been tested through actual simulations in most of the studies. The adjustment of the parameters for genetic operations or the comparison of EA methods like "GA or HC (Hill Climbing method)?" have also been decided experientially or by trial and error. Moreover, in the statistical certainty, the comparison by trial and error needs enough trials. Some of the problems take much time for the evaluation [2], then enough trials are difficult or impossible in actual time. These are general problems in EAs.

This study tries to mathematically quantify the search performance of EAs using the parameters of the operations for the purpose of theoretical investigation of them. That is, the purpose of this study is to derive the formulas that express search performance of EAs using the parameters of the genetic operation. If we can quantify this search performance, the derived formula will enable us to compare and optimize the parameters without actual trials.

This paper defines "times of improved solutions (TIS)" to indicate search performance of EAs and attempts to formulate it. This paper employs hill climbing method to an unimodal increasing function as a basic study for this purpose, and it derives the formula for TIS that consist of the parameters of the genetic operation. This paper shows that appropriate formula is derived and it enables us to compare the performance and optimize the parameters of the operations without actual trials.

2 Problem Setting

We must consider the following three matters to quantify search performance.

(a) Probabilistic behavior of search
(b) Multi-interaction among genes (epistasis)
(c) Influence by shape of landscape of fitness function (local minimum, multimodality, and so on)

However, it is difficult to consider all these matters at the same time, and it can be one of the reasons that theoretical investigation in EAs have kept difficult so far. It may be impossible to formulate completely considering (c), because we use EA methods as we do not know the shape or the peaks of fitness functions. This study tries to formulate the search performance as long as it is possible, and it tries to quantify the impossible parts as parameters presenting the features of fitness functions and put them into the formulas. Then it is expected that we can grasp at least the guideline for the difference of the performance to the parameters or methods.

This paper excludes the factors on (b) and (c) as a basic study of this research. This paper employs an unimodal increasing function which has the problem with no interaction among genes and known fitness function. The fitness value is the number of genes with "1" in a chromosome and each gene is equivalent and independent.

When we quantify the search performance of GA even for only (a), it is thought that a lot of genetic operations (crossover, mutation, selection, and so on) make it difficult. Therefore, this paper employs Hill Climbing method that is the simplest method in evolutionary computation (EC). In the employed HC, the number of chromosomes is 1, a new chromosome is generated per a generation by mutation applied to each gene at the rate of P_m per a gene, and the elite one between them is remained to the next generation, i.e., the parameters used in this paper are the rate of mutation P_m, the chromosome length N, and the number of generations G. This paper theoretically investigates the difference of search performance by that of parameters through this simple problem setting.

3 Times of Improved Solutions: TIS

This paper defines the times of improved solutions (TIS) to indicate search performance of EAs and attempts to formulate it. The TIS is the accumulation values of improved times of the solution within G generations. This paper assumes again that the larger the TIS value is, the higher the search performance is.

3.1 Derivation of TIS

Formulation of TIS

TIS is formulated using the improved probability of solutions ε_n and given by (1). This paper derives ε_n using the decrement ratio of search ability α_n (2). α_n is the probability generating the unimproved solutions.

$$\text{TIS} = \sum_{n=1}^{G} \varepsilon_n, \tag{1}$$

$$\varepsilon_n = 1 - \alpha_n. \tag{2}$$

This paper defines γ_m and γ_{pn} as the decrement factors to the decrement ratio of search ability α_n and derives them.

Decrement Ratio of Search Ability by No Mutation γ_m

γ_m is the decrement rate of search ability by no mutation. It is derived by the probability that the mutation was applied to no genes.

Decrement Ratio of Search Ability by Changing Preimproved Genes γ_{pn}

γ_{pn} is the decrement rate of search ability by changing preimproved genes. When the solution was improved, the chromosome with genes to which mutation was applied was chosen as the elite among these chromosomes. Therefore it can be thought that to change these genes themselves causes an unimproved search. γ_{pn} is derived by the probability changing such genes again. The detail derivation process of γ_m and γ_{pn} was omitted in this paper.

The decrement ratio of search ability α_n is derived by the sum of γ_m and γ_{pn} (3).

$$\alpha_n = \gamma_m + \gamma_{pn}. \tag{3}$$

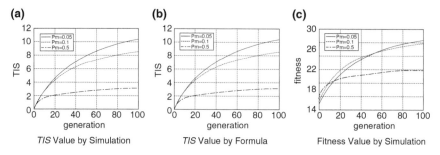

Fig. 1. TIS value by simulation and formula

3.2 Experiment Result

Figure 1a shows the transition of TIS value by actual simulation results. The simulation condition was as follows: the rate of mutation $P_m = (0.05, 0.1, 0.5)$, the chromosome length $N = 30$ and the number of generations $G = 100$. It shows the the average values in 10,000 trials. Figure 1b is the values obtained by the derived TIS in (1). The result of Fig. 1a, b shows that the transitions of them are corresponding and the derived equations for TIS are appropriate.

Moreover, Fig. 1c shows the transition of fitness value by actual simulation results. Figure 1b shows that the TIS can express superiority or inferiority relation of search performance by difference of parameters.

4 Conclusion

This research tries to mathematically quantify the search performance of EAs. As a basic study of this research, this paper quantified the search performance employing hill climbing method to an unimodal increasing function. This paper defined times of improved solutions (TIS) to indicate search performance of EAs and formulated it. The result showed that the derived equations for TIS were appropriate and the proper parameters could be found by using the TIS formula.

References

1. J. Holland, Adaptation in Natural and Artificial Systems, The University of Michigan, 1975/MIT, 1992
2. F. Idia, H. Ayai and F. Hara, Behaviour learning of face robot using human natural instruction. In Proceedings of IEEE International Workshop on RObot and Human Communication (ROMAN 01), pp. 171–176, Bordeaux/Paris, France 2001

Improvement of Search Performance of Genetic Algorithm Through Visualization of Search Process

Daisuke Yamashiro, Tomohiro Yoshikawa, and Takeshi Furuhashi

Summary. Performance in searching solutions by Genetic Algorithm (GA) depends on genetic operators and/or their parameters. For comparison of the performance with some genetic operators and/or parameters, it has been usually employed the transitions of fitness values through actual applications or the number/performance of acquired Pareto solutions in multi-optimization problems. This paper proposes a visualizing method for GA, which can visualize relative distances among chromosomes in search process and give information of not only the performance but also the effects of the genetic operations such as the diversity of chromosomes. This paper shows that the visualized data enables us to interpret the differences in search processes, and to get new information to determine efficient genetic operators and their parameters.

1 Introduction

Genetic algorithm (GA) [1] is one of the most effective methods for searching quasioptimal solutions of optimization problems within practical compilation time. The search performance of GA is heavily dependent on genetic operators and their parameters, e.g., number of individuals, probability of crossover, that of mutation. Comparison of their performance among various genetic operators or parameters has been carried out by trial and error to acquire. In this study, we aim to obtain new information and proper index to determine effective genetic operators and their parameters through visualization of search process of GA.

This paper proposes a visualization method of relative distances among individuals of GA to grasp the search process easily. The proposed method enables us to understand the state of convergence or diversity of individuals, and that of generating or scattering of new individuals generated by crossover and/or mutation. This leads us to tune/select genetic operators in a few trials. This paper investigates the effectiveness of the proposed method by applying it to a simple objective function. This paper compares two different genetic operators both of whose objectives are to keep diversity of individuals and shows that the proposed method helps us to clarify the difference of the features of each search process. Some causes of poor performance are fed back to the selection of the genetic operators and the tuning of parameters. The performance is shown to be improved.

2 Proposed Method

The relative distances among individuals of GA are defined in this paper. The individuals are plotted on a spherical surface with the relative distances being preserved. From the visualized individuals, we can understand the search process of GA. The relative distances can be defined on the distances among genotypes as well as fitness values of phenotypes. In the case of multiobjective optimization problems, relative distances in multidimensional space can be defined. This paper employs Euclidean distance among genotypes as the relative distance.

This method employs self-organizing map (SOM) [2] to visualize the relative relationships among individuals of GA. SOM is one of the unsupervised learning algorithm, and it can map relative relationships among data on two-/three-dimensional plane. This paper uses spherical SOM for avoiding the difficulty of preserving the relative distances on the corner of two-dimensional space. The flow of the proposed method is as follows:

Step 1. Definition of the relative distance among chromosomes suitable to each problem.
Step 2. Selection of genetic operators and setting of their parameters.
Step 3. Mapping of individuals on a spherical surface using SOM in each generation.
Step 4. Individuals are colored to distinguish their roots, i.e., yellow color means the new individual generated by crossover or blue means that of mutation, and so on.

3 Experiment

3.1 Results of Visualization

This section employed Schaffer F1 test function for the objective function [3]. Two different genetic operators to keep the diversity of chromosomes were compared. Genetic operator A employed uniform crossover and genetic operator B employed selection-reproduction process devised in [4]. Figure 1a, b show the visualization results at the 41st generation. In Fig. 1a, most of the individuals had converged to single point. No further increase of the fitness value of the elite was observed. The same chromosomes cannot generate a different chromosome even if the uniform crossover was applied. In this case, the efficiency of search of genetic operator A became poor, because mutation was

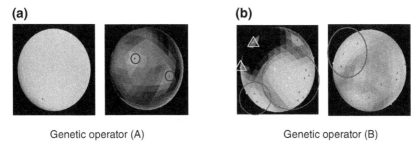

| (a) | (b) |
| Genetic operator (A) | Genetic operator (B) |

Fig. 1. Visualization results 0 at the 41st generation. (**a**) Genetic operator (A) and (**b**) genetic operator (B)

the only way to evolve new chromosomes with a higher fitness value. It might be possible to suppose from fitness transition that the individuals are almost converged. One of the advantages of the proposed method is that the convergence is clearly visualized. On the contrary, in the case of genetic operator B in Fig. 1b, a few diversity of groups were observed and the individuals converged slowly, while the individuals in each group was kept. The color information shows that the yellow squares on sphere mean that the crossover had the effect of local search in their groups and the search relatively far from them, and the blue squares mean the mutation mainly had the effect of local search.

3.2 Feedback

The purpose of this study is to make the selection of genetic operators and tuning of the parameters easy. This section describes the features of the genetic operators which were found from the visualization and the derived conclusions for feedback to selection/tuning of genetic operators. The features of the genetic operators are as follows:

1. The uniform crossover of the genetic operator A generated new individuals quite different from the parent chromosomes. However, this drastic change of chromosomes generated few individuals which survived in the selection process. The performance of the most of offsprings generated by the uniform crossover was poor.
2. The genetic operator B kept the diversity of individuals for a long generations and the diversity and convergence of individuals were well balanced. Then the genetic operators contributed both to global and local searches, which yielded slow but improvement of fitness value for longer generations.

The feedback information to improve search performance for genetic operator A was as follows:

1. The mutation of genetic operator A should be more active to generate new individuals in order to make use of the effect of uniform crossover. Thus, the probability of mutation should be made higher and the simultaneously

changed parts of a chromosomes by the mutation should be enlarged. Then the synergistic effect of the genetic operators, mutation and crossover to keep the diversity is expected.

The above information was fed back to the mutation as follows: the probability of mutation was changed to 0.2 from 0.05 and the number of inverting bits randomly selected for the mutation was increased from one to three. This genetic operator with the new parameters was called "Genetic operator A'". A trial was defined to be a computation for 100 generations. We carried out 100 trials. The t-test showed statistically significant result for $p = 0.01$. The genetic operator A' was superior to the genetic operator A to find out better fitness values. It was confirmed visually that the individuals in the case of genetic operator A' generated several groups of individuals and kept the diversity in later generations.

4 Conclusions

This paper proposed the visualization method of relative distances among individuals of GA to grasp the search process for easy understanding of the effects of the genetic operators. This paper applied the proposed method to a Schaffer F1 test function by employing two types of genetic operators. The proposed method enabled us to understand the convergence or diversity of individuals easily and we could clarify the difference of the features of the two types of genetic operators. The effective feedback was made to the genetic operators and their parameters, and the search performance of GA was improved.

Acknowledgment

This work was supported in part by "Frontiers of Computational Science" at the 21st century COE program.

References

1. Masatoshi Sakawa and Masahiro Tanaka, Genetic Algorithm, Asakura, Tokyo, 1995
2. M. M. V. Hulle, *Self Organization Map – Theory, Design and Application –* (translated by Tokutaka and Fujimura), Kaibundo, Tokyo, 1999
3. Kalyanmoy Deb, *Multi-Objective Optimization Using Evolutionary Algorithms*, Wiley, New York, 2001
4. Daisuke Yamashiro and Tomohiro Yoshikawa and Takeshi Furuhashi, Understanding effects on genetic operators with visualization of search process, Applied FCS/Techo-Sympo/MPS Symposium collected papers, pp. 33–38, 2005

A Proposal for Analysis of SD Evaluation Data by Using Clustering Method Focused on Data Distribution

Shou Kuroda, Kosuke Yamamoto, Tomohiro Yoshikawa, and Takeshi Furuhashi

Summary. In the field of marketing, companies often carry out a questionnaire to consumers to grasp their impressions of products. Analyzing the evaluation data obtained from consumers enables us to grasp the tendency of the market and to find problems and/or to make hypotheses that are useful for the development of products. This paper proposes the clustering method based on Orthogonal Procrustes Analysis (OPA). It shows that this method can investigate the similar relationships among the objects in each group and compare the similarity/difference of impression words used for the evaluation of objects among subjects in the same cluster.

1 Introduction

In the field of marketing with questionnaire survey and analysis, one of the most useful methods for quantifying the humans' impression to the object is semantic differential (SD) method [1]. The SD method uses some pairs of adjectives (impression words) having antithetical meanings with five/seven discrete levels. This paper proposes a clustering method based on orthogonal procrustes analysis (OPA) for SD method using the similarity of impression relationships to objects. The proposed method can extract the subjects among whom the distributed structures of the SD evaluation data are similar by considering the rotation of the distribution, and it has no information loss because it directly treats the data itself. The analysis using this method leads to the discovery of majority/minority groups and/or the characteristics of the groups. In addition, it enables us to analyze the similarity/difference of objects and impression words among clusters and/or subjects by comparing the cluster center and/or transformation matrices. This paper applies the proposed method to an actual SD evaluation data obtained from subjects. It shows that this method can investigate the similar relationships among the objects in each group and compare the similarity/difference of impression words used for the evaluation of objects among subjects in the same cluster.

2 Proposed Method

The proposed method defines the value of objective function of the OPA as the distance between individuals. Matrix $\boldsymbol{A}_i$ represents the ith SD evaluation data set for clustering, where number of data points is n, and number of dimensions is p. Lagrange function for optimizing the objective function in the proposed method is described by (1) using the orthogonalization transformation matrix $\boldsymbol{T}_i$ for transforming each matrix $\boldsymbol{A}_i$ into the kth cluster center matrix $\boldsymbol{B}_k$.

$$F(\boldsymbol{T},\boldsymbol{B},\boldsymbol{L}) = \sum_{k=1}^{C} \sum_{i \in G_k} \text{tr}\left\{(\boldsymbol{A}_i \boldsymbol{T}_i - \boldsymbol{B}_k)^{\text{T}}(\boldsymbol{A}_i \boldsymbol{T}_i - \boldsymbol{B}_k)\right\}$$
$$+ \sum_{j=1}^{N} \text{tr}\left\{\boldsymbol{L}_j\left(\boldsymbol{T}_j^{\text{T}} \boldsymbol{T}_j - \boldsymbol{I}\right)\right\}. \quad (1)$$

The optimal $\boldsymbol{B}_k$ (cluster center matrix) is

$$\boldsymbol{B}_k = \frac{1}{|G_k|} \sum_{i \in G_k} \boldsymbol{A}_i \boldsymbol{T}_i. \quad (2)$$

The optimal $\boldsymbol{T}_i$ (transformation matrix) is solved by

$$\boldsymbol{T}_i = \boldsymbol{V}_i \boldsymbol{W}_i^{\text{T}}, \quad \boldsymbol{M}_i = \boldsymbol{A}_i^{\text{T}}\left(\sum_{k=1}^{C} \boldsymbol{B}_k\right), \quad (3)$$

where $\boldsymbol{V}_i$ and $\boldsymbol{W}_i$ are orthogonal $[p \times p]$ matrices consisting of the eigenvectors of $\boldsymbol{M}_i \boldsymbol{M}_i^{\text{T}}$ and $\boldsymbol{M}_i^{\text{T}} \boldsymbol{M}_i$, respectively.

The algorithm of the proposed method is as follows:

Step 1. Each distribution matrix $\boldsymbol{A}_i$ is preprocessed for conforming the distribution center and the scale, if it is necessary.
Step 2. Initial cluster k ($k \in \{1,\ldots,C\}$) for each $\boldsymbol{A}_i$ is randomly assigned.
Step 3. Initialization of $\boldsymbol{T}_i = \boldsymbol{I}$
Step 4. Each cluster center matrix $\boldsymbol{B}_k$ is calculated by (2).
Step 5. Each distribution matrix $\boldsymbol{A}_i$ is assigned to the cluster that has the smallest evaluation value to $\text{tr}\left\{(\boldsymbol{A}_i \boldsymbol{T}_i - \boldsymbol{B}_k)^T (\boldsymbol{A}_i \boldsymbol{T}_i - \boldsymbol{B}_k)\right\}$.
Step 6. $\boldsymbol{T}_i$ is recalculated by (3).
Step 7. $\boldsymbol{B}_k$ is recalculated by (2).
Step 8. If each $\boldsymbol{B}_k$ and $\boldsymbol{T}_i$ is converged, iteration is end. Otherwise, go to Step 5.

3 Experiment

Thirty laptop photo images were employed as the objects. This experiment requested 22 subjects to evaluate the objects by SD method including 21 impression words.

3.1 Result and Discussion

This section discusses the result by the proposed method when the number of cluster was five. The number of subjects belonging to each cluster was $\{2, 4, 4, 8, 4\}$. A majority (cluster 4) and a minority (cluster 1) group was found in the SD evaluation data. This paper applied the cluster analysis by Ward's method to the distance matrix (similarity matrix) among objects calculated by each cluster center distribution for the cluster 1 and cluster 4. The acquired dendrogram showed that they had some characteristic differences among objects. For example, though the distance in the dendrogram between object 6 and 14 was less than 0.5 in cluster 4 (majority), those in cluster 1 (minority) was approximately 1.5. In addition, though object 20, 23, 18, 29 were same cluster (0.5) in cluster 4, object 20, 23 and object 18, 29 were the farthest distance (3.0) in cluster 1. It is thought that these two groups have different individuality for similarities among objects.

3.2 Analysis of Impression Evaluation

Equation (4) represents the relationship among the SD evaluation data $\boldsymbol{A}_i$, the orthogonal transformation matrix $\boldsymbol{T}_i$ of subject i and the cluster center matrix $\boldsymbol{B}_k$ of cluster k that subject i belongs.

$$\boldsymbol{B}_k \simeq \boldsymbol{A}_i \boldsymbol{T}_i. \tag{4}$$

Principal component analysis (PCA) is applied to $\boldsymbol{B}_k$.

$$\boldsymbol{P}_k = \boldsymbol{B}_k \boldsymbol{C}_k \simeq \boldsymbol{A}_i \boldsymbol{T}_i \boldsymbol{C}_k, \tag{5}$$

where $\boldsymbol{P}_k$ is the matrix of principal component (PC) score, and $\boldsymbol{C}_k$ is that of PC coefficient. When we consider $\boldsymbol{T}_i \boldsymbol{C}_k$ as the matrix of PC coefficient of $\boldsymbol{A}_i$, we can relate the SD evaluation data to the objects.

Figure 1 shows the result of applying (5) to the cluster 1 (subject 11 and 13). The left figure shows the distribution of objects in cluster 1. The right figure shows the relationship of impression words between subject 11 and subject 13 corresponding to the left figure. The numbers plotted in the left figure represent the indexes of objects. Cumulative contributing rate by these two principal components was 60.2%. In the right figure, impression words represented by principal component coefficients of each subject are plotted on principal components of the left figure. In subject 11, IW-02 (Thick–Thin), IW-06 (Unrefined–Neat) and IW-18 (Heavy–Light) have a strong influence to PC1 (horizontal axis). Subject 13 also has a similar tendency. They evaluated Obj-{04, 06, 10, 14, 21, etc.} by using structure impression words of laptop products whose meaning can be easily shared. It is inferred that these objects have stronger structural feature than other objects. In addition, the right figure shows they were impressed "Beautiful" (IW-01) to the objects having "Neat," "Thin," and "Light" structure. Moreover, the bottom-left area in the

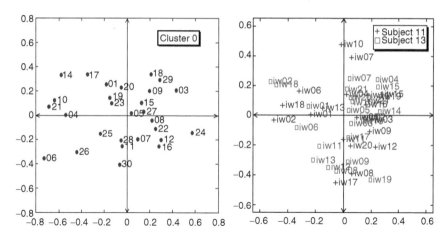

Fig. 1. PC score of objects and impression words of subjects

right figure shows that subject 13 evaluated object 06 using some impression words such as "Neat," "Soft," "Noble," and so on while subject 11 used "Thin" and "Casual." It suggests that the evaluation of subject 13 for these objects in this area was more meticulous than subject 11. In addition, "Thin" for subject 11 was close to "Neat" for subject 13.

4 Conclusion

This paper defined distributed structure of objects in the space of impression words as individuality, and proposed a clustering method based on OPA for clustering of the individuality. This paper applied the proposed method to SD evaluation data obtained from subjects, and showed that this method could investigate the similar relationships among the objects in each group and compare the similarity/difference of impression words used for the evaluation of objects among subjects in the same cluster.

Acknowledgment

A part of this study was supported by the 21st century COE program "Frontiers of Computational Science," Grant-in-Aid for Scientific Research from the Ministry of Education, Culture, Sports, Science and Technology (base research (c) (2) No.16500126) and Toyota Central R&D Labs., Inc.

Reference

1. K. Yamamoto, T. Yoshikawa, and T. Furuhashi, "A proposal on stratification method for SD evaluation data considering individuality," in *FAN Symposium 2005 in Kyoto, Society of Instrument and Control Enginners (in Japanese)*, 2005, pp. 495–500.

Multiobjective Evolutionary RBF Networks and Its Application to Ensemble Learning

Nobuhiko Kondo, Toshiharu Hatanaka, and Katsuji Uosaki

Summary. This paper considers a pattern classification by the ensemble of evolutionary RBF networks. Mathematical models generally have a dilemma about model complexity, so the structure determination of RBF network can be considered as the multi-objective optimization problem concerning with accuracy, complexity, and smoothness of the model. The set of RBF network are obtained by multi-objective evolutionary computation, and then RBF network ensemble is constructed of all or some RBF networks at the final generation. Some experiments on the benchmark problem of the pattern classification demonstrate that the RBF network ensemble has comparable generalization ability to conventional ensemble methods.

1 Introduction

Over the past decade, the ensemble learning in which outputs of weak learners are combined has taken much attention in the fields of the machine learning and the artificial neural networks. It is a promising approach for giving a high generalization ability and solving complex problems. The ensemble adopting neural networks (NNs) as members is called NN ensemble. In NN ensemble, the problems are how to construct individual NNs and how to combine outputs.

In this study, the construction of the ensemble from the set of RBF (radial basis function) networks, which is a kind of NNs, obtained from the perspective of the multiobjective optimization is considered. At first the set of Pareto-optimal RBF networks based on evaluation of approximation ability and structure complexity using evolutionary computations which show good results on multiobjective optimization is obtained. The RBF network ensemble is constructed from the Pareto set since the obtained Pareto-optimal RBF networks are diverse on their structure and the diversity makes them suitable to members of the ensemble. The RBF network ensemble is applied to the pattern classification problem and evaluated by numerical experiments.

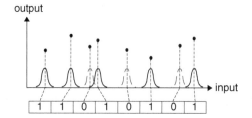

Fig. 1. Genetic representation of RBF network

2 Ensemble of Pareto RBF Networks

Let $\phi_j(\boldsymbol{x})$ be the jth basis function, the output of RBF network $\hat{y}$ is represented as follows :

$$\hat{y} = w_0 + \sum_{j=1}^{m} w_j \phi_j(\boldsymbol{x}), \quad \phi_j(\boldsymbol{x}) = \exp\left(-\frac{(\boldsymbol{x} - \boldsymbol{c}_j)^{\mathrm{T}}(\boldsymbol{x} - \boldsymbol{c}_j)}{2\sigma_j^2}\right) \quad (1)$$

Here, $\boldsymbol{x} = (x_1, x_2, \ldots, x_d)^{\mathrm{T}}$ is the input vector, $\boldsymbol{c}_j = (c_{1j}, c_{2j}, \ldots, c_{dj})^{\mathrm{T}}$ and σ_j^2 are the jth center vector and the width parameter, and superscript T represents the transposition of matrix. m indicates the number of the hidden layer neuron which is corresponding to the structure of the networks. w_j is the weight from jth neuron to the output layer, which is generally estimated by a numerical optimization algorithms.

In this paper, the number of the hidden layer neuron m and the centers $\boldsymbol{c}_j$ are estimated by multiobjective GA, and the weights w_j are estimated by real-coded GA. The widths σ_j is assumed to be a constant value for simplicity.

The binary string whose length is same to the number of training data is used as the chromosome of multiobjective GA. Each gene corresponds to the position of one of training data. If a gene value is 1, the center of basis

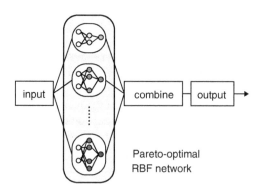

Fig. 2. RBF network ensemble

function is placed at the corresponding position. Conversely if a gene value is 0, basis function is not placed. This coding can represent the number of basis functions and the centers of them.

Since there generally exists a dilemma about model complexity, we consider the learning of RBF networks as the multiobjective optimization problem. In particular, the following three values are considered as fitness values on multiobjective GA. The first is the number of hidden layer neuron, which represents the model complexity. The second is the value of half-adjust of log of MSE (mean squared error) in the third place of decimal point, which represents the representation ability of the model. The third is the sum of absolute weights $\sum_{j=0}^{m} |w_j|$, which is based on the concept of regularization and represents the smoothness of the model. The relation of domination among individuals are investigated based on the minimization of these three fitness value. MSE is used as the fitness of the real-coded GA which estimates connection weights.

An algorithm for the construction of Pareto-optimal RBF networks is constructed based on above. It is the algorithm of double structure in which the multiobjective GA which determine the number and the centers of basis functions includes the real-coded GA which estimates the connection weights. NSGA-II [1] is used as the multiobjective GA. When the offsprings are generated, the connection weights of each individual are estimated by real-coded GA. UNDX [2] and MGG [3] are used for the crossover and the generation change model, respectively.

As stated in Sect. 1, the ensemble learning has taken much attention in the fields of the machine learning and the artificial neural networks over the past decade. It is known that the members should be as accurate as possible and be diverse at the same time to construct the ensemble having good generalization ability. In this study, we consider the construction of the RBF network ensemble from the set of RBF network obtained by multiobjective evolutionary algorithms, since the obtained Pareto-optimal RBF networks are diverse on their structure and the diversity makes them suitable to members of the ensemble.

As the members of the ensemble, following three patterns are considered:

1. All RBF networks at the final generation.
2. All Pareto-optimal RBF networks at the final generation.
3. Among all RBF networks at the final generation, RBF networks whose accuracy rate for training data is higher than the average of the set.

For descriptive purposes these three patterns are named *member 1*, *member 2*, and *member 3*. The outputs of ensemble is calculated by combining the outputs of these members.

As the combination method, following three patterns are considered :

- *Simple averaging.* The output of the ensemble is formed by a simple averaging of output of members.

Table 1. Parameters of GAs

multiobjective GA	
number of generation	10
population size	30
crossover probability	0.7
mutation probability	0.1
real-coded GA	
number of generation	20
population size	30
number of crossover	30

- *Majority voting.* The output of the greatest number of members is the output of the ensemble.
- *Winner-takes-all.* The output of the ensemble is the output of the member whose output has the highest activation.

3 Numerical Simulations

We dealt with the Australian credit card assessment problem as a benchmark of the pattern classification, whose data set is provided by UCI Machine Learning Repository [4]. The data set contains 690 patterns. The patterns have 14 attributes which include six numeric and eight discrete ones. The output has two classes. The proposed algorithm is evaluated by 10-fold cross-validation. The parameters of GAs are shown in Table 1. The widths of basis functions σ^2 are set to 5.0. Table 2 shows the results of cross-validation

Table 2. Accuracy rates for test data

	*member*1	*member*2	*member*3	MPANN	EENCL
simple averaging					
mean	0.865	0.862	0.865	0.865	0.855
SD	0.037	0.042	0.039	0.043	0.039
max	0.942	0.942	0.942	0.928	0.913
min	0.826	0.812	0.812	0.797	0.797
majority voting					
mean	0.862	0.865	0.865	0.862	0.857
SD	0.043	0.039	0.037	0.049	0.039
max	0.957	0.942	0.942	0.928	0.913
min	0.812	0.812	0.826	0.783	0.812
winner-takes-all					
mean	0.862	0.861	0.865	0.858	0.865
SD	0.045	0.043	0.037	0.044	0.028
max	0.957	0.942	0.942	0.913	0.913
min	0.812	0.797	0.826	0.797	0.812

about the accuracy rates and compares results of our proposed method with MPANN by Abbass [5] and EENCL by Liu et al. [6], which are NN ensemble methods.

The accuracy rates of the ensembles constructed of *member 1* or *member 3* are more stable than those of *member 2*. The number of Pareto individuals of the final generation is not very large, so there is a possibility that the members of ensemble of *member 2* are not so diverse.

The mean of accuracy rate of our proposed method with and *member 3* are equivalent to those of MPANN and EENCL. Moreover, the variance is smaller and the maximum and the minimum are higher than those of these two methods. These results indicate our proposed method is more accurate and stable than conventional ensemble methods.

4 Conclusions

In this paper, the method in which RBF network ensemble is constructed from Pareto-optimal set obtained by multiobjective evolutionary computation is proposed and its behavior is evaluated by numerical experiments. The benchmark tests of pattern classification indicate that our proposed method can get better or comparable results to conventional evolutionary ensemble methods by selecting the member selection method, the ensemble method and the parameters appropriately, in other words this result indicates there is a issue of how to select such parameters, etc.

References

1. Deb K, Pratap A, Agarwal S, and Meyarivan T (2002). A fast and elitist multi-objective genetic algorithm: NSGA-II, IEEE Transactions on Evolutionary Computation, 6(2), pp. 182–197
2. Ono I and Kobayashi S (1997). A real-coded genetic algorithm for function optimization using unimodal normal distribution crossover, Proceedings of Seventh International Conference on Genetic Algorithms, pp. 246–253
3. Sato H, Ono I, and Kobayashi S (1997). A new generation alternation model of genetic algorithms and its assessment, Journal of Japanese Society for Artificial Intelligence, 12(5), pp. 734–744
4. http://www.ics.uci.edu/~mlearn/
5. Abbass HA (2003). Pareto neuro-evolution : constructing ensemble of neural networks using multi-objective optimization, The IEEE Congress on Evolutionary Computation 2003, 3, pp. 2074–2080
6. Liu Y, Yao X, and Higuchi T (2000). Evolutionary ensembles with negative correlation learning, IEEE Transactions on Evolutionary Computation, 4(4), pp. 380–387

A New Concept of Cerebral Autoregulation Using Mathematical Expressions

Tomoki Youkey Shiozawa, Hiroki Takada, and Masaru Miyao

Summary. Stress was induced by graded head-up tilt (HUT) in 13 healthy subjects, and their beat-by-beat values of cerebral blood flow velocity (CBV) $[\mathrm{cm\,s^{-1}}]$ and arterial blood pressure (ABP) [mmHg] were measured. Low frequency (LF: 0.07–0.2 Hz) and high frequency (HF: 0.2–0.4 Hz) transfer function gains (TFGs) $[(\mathrm{cm\,s^{-1}})/\mathrm{mmHg}]$ were calculated to assess the dynamic cerebral autoregulation (DCA). By mathematically expressing the transition of DCA during the HUT, a new concept of cerebral autoregulation was discovered. This type of cerebral autoregulation rapidly leads DCA to a different stationary state after exposure to a sudden change in the gravitational load.

1 Introduction

It is classically believed that when the changes in the arterial blood pressure (ABP) are kept within a regular or permissible range, the cerebral blood flow volume (CBFV) remains constant. This autoregulatory system of the cerebral hemodynamic system is well known as "cerebral autoregulation" [1]. However, it is also known that CBFV can sometimes temporarily respond to spontaneous but rapid and transient changes in the ABP [2]. This means there is another type of autoregulatory system called "dynamic cerebral autoregulation" (DCA) that suppresses this transient change in the CBFV [3]. DCA has been reported to be generally frequency dependent, and it can be quantified by the transfer function analysis of the spontaneous beat-by-beat fluctuations in the ABP (input) and CBFV (output) [2, 3].

The purpose of the present study is to express the transition in DCA during orthostatic stress in humans by using mathematical equations.

2 Materials and Methods of Experimental Study

Stress was induced in thirteen healthy subjects (10 males and 3 females; age, 28.8 ± 4.6 years; height, 168.8 ± 7.5 cm; weight, 65.8 ± 12.0 kg, (mean $\pm$ SD)) by graded head-up tilt (HUT; 0, 15, 30, 45, 60, 90 degrees of passive standing for 6 min each). The blood flow velocities of middle cerebral arteries (CBVs) were measured by transcranial Doppler ultrasound (TCD; Waki, Atys Medical Co. Ltd.), and the ABPs were measured by using photoplethysmography (Finapres, Ohmeda Co. Ltd.). Beat-by-beat mean ABP and mean CBV were obtained by integrating the signals within each cardiac cycle. Transfer function gain (TFG) [$(cm\,s^{-1})$/mmHg] between the fluctuations or changes in the mean ABP (input) and CBV (output) in the low frequency (LF: 0.07–0.2 Hz) and high frequency (HF: 0.2–0.4 Hz) ranges [4] was calculated to assess DCA [3]. The coherence function was also calculated in the same frequency ranges to assess the linear relationship between these two variables and the reliability of the TFG [5].

Variables were compared by using one-way repeated-measures ANOVA for each HUT (the control was supine position and the HUT postures were at each 15–90 degrees). This was done in conjunction with Dunett's post hoc tests for comparisons with a control value. A p-value of <0.05 was accepted as statistically significant, and all the data were represented as mean ±SE.

3 Results of Experimental Study

All the values of the coherence function were more than 0.5, and each datum of TFG was considered to be reliable from this result.

By comparison with the control TFG (supine position), it was found that the TFG tended to decrease with an increase in the orthostatic stress and the decrease in TFG was statistically significant at 45- and 90-degree HUT in the LF range and at 60- and 90-degree HUT in the HF range (Fig. 1).

4 Discussion

The TFG tended to decrease due to an increase in the orthostatic stress. This indicates that the operation of DCA tends to be reinforced by an increase in

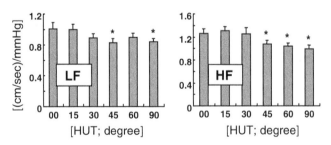

Fig. 1. Transition in transfer function gain at LF and HF ranges; $n = 13$, mean ± SE, $^{*}p < 0.05$, $^{**}p < 0.01$ (vs. 0-degree HUT)

the gravitational stress along the downward vector of the axis of the human body (Gz).

Mathematical Expressions. Although changes in some inputs or control variables (orthostatic stress in this study) influence some outputs or measured values (e.g., ABP, CBV, or DCA) due to some interactions of several physiological functions, these effects can usually be observed through some "black boxes" because the real processes of these effects cannot easily be analyzed. However, we can express these processes by using some simple dynamic models. To achieve this, we attempt to use nonhomogeneous second-order differential equations with constant coefficients.

The concept of time course is adopted during the process from the input to output. Let us suppose that the DCA, i.e., TFG, will be analyzed as the output "T." Let us also suppose the following control variables (I) a change in the Gz denoted by "u"; (II) a parameter that indicates a difference in the frequency range of the fluctuations in ABP or CBV, denoted by "α"; (III) the monitoring duration time during which the control stress "τ" is applied; and (IV) the time, during which the input u changes to output T, denoted by "t." This output T can be expressed as:

$$T = F(u(\tau), \alpha, t). \quad (1)$$

A dynamic model that expresses an operation of a control variable (input u) toward TFG (output T), can be written as:

$$a \cdot \frac{d^2 T(t)}{dt^2} + b \cdot \frac{dT(t)}{dt} + c \cdot T(t) = u(t). \quad (2)$$

Since, in this case, $u(t)$ is a stepped equation (2) can be expressed as:

$$a \cdot \frac{d^2 T(t)}{dt^2} + b \cdot \frac{dT(t)}{dt} + c \cdot T(t) = u_1(t_0 < t \leqslant t_1) \quad (3)$$

$$u_2(t_1 < t \leqslant t_2).$$

$$\vdots \quad \vdots$$

$$u_n(t_{n-1} < t \leqslant t_n)$$

Here the coefficients "a," "b," and "c" are functions that include several known parameters, for example, external conditions (LF, HF), and so on. They can be expressed as:

$$a = a(\alpha, \beta, \ldots), \ b = b(\alpha, \beta, \ldots), \ c = c(\alpha, \beta, \ldots) \quad (4)$$

A solution of the dynamic equation must be the same as (1); it should consist of a stationary solution $T_i^*(u_i)$ and a complimentary solution T_i^C. According to (2), the stationary solution $T_i^*(u_i)$ can be expressed as:

$$a \cdot \frac{d^2 T_i^*}{dt^2} + b \cdot \frac{dT_i^*}{dt} + c \cdot T_i^* = u_i. \quad (5)$$

According to the definition, $d^2T_i^*/dt^2 = 0$ and $dT_i^*/dt = 0$ at the stationary state.

$$\therefore T_i^* = \frac{u_i}{c}. \tag{6}$$

On the other hand, the complimentary solution T_i^C can be expressed as:

$$a \cdot \frac{d^2T_i^C}{dt^2} + b \cdot \frac{dT_i^C}{dt} + c \cdot T_i^C = 0 \tag{7}$$

$$\therefore T_i^C = C_1 \cdot e^{r_1 \cdot t} + C_2 \cdot e^{r_2 \cdot t} \tag{8}$$

(C_1 and C_2 are settled by initial conditions.

$$r_1 = \frac{-b + \sqrt{b^2 - 4ac}}{2a}, r_2 = \frac{-b - \sqrt{b^2 - 4ac}}{2a}, \frac{b}{a} > 0.\Big)$$

Since the human body is an equilibrium and conservative system, both r_1 and r_2 can be supposed to be negative values. TFG (DCA) is considered to be the sum of the stationary solution and the complimentary solution expressed as:

$$T_i = T_i^* + T_i^C = \frac{u_i}{c} + C_1 \cdot e^{r_1 \cdot t} + C_2 \cdot e^{r_2 \cdot t} \tag{9}$$

$$\therefore \lim_{t \to \infty} T_i = T_i^* = \frac{u_i}{c}. \tag{10}$$

In the experimental study, the measurement of "T_i" was to be started 2–3 min after u_i changed (t = 2–3 min). It has already been strictly confirmed in a previous study that TFG did not change in a stationary condition for several hours [6]. A schematic diagram of this process is shown in Fig. 2-a. The process that was repeated from "u_1" to "u_n" is also shown in Fig. 2-b.

On the other hand, the values of the coefficient c in both the LF and HF ranges increased almost linearly with an increase in the gravitational stress (Fig. 3). These results imply that a gravitational load reinforces a type of function that leads DCA into a different stationary state as quickly as possible after

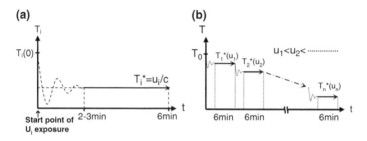

Fig. 2. Schematic diagrams of the transitions in TFG

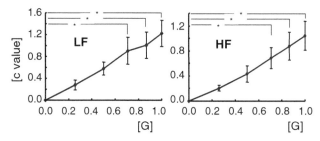

Fig. 3. Relationship between the Gz [G] and c value [c: $T_i^* = u_i/c$], $n = 13$, mean $\pm$ SE, $^*p < 0.05$ (vs. 0-degree HUT)

a change in Gz. This type of a physiological function has not been reported before. In this study, we propose the new concept of cerebral autoregulatory system.

References

1. Paulson OB et al. (1990) Cerebrovasc Brain Rev 2: 161–192
2. Giller CA (1990) Neurosurgery 27: 362–368
3. Zhang R et al. (1998) Am J Physiol 274: H223–241
4. Welch PD (1967) IEEE Trans Audio Electroacoust AU-15: 70–73
5. Marmarelis, VZ (1988) Ann Biomed Eng 16: 143–157
6. Zhang R et al. (2000) Am J Physiol 278: H1848–1855

Evolution Strategies on Desktop Grid Environment Using Scheduling Algorithms RR for Same Length Tasks

Yoshiyuki Matsumura, Noriyuki Fujimoto, Xin Yao, Yoshikazu Murayama, and Kenichi Hagihara

Summary. This chapter empirically investigates the effect of List Scheduling with Round-robin order Replication (RR) on the computational cost of Evolution Strategies (ES). Computer experiments using a desktop grid environment in a local network show that RR cuts the computational time of ES and verify that the actual energy loss of RR is less than the theoretical maximum value.

1 Introduction

In [2], we have been investigating the use of evolution strategies (ES) for the automatic design of robot controllers. We have made a number of advances, including the invention of more efficient evolutionary methods. Despite this the robot controllers still take considerable amounts of computational time to evolve. This is a very time consuming process. Then, we plan to improve the approach by distributing the tasks over many computers at Shinshu University, the university of Birmingham and Osaka University using Grid Computing.

The most common objective function of grid task scheduling problems (both for a grid and for a nongrid parallel machine) is the makespan. However, on a grid, the makespan of a nonoptimal schedule may be much longer than the optimal makespan because the computing power of a grid varies over time. For example, consider an optimal schedule with makespan OPT. If a grid is suddenly slowed down at time OPT and the slow speed situation continues for a long period, then the makespan of the second optimal schedule is far from OPT. So, if the criterion of a schedule is the makespan, there is no approximation algorithm in general for scheduling onto a grid [1].

In contrast to this, recently the authors [1] proposed a novel criterion for a schedule called TPCC and gave a $(1 + [m(\log_e(m-1)+1)]/n)$-approximation algorithm for minimum TPCC scheduling of a coarse-grained parameter-sweep application with the same length tasks where n is the number of tasks and m is the number of processors. TPCC represents the total computing power consumed by the parameter-sweep application. This approximation algorithm called List Scheduling with round-robin order replication (RR) does not use any prediction information on the performance of underlying resources. Hence, this result implies that, regardless how the speed of each processor varies over time, the consumed computing power can be limited within $(1 + [m(\log_e(m - 1) + 1)]/n)$ times the optimal one in such a case.

This paper empirically investigates the effect of RR on the computational cost of ES. Computer simulations are conducted which compare the theoretical maximum energy loss with the actual measurement for a given scheduling task.

2 Classical Evolution Strategies

The Classical ES (CES) [3] algorithm adopted in this paper is described as follows:

1. Generate an initial population of μ individuals, and set $g = 1$. Each individual is taken as a pair of real-valued vectors $(\boldsymbol{x}_i, \boldsymbol{\eta}_i), \forall i \in \{1, \ldots, \mu\}$, where $\boldsymbol{x}_i$ and $\boldsymbol{\eta}_i$ are the i-th coordinate value in the object and the strategy parameters (larger than zero), respectively.
2. Evaluate the objective value for each individual $(\boldsymbol{x}_i, \boldsymbol{\eta}_i), \forall i \in \{1, \ldots, \mu\}$ in the population, based on the objective function $f(\boldsymbol{x}_i)$.
3. Each parent $(\boldsymbol{x}_i, \boldsymbol{\eta}_i), i = 1, \ldots, \mu$, creates λ/μ offspring on average, so that a total of λ offspring are generated. The offspring are generated as follows: for $i = 1, \ldots, \mu$, $j = 1, \ldots, n$, and $p = 1, \ldots, \lambda$,

$$\eta'_p(j) = \eta_i(j)\exp\{\tau' N(0,1) + \tau N_j(0,1)\}, \quad (1)$$

$$x'_p(j) = x_i(j) + \eta'_p(j) N_j(0,1), \quad (2)$$

where $x_i(j), x'_p(j), \eta_i(j)$, and $\eta'_p(j)$ denote the j-th component of the vectors $\boldsymbol{x}_i, \boldsymbol{x'}_p, \boldsymbol{\eta}_i$, and $\boldsymbol{\eta'}_p$, respectively. $N(0,1)$ denotes a normally distributed one-dimensional random number with mean zero and standard deviation one. $N_j(0,1)$ indicates that the random number is generated anew for each value of j. The factors τ and τ' are commonly set to

$$\left(\sqrt{2\sqrt{n}}\right)^{-1}$$

and

$$\left(\sqrt{2n}\right)^{-1}.$$

Various types of recombination operators can also be applied before calculating equations (1) and (2).

4. Calculate the fitness of each offspring $(\boldsymbol{x'}_i, \boldsymbol{\eta'}_i), \forall i \in \{1, \ldots, \lambda\}$, according to $f(\boldsymbol{x'}_i)$.

5. Sort offspring $(\boldsymbol{x}'_i, \boldsymbol{\eta}'_i)$, $\forall i \in \{1, \ldots, \lambda\}$ according to their fitness values, and select the μ best offspring out of λ to be parents of the next generation.
6. Stop if the halting criterion is satisfied; otherwise, $g = g + 1$ and go to step 3.

3 The Dynamic Scheduling Algorithm RR

The dynamic scheduling algorithm RR [1] in this paper is described as follows: First of all, a data structure called a ring is defined. Then, using a ring of tasks, RR is described. A ring of tasks is a data structure which manages a set of tasks. The tasks in a ring have a total order such that no task has the same order as any other task. A ring has a head which points to a task in the ring. The head in a ring is initialized to point to the task with the lowest order in the ring. The task pointed to by the head is called the current task. The next task in a ring is defined as follows. If the current task is the task with the highest order in the ring, then the next task in the ring is the task with the lowest order in the ring. Otherwise, the next task in a ring is the task with the minimum order of those tasks that are of higher order than the current task. A head can be moved so that the head points to the next task. Hence, using a head, the tasks in a ring can be scanned in the round-robin fashion. Arbitrary tasks in a ring can be removed. If the current task is removed, then a head is moved so that the next task is pointed to.

RR runs as follows. At the beginning of the dynamic scheduling by RR, every processor is assigned exactly one task, respectively. If some task of the assigned tasks is completed, then RR receives the result of the task and assigns one of the yet unassigned tasks to the processor. RR repeats this process until all the tasks are assigned. At this point in time, exactly m tasks remain uncompleted. RR manages these m tasks using a ring of tasks. Then, RR repeats the following process until all the remaining m results are received: If the task instance of some task v is completed on processor p, then RR receives the result of v from p, kills all the task instances of v running on processors except p, removes v from the ring, selects task u in the ring in the round-robin fashion, and replicates the task u onto the processor p.

The original of task v is the task instance which is assigned earliest of all the task instances of v. The other task instances of v are called the replicas of v. Notice that the original of every task v is unique in a schedule generated by RR.

4 Computer Simulations

We conduct a series of computer simulations using the Web service. In this paper, the results for the sphere function are presented. This test function defines a 1,000 dimensional problem. The experimental setup is based on Yao

Table 1. The actual measurements and theoretical value on homogeneous machines (RedHat 9.0 & Xeon 2.4 GHz) in the local network (1,000-dimension)

the number of CPU	averaged time(s) (measure)	max loss(%) (measure)	max loss(%) (theory)
1	30,511.31	0	0
2	15,255.62	1.09	4.00
3	10,375.97	2.23	10.16
4	7,904.32	6.88	16.79
5	6,175.05	11.36	23.86
6	5,523.81	11.22	31.31

and Liu [3]: $(\mu, \lambda) = (30, 200)$ with Gaussian mutation, no recombination, and no correlated mutations. Last generation is 2,000 in one task. The number of independent tasks is 50. All simulations are independently calculated for 10 times.

In [1], we found that theoretical maximum loss is $\{(1 + (m \log_e(m-1) + m)/n) - 1\} \times 100$, where n is the number of independent tasks and m is the number of processors. Table 1 shows that RR cuts the computational time and verifies that the maximum real loss energy of RR is less than the theoretical loss energy.

5 Conclusions

This paper empirically investigated the effect of grid computing on the computational cost of ES. We showed that RR cuts the computational time and verified that the maximum real loss energy of RR is less than the theoretical loss energy.

Acknowledgment

The authors acknowledges financial support in part through Grant-in-Aid for Scientific Research (16016262) and (17700233), and also Grant-in-Aid for 21COE from the Ministry of Education, Culture, Sports, Science and Technology of Japan.

References

1. N. Fujimoto and K. Hagihara (2003), "Near-Optimal Dynamic Task Scheduling of Independent Coarse-Grained Tasks onto a Computational Grid," Proceedings of the 32nd Annual International Conference on Parallel Processing (ICPP-03)

2. Y. Matsumura, X. Yao, J. Wyatt, K. Ohkura and K. Ueda (2003), "Robust Evolution Strategies Use Adaptive Search Strategies to Design the Continuous-Time Recurrent Neural Networks," Proceedings of 2nd International Conference on Computational Intelligence, Robotics and Autonomous Systems (CIRAS 2003)
3. X. Yao and Y. Liu (1997), "Fast Evolution Strategies," *Control and Cybernetics*, Vol. 26, No. 3, pp.467–496

Solution of Black-Scholes Equation By Using RBF Approximation

Zhai Fei, Yumi Goto, and Eisuke Kita

Summary. This article described the evaluation of European option price by using radial basis functions. The governing equation is discretized by Crank-Nicholson scheme and radial basis function approximation. Finally, numerical results are compared with analytical ones.

1 Introduction

Recently, the billing of financial derivatives has been increasing drastically for hedging the financial risk. The option transaction is one of the most important financial derivatives and therefore, several schemes have been presented for their pricing [1, 2].

This paper focuses on the evaluation of European option. Since the European option is the most fundamental one, the evaluation scheme can be extended easily to the others. The European option price is governed with the Black-Scholes differential equation. The finite difference method (FDM) is a typical numerical solution for the equation. The domain discretization is necessary in the FDM and the value is appreciable only by a specific lattice point. In this paper, we will apply the radial basis function (RBF) approximation. It has fast convergence property although the domain discretization is not necessary. The present formulation is summarized as follows. The Black-Scholes equation is formulated with the Crank-Nicholson scheme on the time axis and the option price is approximated with radial basis functions with unknown parameters at each time step. The initial values of the parameters are determined from the payoff condition on an expiration date. Then, the option price at the day of purchase are evaluated according to the backward algorithm from the maturity day. The numerical solutions are compared with the analytical.

2 Evaluation of European Option

2.1 Black-Scholes Equation [1, 2]

When the option is defined from only one asset, the price of the European option V is governed with

$$\frac{\partial}{\partial t}V(S,t) + \frac{1}{2}\sigma^2 S^2 \frac{\partial^2}{\partial t^2}V(S,t) + r\frac{\partial}{\partial S}V(S,t) - rV(S,t) = 0, \quad (1)$$

where the parameters t, S, μ, and σ denote the time, the stock price, the drift, and the volatility, respectively.

Besides, the boundary condition for the European put-option is given as

$$V = \max(E - S, 0) \quad (2)$$

and the boundary condition for the European call-option is given as

$$V = \max(S - E, 0), \quad (3)$$

where E denotes the expiration price for the option and the function $\max(x, 0)$ gives the larger value between x and 0.

2.2 Solution Procedure Using RBF Approximation

Discretizing (1) with Crank-Nicholson Scheme, we have

$$HV^{(t+\Delta t)} = GV^{(t)}, \quad (4)$$

where

$$V(S,t) = V^{(t)}, \quad V(S, t+\Delta t) = V^{(t+\Delta t)} \quad (5)$$

$$H = 1 + (1-\theta)\Delta t \left(\frac{1}{2}\sigma^2 S^2 \frac{\partial^2}{\partial t^2} + r\frac{\partial}{\partial S} - r\right)$$

$$G = 1 - \theta\Delta t \left(\frac{1}{2}\sigma^2 S^2 \frac{\partial^2}{\partial t^2} + r\frac{\partial}{\partial S} - r\right),$$

where Δt means the time-step size and the parameter θ is taken in the range of $0 \leq \theta \leq 1$.

In this study, the following radial basis function is adopted:

$$\phi(S, S_j) = \sqrt{c^2 + r_j^2}, \quad (6)$$

where d is the parameter and $r_j = \|S - S_j\|$.

The derivative price V governed with (1) is approximated with the radial basis function as follows:

$$V^{(t)} = \sum_{n=1}^{N} \lambda_n \phi_n, \qquad (7)$$

where N and λ_n denote the total number of data points and the unknown parameters, respectively.

Substituting (7) into (4), we have

$$\sum_{n=1}^{N} H\lambda_n^{(t+\Delta t)} \phi_n = \sum_{n=1}^{N} G\lambda_n^{(t)} \phi_n. \qquad (8)$$

2.3 Solution Algorithm

The algorithm of the present scheme is as follows:

1. Discretize from $t = 0$ to $t = T$ with time-step $\Delta t = T/M$.
2. Specify V^T at the expire date T from the boundary condition and calculate λ_n^T at the time T.
3. $t \leftarrow T - \Delta t$
4. Calculate λ_n^m from (8).
5. $t \leftarrow t - \Delta t$
6. Repeat step 4 and 5 if $n \geq 0$.
7. Evaluate $V^0 = \sum_{n=1}^{N} \lambda_n^0 \phi_n$.

3 Numerical Example

First, we will discuss the effect of the parameter c in RBFs to the accuracy. The parameters are specified in Table 1. The time step is $M = 100(\Delta t = 0.005)$ and the number of data points is $N = 121$. For different values of the parameter c, the solution error and the condition number of the coefficient matrix $H\phi_j$ are shown in Table 2. According to the increase of the parameter

Table 1. Parameters for numerical result

expiration date	$T = 0.5$ [year]
expiration price	$E = 10.0$
risk free interest rate	$r = 0.05$
volatility	$\sigma = 0.2$
Crank-Nicholson method	$\theta = 0.5$
maximum stock value	$S_{\max} = 30$

Table 2. The condition number and the error for c

c	condition number	ε
0.5	1.81×10^6	0.000291533
0.8	6.01×10^7	0.000290701
1.0	6.45×10^8	0.000289184
1.2	7.05×10^9	0.000287901
1.5	2.61×10^{11}	0.000286333
1.8	9.81×10^{12}	0.000284856

c, the error decreases and the condition number increases gradually. From the numerical experiments, we notice that the condition number should be smaller than 6.7×10^8 for the accurate solution. Therefore, in this case, we will take $c = 1.0$.

In the European put option, the numerical solution V_{RBF} is compared with the theoretical one $V_{Analytical}$ in Table 3. We notice that the numerical solutions agree well with the theoretical ones.

4 Conclusions

This paper describes the numerical solution for the evaluation of the European option using radial bases function approximation. In the formulation, the Black-Scholes equation is expanded with Crank-Nicholson scheme and the stock option price is approximated with radial basis functions. The present scheme is applied to the evaluation of the European stock option. The numerical solutions well agree with the theoretical ones. In the future, we are going to extend the present formulation for the evaluating the other more complicated options such as Look-Back and Asian options.

Table 3. Results of European put option

stock S	V_{RBF}	$V_{Analytical}$
0.0	9.7531	9.7531
2.0	7.7531	7.7531
4.0	5.7531	5.7531
6.0	3.75318	3.75318
8.0	1.79823	1.79871
10.0	0.44055	0.44197
12.0	0.04780	0.04834
14.0	0.00271	0.00277
16.0	0.00010	0.00010
18.0	0.00008	0.00000
	$\varepsilon = 0.000288675$	

References

1. G. Courtadon. A more accurate finite difference approximation for valuation of options. *Journal of Financial and Quantitative Analysis*, Vol. 17, pp. 697–703, 1982
2. P. Wilmott, J. Dewynne, and S. Howison. *Option Pricing: Mathematical Models and Computation*. Oxford Financial Press, Oxford, 1993

Visualization of Differences on Two Subjects' Impression Words Based on Correlation of *SD Data*

Teruyuki Kojima, Takeshi Furuhashi, Tomohiro Yoshikawa, and Kosuke Yamamoto

Summary. The purpose of this study is to visualize the impression differential between 2 subjects. We employ semantic differential (SD) method which is one of the most popular methods to quantify individual subjectivity. The number of dimensions of each impression word is same to that of objects in SD data. It is necessary to reduce the dimension to less than three-dimensions for visualization. This paper proposes the visualizing method which focuses on correlation of SD data between two subjects. The impression words are visualized on three-dimensional space where impression words having high correlation between two subjects' SD data are put close each other. We can investigate and discuss the similarities/differences of impression between two subjects through the visualized space.

1 Introduction

The semantic differential (SD) method is well known for one of the methods to quantify individual impression to objects. In the SD methods, a subject evaluates the objects by some pairs of antonymous adjectives with some discrete levels, and it clarifies the semantic structure to objects. The SD method is widely used to get subjective evaluation easily. A latent factor is often extracted from some pairs of impression words by employing factor analysis (FA) to SD evaluation data, because even the same subject does not always evaluate identically. In some conventional studies, the FA has been employed to averaged SD data gained from a lot of subjects [1, 2]. However, the averaging subjects' SD data means the disappearance of the individuality, because an individual differential exists on impression to even a same object.

This paper proposes the visualizing method to an individual differential of impression words, which employs the correlation of SD data on each impression word between two subjects. There is an individual differential of impression on a same impression word and that of evaluation scale. Then, comparing a same impression word such as "cute" or an absolute evaluation value between subjects does not mean considering individuality. This study focuses on correlation of SD data. It assumes that if fluctuation of evaluation

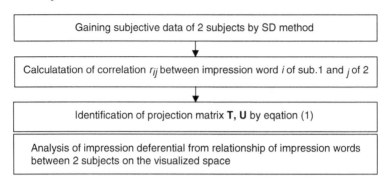

Fig. 1. Flow of the proposed method

data on an impression word to objects is similar between two subjects, even though the impression word or the value itself is different, it shows the similar subjective expression impressed by the objects between them.

2 The Proposed Method

Figure 1 shows the flow of the proposed method.

The number of impression words used in the SD is p and that of objects is q, respectively. Each SD data of two subjects is represented by a matrix $\mathbf{A}$ or $\mathbf{B}$ which has $p \times q$ elements. Then, each matrix of coordinates of impression words in three-dimensional space is represented by a matrix $\mathbf{X}$ or $\mathbf{Y}$ with $p \times 3$ elements. In the case of constructing visualized space by linear projection, $\mathbf{X}$ and $\mathbf{Y}$ are calculated by equation $\mathbf{X} = \mathbf{AT}$ and $\mathbf{Y} = \mathbf{BU}$, respectively. $\mathbf{T}/\mathbf{U}$ is a matrix for linear projection which has $q \times 3$ elements.

This study focuses on correlation relationship among impression words of two subjects' SD data. $\mathbf{T}$ and $\mathbf{U}$ are identified to realize that impression words with high correlation in SD data are put close one another in three-dimensional visualized space, and vice versa. The perspective of impression words expresses similarity of tendencies for impression words and difference of same impression words between two subjects. Then, they can be grasped intuitively by this method. Equation (1) shows the objective function employed in this paper.

$$E = \sum_{i}^{p}\sum_{j}^{p} \{\alpha_R \times (1 - r_{ij}) - \overline{x_i y_j}\}^2$$

$$\left(x_i = \left(\sum_{k=1}^{q} a_{ik} t_{k1}, \sum_{k=1}^{q} a_{ik} t_{k2}, \sum_{k=1}^{q} a_{ik} t_{k3} \right), \right. \tag{1}$$

$$y_j = \left(\sum_{k=1}^{q} b_{jk} u_{k1}, \sum_{k=1}^{q} b_{jk} u_{k2}, \sum_{k=1}^{q} b_{jk} u_{k3} \right),$$

$a_{ik}, b_{jk}, t_{k1} \sim t_{k3}, u_{k1} \sim u_{k3}$ represent elements of $\mathbf{A}, \mathbf{B}, \mathbf{T}, \mathbf{U}$.)

Here, α_R represents a constant value to adjust the scale, r_{ij} represents a correlation value for each element of SD evaluation data between impression word i of **A** and impression word j of **B**, and $\overline{x_i y_j}$ represents Euclid distance between x_i and y_j in visualized space. In the proposed method, **T** and **U** are identified by repetitive optimization to minimize E in (1).

3 Experiment

The proposed method is applied to two subjects' SD evaluation data with 42 pairs of impression words and 24 objects of car figures. Figure 2a–c shows acquired visualized space by the proposed method. Figure 2b–c are the magnified figures of Part (A) and (B) in Fig. 2a, respectively.

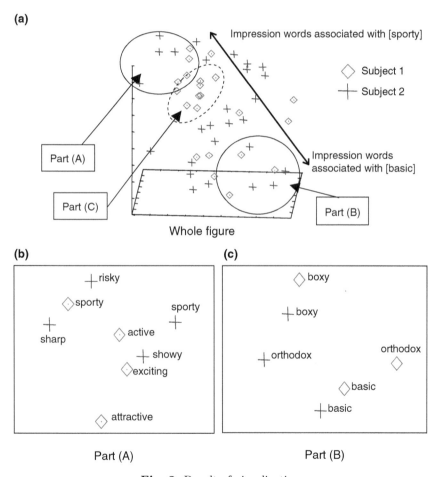

Fig. 2. Result of visualization

In Fig. 2a, it shows the tendency of distributed impression words corresponding to sportiness and basicness oppositely. It is supposed that each subject generally gave similar evaluation to the car designs. In Fig. 2b, different impression words are closer one another such as [1: exciting − 2: showy] and [1: sporty − 2: sharp/riskey] than the same impression words. Therefore, it is supposed that two subjects expressed similar impression on the objects by using different impression words in this part. In Fig. 2c, the same impression words are put close one another such as [boxy] and [basic]. Similar impression words were used to evaluate the objects in this part. Moreover, the part (C) in Fig. 2a does not have the impression words of subject 2. The impression words "sharp, luxurious, pleasant, opened, showy, urban, advanced, exclusive, fashionable" or impression itself expressed by subject 1 do not have the corresponding words for subject 2 employed in this experiment.

4 Conclusion

This paper proposed the visualization method to grasp individual differential of impression words, which employed the correlation of 2 subjects' SD data. This paper applied the proposed method to the SD evaluation data, and it showed that the visualized space which represented the relationship of impression words between two subjects could be constructed. This paper investigated the individual differential of them through the experimental result.

Acknowledgment

This work was supported in part by Frontiers of Computational Science at the 21st century COE program and Grant-in-Aid for Scientific research 16500126 from the Ministry of Education, Culture, Sports, Science and Technology.

References

1. T. Ishizaka, S. Kawaji, T. Tajima, M. Arao, "Comparison of the impression by motion of Pet Robot," Proceedings of the 3rd Annual Conference of JSKE 2001(in Japanese), pp. 124
2. M. Nantani, K. Satoh, H. Hayashi, Y. Nakamori, J. Nomura, K. Imamura, M. Nagamachi, "Fuzzy Modeling for Products Subjective Evaluation," Proceedings of 11th Fuzzy System Symposium (Okinawa, July 12–14, 1995) (in Japanese), pp. 279–282

Evaluation of High Imaging Quality LCDs Which Features Moving Picture: Comparison using Stabilometers

Kazuhiro Fujikake, Koichi Miura, Takahiro Sakurai, Hiroki Takada, Satoshi Hasegawa, Masako Omori, Ryumon Honda, and Masaru Miyao

Summary. Optically compensated bend (OCB) mode liquid crystal display (LCD) panels are newly developed displays that have excellent moving picture quality, almost equivalent to that of a cathode ray tube (CRT). High performance, with a brightness of $600\,\text{cd}\,\text{m}^{-2}$ and a contrast ratio of 600:1, was obtained by using the pseudoimpulse driving method to insert a black period between two continuous frames, and by using the blinking backlight method. Measuring center of gravity of the human body, we compared statokinesigrams while viewing moving maps on OCBs with those on conventional model displays. The results suggest that OCBs are better than conventional displays for scrolling map images like those in car-navigation systems.

1 Introduction

Liquid crystal displays (LCDs) are come into wide use as general visual display terminals. They have features such as larger display size, miniaturization and lighter weight, space-saving size, and low power consumption. However, users viewing movie displays on the LCDs complain of blurring and bleeding of the images.

When users viewed moving pictures on the LCDs they experienced visually induced motion sickness which was caused by disagreement between visual stimulation and that of the inner ear. The blurred images on the LCDs could induce "image sickness" in viewers, which was an unpleasant feeling similar to that of motion sickness [2].

In this study, a newly developed LCD was compared with a previous model LCD. Using stabilometers, we compared the sway of center of gravity with two different displays mentioned above. The center of gravity of subjects was compared as they viewed a moving map.

2 Method

We adjusted the temperature of 25°C in the experiment room, which was dark. The photograph shows the experimental setting.

Subjects stood quietly on the detection stand of a stabilometer (G5500, Anima Co., Ltd) in the Romberg posture with their feet together, for 1 min before the recording of the sway. The sway of each center of gravity was then recorded with a sampling frequency of 20 Hz.

Subjects then stood with their eyes opened for 1 min (resting state), and looked at each display with a moving map task for the following 1 min (testing state). Afterward, subjects closed their eyes for 1 min. The visual distance was 1 mater.

2.1 Display Device

We prepared two types of displays that are a previous model display and an optically compensated bend (OCB) display. The former is a hold type display, which is a general LCD model. The OCB display is a newly developed pseudoimpulse drive type.

2.2 Subject

The subjects were six persons from 20 to 27 years old, with no history of equilibrium function problems.

2.3 Moving Map Task

The map of a fictitious city scrolled from left to right. The name of a place was read out from the moving map as a moving map task. Scroll speed was 20 dot/s as the moving map.

2.4 Evaluation

We calculated indices for the statokinesigrams such as are "Area of sway," "Total trajectory length," and "Trajectory length per unit area" that are commonly used in the clinical field [3]. In addition, new quantification indices called "Sparse density: S_2," "Sparse density: S_3," "Trajectory length of chain 1," and "Trajectory length of chain 2" were also estimated [4].

3 Results

Typical statokinesigrams were shown in Fig. 1, which shows an example of the results for stabilometry.

The left one was observed when a subject watched the previous model display, and another figure was observed when a subject watched the OCB display. In these figures, the vertical axis shows the anterior and posterior movement in the center of gravity, and the horizontal axis shows the right and

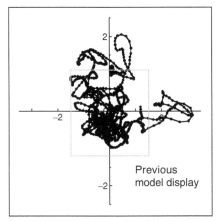 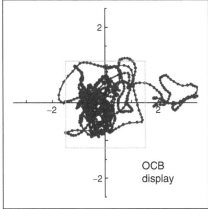

Fig. 1. Center of gravity (extraction example)

left movement in the center of pressure. The red squares in the two figures indicate the range recorded in the resting state. Amplitudes of the sway were larger when the subjects viewed the previous model display than when they were in the resting state. Right–left movement was larger when the subjects viewed the displays than when they were in the resting state.

The measured values are shown in the Table 1. Most values were the largest with the old model display, followed by the values with the OCB display.

4 Discussion and Conclusion

We statistically compared the resting state among the testing states using values of the indices evaluating the statokinesigrams. Among these states, the results of one-way analysis of variance (ANOVA) showed a significant difference in the indices except for the total locus of chains ($p < 0.01$). According to multiple comparisons, values of indices except for trajectory length per unit area were significantly larger when the subjects viewed the previous model

Table 1. Result of stabilometry (mean ± standard deviation)

	pre	old model	OCB
area of sway	2.3 ± 1.0	6.8 ± 3.4	4.9 ± 2.7
total trajectory length	53.8 ± 16.4	96.5 ± 21.4	96.3 ± 22.0
trajectory length per unit area	38.8 ± 9.1	52.2 ± 10.4	56.3 ± 11.6
sparse density S_2	1.3 ± 0.1	1.9 ± 0.2	1.7 ± 0.3
sparse density S_3	2.0 ± 0.4	3.3 ± 0.4	2.7 ± 0.7
total of chain 1	1.9 ± 1.4	3.3 ± 1.6	3.0 ± 1.5
total of chain 2	1.7 ± 1.4	2.9 ± 1.5	2.4 ± 1.5

display than when the subjects were in the resting state ($p < 0.05$). These statistical results indicated that visually induced motion sickness was affected by the previous model display. In contrast, there was no significant difference between resting state and testing state for the OCB display although values of the indices tended to be larger than when the subjects viewed the OCB display than when the subjects were in the resting state. This result indicated that the use of the OCB display suppressed visually induced motion sickness, which was affected by the previous model display.

Comparing testing states for the previous model display with the OCB display, no significant difference was found in those indices except for the sparse density S_3. This index S_3 might be convenient with the evaluation of displays. According to the previous researches, the aging does not produce an effect on the sparse density but on the chains [4]. However, the sparse density in the stabilometry could judge whether subjects had drunk an hour before [2]. It is well known that the equilibrium function in vestibule–cerebellum system is deteriorated by ingestion of alcohol as a medical action [5]. Therefore, the image sickness induced by the blurred images on the LCDs might exert influence on the vestibule–cerebellum system.

OCB displays have excellent moving picture quality. The present results suggest that OCB displays are better than conventional displays for many uses.

References

1. Omori M., Watanabe T., Takada H., Miyao M. (2002) Behav Inf Technol. **21**: 313–316
2. Takada, H., Kitaoka Y., Ichikawa, S., Miyao, M. (2003) Equilibrium Res. **62**(3):168–180
3. Suzuki J., Matsunaga, T., Tokumatsu, K. et al. (1996). Q&A and a manual in Stabilometry. Equilibrium Res: **55**(1):64–77
4. Takada, H., Kitaoka,Y., Iwase, S., Shimizu, Y., Watanabe, T., Nakayama, M., Miyao, M., Mihashi, K. (2003) Characteristic changes of sway of center of gravity with advancing age. Env. Med. **47**:85–89
5. Kaga, K. (1992). Memaino Kouzo: Structure of vertigo. Tokyo: Kanehara: 23–26, 95–100 (in Japanese)

Template Translation for Multilingual Disaster Information System

Shozo Tanaka, Masaru Miyao, Kohei Okamoto, and Satoshi Hasegawa

Summary. Multilingual disaster information system (MLDI) [1] has been developed to overcome the language barrier during times of natural disaster. MLDI is a Web-based system that includes templates in five languages so that translated texts can be made available immediately. This paper also describes the necessity of multilingual information.

1 Introduction

Disaster information or other important information must be made available to all people in a country, including foreign residents and travelers who cannot understand the native language of the country. The Great Hanshin-Awaji Earthquake occurred in Japan on 17 January, 1995, and killed 6,433 persons including 910 people who died afterward as a result of the earthquake (Great Hanshin-Awaji Earthquake Memorial Research Institute 2004) [2]. In the period following the earthquake, many foreigners who could not understand the Japanese language experienced great difficulty in finding refuge or in obtaining goods for survival in the disaster area.

2 System Aims of MLDI

MLDI system aims for simple operation and production of accurate information for communication to all people in disaster period. To achieve the aims, we propose template translation and distributed Web system. Template translation promotes exact translation filling the blanks in sentences prepared in advance. MLDI system constructed by a number of Web server realize robust system doing load and function distribution.

3 System Architecture of MLDI

MLDI system can be built from Web server and xml Web services server in order to elicit high performance in distributed environments at disaster

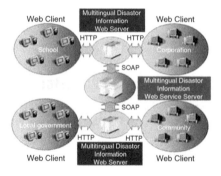

Fig. 1. System architecture of MLDI

period (Fig. 1). Web client connects to Web server using HTTP. Simple object access protocol (SOAP) is available for exchanging the information among the servers. Client applications as well as servers can also take advantage of XML Web services over SOAP.

4 Operation of MLDI System

The operation of MLDI system which aims simple operation to realize exact translation is summarized later.

4.1 Select the Templates

The seven categories of the templates (Fig. 2) can be selected by pushing a button. Rapidly a user can easily choose the appropriate template.

4.2 Fill with the Blanks

The blanks in the Japanese text can be filled with the numbers, Roman letters or Japanese characters representing the name of places, dates, times, telephone numbers, or other information (Fig. 3).

Fig. 2. Select the templates of the seven categories

Fig. 3. Fill with the blanks in the Japanese text

4.3 Confirm the Translated Text

The edited Japanese templates are translated into complete text in four languages, English, Korea, Chinese and Portuguese (Fig. 4). We can confirm the translated text. If translated texts are incorrect, these texts can be editing.

4.4 Display all Translated Text List

The Japanese, English, Korean, Chinese and Portuguese text list (Fig. 5) translated into complete text in four languages are displayed by pushing the "View translated list" button.

4.5 Make Use of Multilingual Translated Text after Downloaded

After multilingual translated text is downloaded, several applications can make use of it. An unicode compliant word processor can edit translated text and paste that text to e_mail, and send multilingual e_mail for many people who lives in disaster area.

Fig. 4. Confirm the translated text

Fig. 5. Display all translated text list

5 Conclusion

Now, various disasters, earthquake, seismic sea wave, tropical storm and so on, occur in the world, and disaster information is becoming a common concern of all countries. But it's very difficult to communicate disaster information for all people including foreigners who live in distressed area.

MLDI system has been developed to support most vulnerable people since 2002. The multilingual translation text which makes use of this system is effectively utilized as shared disaster information, for example printing translated text by word processor and sending e-mail. Currently MLDI system is mush interested by disaster manager of prefecture, city governments, corporations, and schools. MLDI system is necessary to expand function with operability and robustness.

References

1. Miyao M, Okamoto K (Representatives) (2004) Multi lingual Disaster Information System Consortium, http://tagengo.seis.nagoya-u.ac.jp/
2. Great Hanshin-Awaji Earthquake Memorial Research Institute (2004) Great Hanshin-Awaji Earthquake Research Paper, http://www.hanshin-awaji.or.jp/kyoukun/eng/

An Application of Double-Wayland Algorithm to Detect Anomalous Signals

Hiroki Takada, Hitoshi Tsunashima, Daishi Yamazaki, Tomoki Shiozawa, Meiho Nakayama, Takayuki Morimoto, and Masaru Miyao

Summary. The Wayland algorithm has been improved in order to evaluate the degree of visible determinism for dynamics that generate a time series simply and accurately. The Double-Wayland algorithm that we proposed could also detect phase transitions among multistates and nonstationarity in the dynamics. We herein apply this algorithm to detect defects in railways as anomalous signals.

1 Introduction

Dynamical equation systems (DESs) were obtained as mathematical models that regenerated time series data. Anomalous signals can either be given by the degeneration of the potential function in the DESs or they essentially change, for instance, their degree of freedom increases or stochastic factors are added to them. Visible determinism in the latter case would be different from that in the case where random variables do not exist. There is no common method in calculus to distinguish deterministic models from stochastic ones; however, the Grassberger–Procaccia and the Wayland algorithm [1] might be used as mathematical methods to evaluate their degree of freedom. The dimension of the DES can be estimated as a fractional number by using the Grassberger–Procaccia algorithm, which is considered to be beneficial for ensuring accuracy. However, it is difficult to examine whether a stochastic process is suitable for the mathematical model of time series data. It would not be easy to complete the former computation because the fractal dimension of the DES was derived from the calculations of all the points in the embedding space. In contrast, visible determinism could be estimated statistically in the case of the latter algorithm. The statistics shortens computation time. It is well known that 0.5 is the empirical threshold of the translation error to classify mathematical models as deterministic and stochastic generators; however, the translation error was not commonly estimated at the same value as that in the case of a larger signal to noise (S/N) ratio [2]. The authors then compared the various translation errors involved in the time series along with their differences [3]. In this research, an application of this Double-Wayland algorithm is presented in order to demonstrate its usefulness. There are several types of defects such as irregularity of cross level, alignment and leveling defects, improper distance between two rails, twist, and corrugation. Railway accidents might occur due to these defects that can be identified either by conducting visual checks or by using expensive detectors in the track inspection cars. We examine whether corrugation can be easily detected based on the measurements of accelerations on general vehicles and apply the Double-Wayland algorithm to the time series data of the accelerations.

2 Method (Double-Wayland Algorithm)

Delay coordinates $\{x_t\}$ can reconstruct a continuous trajectory without crossings in the embedding space that has a high dimension. The latter algorithm supposes that the difference vectors $\mathbf{v}_t = \mathbf{x}_{t+\tau} - \mathbf{x}_t$ in this space are approximated to temporal variations of the trajectories (Fig. 1) and estimates the translation error. This translation error is a statistical index that measures the complexity of the dynamics that generate the time series as follows. First, M onset periods t_0 are chosen at random. Second, the values of

$$E_{\text{trans}}(t_0) = \frac{1}{K+1} \sum_{i=0}^{K} \frac{|\mathbf{v}(t_i) - \bar{\mathbf{v}}|}{|\bar{\mathbf{v}}|} \quad (1)$$

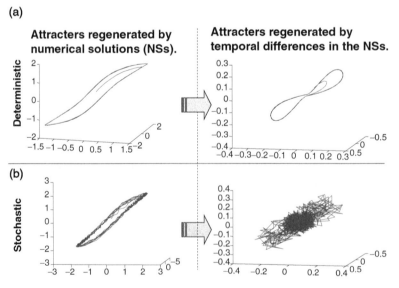

Fig. 1. This figure shows attracters regenerated by numerical solutions (NSs) to van der Pol equations without (**a**) and with random noise terms (**b**)

standardized by the average of the difference vectors at $K+1$ points $\{x_{t_i}\}_{i=0}^{K}$,

$$\bar{\mathbf{v}} = \frac{1}{K+1}\sum_{i=0}^{K} \mathbf{v}(t_i), \qquad (2)$$

are obtained at every onset period, where K points nearest to x_{t_0} are selected as $\{x_{t_i}\}_{i=1}^{K}$. Third, a median of M values with regard to (1) is extracted. Finally, Q medians are obtained with repeated measures. The translation error is estimated by the expectation value of Q medians. In addition, translation errors involved in the time series and their differences are estimated by the abovementioned Wayland algorithm. If the DES that includes stochastic factors were the generator of the time series, the flow would not be smooth (Fig. 1). In such a case, the latter translation errors would be much higher than the former.

We observed horizontal/vertical accelerations on a general vehicle and applied this Double-Wayland algorithm to time series data of the accelerations.

3 Results

Figure 2 shows the vertical acceleration, which was measured during business hours. One of these accelerations was recorded in the operating section; the visual checks revealed no corrugation (Fig. 2a). On the other hand, in another section the corrugation was confirmed (Fig. 2b). In this paper, the former section is termed as a normal section, and the latter is referred to as an anomalous section. The Double-Wayland algorithm was employed in every interval for 1.5 s; the total number of sequences for each interval was 3,000. The translation errors in the normal section were estimated to be approximately 0.5 (Fig. 3). The translation error displayed a weak dependence on the embedding dimension and decreased with leniency (Fig. 3a). Although these values were lower than any translation error values derived from the temporal differences of the time series in the accelerations, the order was reversed in the anomalous section (Fig. 3b).

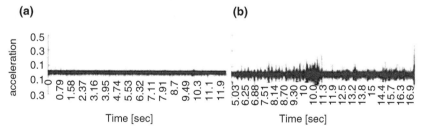

Fig. 2. Typical time series of the vertical acceleration. The acceleration was observed in the normal section (**a**) and the anomalous section (**b**)

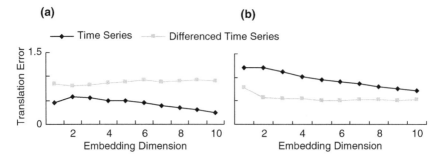

Fig. 3. Translation errors were derived from the time series of the acceleration in Fig. 2 for each embedding space. The accelerations were observed in the normal section (**a**) and the anomalous section (**b**)

4 Discussion

According to the Double-Wayland algorithm, the process that generated accelerations was generally regarded as a stochastic one. Moreover, with regard to the anomalous section, the mathematical model for the accelerations was changed into a deterministic one (Fig. 3). The regularity in acceleration might be physically enhanced by undertaking periodic structural changes in the railways. Thus, we recognized the order in the translation errors derived from the time series of horizontal/vertical accelerations and the temporal differences. We succeeded in detecting the anomalous sections in which corrugation was observed by conducting visual checks. These sections could also be detected by a complex multiresolution analysis, which indicated that the time series of horizontal/vertical accelerations contained a certain frequency band [4].

This result suggested the availability of a mathematical approach to detect the defects in the railways. The defects were previously identified by conducting visual checks or by using expensive detectors in the track inspection cars. However, due to hectic train schedules, it is not always possible to identify these defects. The proposed method of detecting defects based on the measurements of the accelerations of general vehicles is considered to be convenient since it enables the identification of the defects irrespective of the situation.

Currently, we are applying the Double-Wayland algorithm to several fields such as stock price and stabilometry; however, due to space limitations, we do not report this data for targets related to the fields of medicine and economics. These applications will be reported at the next symposium.

Acknowledgment

A part of this study was supported by the 21st Century Center of Excellence (COE) Program for Frontiers of Computational Science, the Japan Science and Technology Agency (JST), and the Program for Promoting Fundamental Transport Technology Research from the Japan Railway Construction, Transport and Technology Agency (JRTT).

References

1. Wayland R et al. (1993) Phys. Rev. Lett. 70: 530–582
2. Matsumoto T et al. (2002) Chaos and time series. Baihukan, Tokyo
3. Takada H et al. (2005) Bulletin of Society for Science on Form, 17 (3): 301–310
4. http://www.ntsel.go.jp/ronbun/happyoukai/17files/poster11.pdf

CPSIA information can be obtained
at www.ICGtesting.com
Printed in the USA
LVHW081936210620
658633LV00003B/108